Psychological Knowledge

A social history and philosophy

Martin Kusch

London and New York

First published 1999
by Routledge
First published in paperback 2006
2 Park Square, Milton Park, Abingdon, Oxon, OX14 4RN

Simultaneously published in the USA and Canada
by Routledge
270 Madison Avenue, New York, NY 10016

Reprinted 2006

Typeset in Palatino by Routledge
Printed and bound in Great Britain by MPG Books Ltd,
Bodmin, Cornwall

British Library Cataloguing in Publication Data
A catalogue record for this book is available from the British Library

Library of Congress Cataloguing in Publication Data
Kusch, Martin.
Psychological Knowledge: A social history and philosophy /
Martin Kusch.
(Philosophical issues in science)
1. Philosophical – Philosophy. 2. Psychology – History.
3. Philosophy of mind. I. Title. II. Series
BF38.K87 1999 98-35091
150'.1-dc21 CIP

ISBN 10: 0-415-19253-6 (Hardback)
ISBN 10: 0-415-37931-8 (Paperback)

ISBN 13: 978-0-415-19253-8 (Hardback)
ISBN 13: 978-0-415-37931-1 (Paperback)

And pardon that thy secrets should be sung…

John Keats, *Ode to Psyche*

Contents

List of illustrations ix
Acknowledgements xi

Introduction 1

PART I
A social history of psychological knowledge: the controversy over thought psychology in Germany, 1900–20 7

Introduction to Part I 9

1 The Würzburgers 18

2 Friends and foes 71

3 Recluse or drillmaster versus interlocutor and interrogator 95

4 Purist versus promiscuist 131

5 Collectivist versus individualist 168

6 Protestant versus Catholic 194

7 Conclusions 232

Interlude 255

PART II
The sociophilosophy of folk psychology **277**

Introduction to Part II 279

8 The folk psychology debate 282

9 Folk psychology as a social institution 321

Notes 369
Bibliography 375
Index 405

Illustrations

Tables

1.1 The five main groups of reaction experiments 24

Figures

1.1 Ach's theory of awareness 35
1.2 Marbe's externalist theory of judgement 46
1.3 The share of visual presentations across Watt's tasks for experimental subjects 50
1.4 Time used by experimental subjects across Watt's tasks 51
1.5 Familiarity across Watt's tasks 52
1.6 Ach's theory of determining tendencies 58
1.7 The creation of new associations in Ach's theory 60
1.8 Bühler's analysis of thought recall 64
7.1 Summary of the analysis: Wundt 233
7.2 Summary of the analysis: the Würzburgers 234

Acknowledgements

The story of writing this book is a tale of three cities. I started the research reported here as a member of the Science Studies Unit of the University of Edinburgh in 1994, and I could never have completed this study without the support of my friends, colleagues, and students in the Scottish capital. David Bloor's advice and encouragement have been crucial throughout: David and I have discussed central themes, and various versions, of this book for many hours, and there is hardly an idea here that was not first tested by him. Other key Edinburghers were Celia Bloor, Claude Charpentier, Sarah Gore-Cortes, Heini Hakosalo, John Henry, Peter Imhof, Matthias Klaes, Donald MacKenzie, Pauline Padfield, Irene Rafanell, Steve Sturdy, Carole Tansley (who kindly corrected the English of some early versions of parts of the manuscript, and who unsuccessfully tried to teach me the British sense of humour), and the students of my courses on 'The History of Experimental Psychology', 'The Sociology of Scientific Knowledge', 'The Continental Tradition in the Philosophy of the Social Sciences', and 'Science and Society'.

The main bulk of this text was written during the academic year 1996–7, while I was a visitor to Gerd Gigerenzer's research group at the Max-Planck Institute for Psychological Research in Munich, and holder of a fellowship of the Alexander-von-Humboldt Foundation. I gratefully acknowledge Gerd's help and the generous support of both institutions. All members of Gerd's 'ABC-group' were, in one way or another, supportive of my work, and thus it seems only right and proper that I list them here: Ellen Bein, Marina von Bernhardi, Patricia Berretty, Bernhard Borges, Valerie Chase, Jean Czerlinski, Jennifer Davis, Gerd Gigerenzer, Dan Goldstein, Adam Goodie, Ralph Hertwig, Ulrich Hoffrage, Tim Ketelaar, Nicola Korherr, Alejandro Lopez, Laura Martignon, Franz Mechsner,

Andreas Ortmann, Arne Schwarz, Anita Todd, Peter Todd, Angelika Weber, Greg Werner, and Maria Zumbeel. In particular, I am indebted to Anita for checking the style and grammar of the whole manuscript; to Angelika for breakfasts, conversations, and insomnia treatments at the 'Extrablatt'; to Laura for numerous joint explorations into social institutions and Italian food; to Marina for 'Kalbszunge', political education, wonderful outings, and assistance in a million-and-one practical matters; to Trish for tours around Bavaria, Austria, and the Czech Republic, spiced with arguments about 2 + 2 = 4; and to Adam, Arne, Franz, Gerd, Jean, Jennifer, Ralph, and Tim for several useful exchanges. Outside Gerd's research group, I profited from conversations with Fiorello Banci, Harriet Murphy, Ulla Mitzdorf, and Gertrud Nunner-Winkler.

I put the finishing touches to this book at my most recent academic home, the Department of History and Philosophy of Science at the University of Cambridge. Although I have only been here since the autumn of 1997, the HPS department has had an impact on the shaping of this book for the past three years. I would not have written Part II at all had it not been for Peter Lipton's friendly insistence in 1995 that I present a 'philosophical' paper in the departmental seminar. And I first formulated some of the conclusions of Part I for my 'job-talk' in the spring of 1997. A number of my Cambridge colleagues have made invaluable suggestions on these and other occasions. I particularly wish to thank John Forrester and Simon Schaffer for reading the whole manuscript and for helping me to improve both style and content. I am also obliged to Soraya de Chadarevian, Marina Frasca-Spada, Nick Jardine, Peter Lipton, Jim Secord, Liba Taub, Paul White, and, from outside the department, Akiko Saito.

Although this book was written in Edinburgh, Munich, and Cambridge, I am also indebted to friends and colleagues elsewhere. My mother, Erna Kusch, has again kept me updated on literature published in German; Mitchell Ash wrote a very valuable referee's report on a first version of Chapter 3; and Ed Haupt assisted me in finding my way around Müller's writings. Encouragement and comments also came from Jouko Aho, Jim Brown, Harry Collins, Jeff Coulter, York Gunther, Esa Itkonen, Jonathan Harwood, Jaakko Hintikka, Marja-Liisa Kakkuri-Knuuttila, Lisa Kanerva, Simo Knuuttila, Riitta Korhonen, Eerik Lagerspetz, Gerd Lüer, Juha Manninen, Ilkka Niiniluoto, Sybilla Nikolow, Volker Peckhaus, Frederick Stoutland, Raimo Tuomela, Erkki Urpilainen, Elizabeth

Valentine and Georg-Henrik von Wright. I also wish to thank the many helpful archivists and librarians in Edinburgh, Glasgow, Göttingen, Leipzig, Munich, and Würzburg.

Various parts of this book were first presented as lectures to academic audiences in Bolzano, Boston, Cambridge, Edinburgh, Helsinki, Munich, Oulu, Stockholm, and Toronto. I am obliged to all those kind people who attended these talks, especially to those who raised critical questions.

Finally, thanks are due to Bill Newton-Smith and Adrian Driscoll for accepting my manuscript for publication with Routledge.

An early version of Chapter 3 was published as 'Recluse, Interlocutor, Interrogator: Natural and Social Order in Turn-of-the-Century Psychological Research Schools', *ISIS*, vol. 86, 1995, pp. 419–39, © 1995 by the History of Science Society, Inc. All rights reserved. An early version of Part II was published as 'The Sociophilosophy of Folk Psychology', *Studies in History and Philosophy of Science*, vol. 28, 1997, pp. 1–25. © 1997 by Elsevier Science. I am grateful to both publishers for permission to reproduce parts of these writings in what follows.

I dedicate this book to my four god-children, Ella-Onerva, Henri, Samuel, and Santtu. My dear young friends, please accept this dedication as a small compensation for my having forgotten so many of your birthdays.

Introduction

The main thesis of this study can be stated in one short sentence: *Bodies of psychological knowledge are social institutions*. Or, in a slightly longer formulation: Both *scientific*-psychological theories and *folk*-psychological bodies of knowledge have the same form of existence as do marriage, money, or the monarchy. They are social; they are conventional; and they are theories and bodies of knowledge only because they are *taken* to be theories and bodies of knowledge. Why are some pieces of metal coins – that is, money? Because there are collectives that *take* these pieces of metal to be coins. Why are some sets of sentences theories about the mind? Because there are collectives that *take* these sets of sentences to be theories about the mind.

To analyse something as a social institution calls for more, however, than just the pinpointing of such self-fulfilling, or self-referential, 'takings'. Social institutions do not exist in isolation from one another; they are not independent from the individuals that sustain them; and they involve characteristic actions and artefacts. To analyse a given body of psychological knowledge as a social institution therefore involves displaying its links to other institutions, describing its relations to individuals, and depicting its typical actions and artefacts. I shall now explain in more detail what this involves.

To begin with the links between social institutions, the obvious point to make is that social life is holistic. Social institutions are interconnected and often embedded in one another, just as the institution of promising is partly embedded in the institution of marriage. Social institutions overlap, crisscross, compete, strengthen, and weaken one another. How do bodies of psychological knowledge (as social institutions) stand in this 'field of force' of social institutions? In this book I shall suggest the following answer to this question. First, individual bodies of *scientific*-psychological knowledge, that is theories of scientific psychology, are, variously but closely, intertwined with particular – and thus differently organised – instances of social institutions such as the following: the psychological experiment, the research school, the structure of subdisciplines within psychology, the system of academic disciplines, the churches, and the polity at large. Second, bodies of *folk*-psychological knowledge also have some such links; for the most part, however, they are not so easily influenced and shaped by other institutions. This is because folk psychology is not only *a* social institution, but *the most fundamental* social institution. Folk psychology is the basis for all other social institutions, and it is the social institution that most of us have the least interest in destabilising. Therefore, folk psychology changes much more slowly than do theories of scientific psychology.

Institutions relate to individuals in more than one way. One particularly important way is that individuals – usually in cooperation with others – seek to build or destroy, change or preserve social institutions. They do not engage in institution building from some completely pre-social vantage point, of course; to be able to convince others of changes to, say, the institution of marriage demands participation in the social institution called 'language'. Why do individuals and groups build or destroy, change or preserve institutions? The answer is not hard to find. Much of the social world is a realm of struggles over power between social actors – individuals and groups. Social institutions are the resources, the prizes, the tools, the objects, the victims, and the battlegrounds for such struggles. Actors try to change or destroy, preserve or protect a given institution because they perceive its existence as good or bad for the survival, or the coming-to-be, of other, more highly cherished, institutions.

These actions and motives are easily observed in the case of struggles over *scientific*-psychological theories. For instance, in their theory building and theory selecting, scientific psychologists are influenced by their beliefs and desires *vis-à-vis* their factual or their desired social worlds. One psychologist might favour one theory over its competitors because that theory allows the psychologist to improve his or her standing within his or her profession, speciality, and department or institute. Such might be the case if the successful use of the theory in question presupposes some special, say mathematical or experimental, skills that the psychologist possesses to an unusually high degree. Another psychologist might prefer another theory because it uses social imagery that benefits that psychologist, his or her generation, class, political party, or religious community. And such benefit might of course also result from the general acceptance of a theory that does not involve any transparent social imagery.

Individuals are not only creators of social institutions; individuals – their actions, behaviour, and mental states – are also the products or artefacts of social institutions. Some versions of this claim will hardly be controversial; of course we all are the products of our families, schools, or other educational institutions. Other versions of the claim that individuals are the products of institutions will meet, or have met, with more resistance. Thus not everyone would agree that criminals are products of penal institutions. For our present concerns, we need to apply the general claim to bodies of scientific-psychological and folk-psychological knowledge. Leaving aside the training of psychologists, the most provocative application of the general claim to psychology is to ask whether psychological knowledge – *qua* social institution – creates or shapes its human referents. Such would be the case if psychological knowledge became constitutive of the self-understanding and mental life of its human referents, or if it changed behaviours in ways that made its originally false claims come out true. Psychological knowledge would then have the character of a self-fulfilling prophesy. I shall argue in this book that this is a plausible thesis in at least some important cases of psychological knowledge.

We now know – at least on a general level – what is required to defend the thesis that bodies of psychological knowledge are social institutions. I shall now proceed to motivate the specific lines of enquiry chosen in this book.

I have suggested above that, as social institutions, theories of scientific psychology and bodies of folk-psychological knowledge are rather different. In the network or web of social institutions, theories of scientific psychology are phenomena of the short term, and highly sensitive to influences of other institutions. Bodies of folk psychology, on the other hand, are fundamental and phenomena of the *durée longue*. This difference cannot but influence how I shall go about establishing my general claim that *all* bodies of psychological knowledge are social institutions. *Vis-à-vis* scientific-psychological theories, the challenge is to show in detail how psychological theories are linked to other institutions; how they are tools, targets, and victims of power struggles; and how they can become constitutive of the phenomena they concern. Obviously, such an argument cannot be fruitfully made in the philosophical abstract – it calls for a case study of a key episode in the history of psychology. Part I of this book will therefore be a social history of the debate over thought psychology ('imageless thought') in Germany during the first two decades of the twentieth century. I have chosen this, rather than any other, episode because it is pretty much the only controversy in the history of psychology familiar to historians, sociologists, and philosophers alike.

The situation is different for folk psychology. Because folk psychology is a very slowly changing body of knowledge, establishing links between it and specific historical events and institutions is well nigh impossible – at least it is close to impossible for the core concepts of (our) folk psychology, for 'belief', 'action', and 'desire'. Moreover, in the case of folk psychology – though not in the case of scientific psychology – there exists a substantial amount of philosophical theorising, a body of work that tries to determine the nature and structure of folk psychology. Most of this work takes its starting point from premises that run directly counter to my claim that folk psychology is a social institution. It is therefore imperative that my treatment of folk psychology is philosophical rather than historical. I need to take on the arguments of alternative proposals, show what is wrong with them, and demonstrate how my own social–philosophical approach constitutes an advance. I shall do so in Part II of this book, under the title of 'a sociophilosophy of folk psychology'.

Thus what motivates me to deal with folk psychology and scientific psychology separately and distinctly are two factors: their differences *qua* social institutions, on the one hand, and the different state of research and scholarship with respect to them, on the other.

Having made my case for a separate and distinct treatment of folk and scientific psychology, I hasten to add that there are numerous important connections between Parts I and II of this book. And there have to be, too; after all, there are also many parallels and links between scientific and folk psychology. For instance, many philosophers of mind see bodies of folk-psychological knowledge as theory-like entities, that is as entities similar to the theories of scientific psychology. Others make claims about the influence of folk psychology upon scientific psychology. Still others wonder whether scientific psychology – or else cognitive science or neuroscience – will ever come to replace, or 'eliminate', folk psychology. Obviously, my sociophilosophical treatment of folk psychology will not be satisfactory unless I am able to say something about these difficult issues. In doing so in Part II, I shall draw on and develop what I take to be the philosophical lessons of Part I.

My book is strongly influenced by, and meant as a contribution to, the sociology of scientific knowledge (SSK). Particularly important for my project is recent work in SSK on the 'performative theory' of social institutions. I have already introduced the basic idea of this theory in the first few paragraphs of this introduction. For the social history of Part I, we need no more than this basic idea. Indeed, as I shall argue in the conclusion of Part I, even in this, its most simple form, the performative theory of social institutions allows for the formulation and defence of a number of philosophically important theses. Such theses are that scientific experiments and instruments are best thought of as social institutions; that the traditional distinction between 'the rational' and 'the social' is untenable; and that interpreting experiments and theories as social institutions does not imply any idealistic denial of a mind-independent natural world.

For the purposes of Part II, however, we need a richer version of the performative theory of social institutions. I have therefore placed a more detailed account as an 'Interlude' between the two main parts of this book.

Part I

A social history of psychological knowledge

The controversy over thought psychology in Germany, 1900–20

Introduction to Part I

Almost all histories of psychology have a chapter on the early twentieth-century German controversy over thought psychology. The prominence given to this debate – usually under the slightly misleading title 'imageless thought controversy'[1] – attests to the fact that it was one of *the* most decisive junctures in the history of psychology, and, thus, one of the key events in twentieth-century science. In the exchanges between the antagonists in this controversy many central parameters of the twentieth-century scientific study of human beings were identified, and some were permanently fixed. Topics covered in this debate were the nature of thought, its accessibility to observation and introspection, the role of experimentation in psychology, the relation between logic and psychology, and the differences between the natural and the social sciences.

The debate over thought psychology was first provoked by a number of methodological and theoretical writings coming out of the Psychological Institute of the University of Würzburg between, roughly, 1900 and 1907. As the members of that institute moved on to positions elsewhere, thought psychology spread, too. Soon thought-psychological studies as well as their theoretical interpretations and philosophical defences were also produced at the Universities of Bonn, Munich, and Königsberg. To avoid clumsy circumlocation, I shall speak of the thought psychologists as 'the Würzburgers' even when their institutional location was Bonn, Munich, or Königsberg. Key Würzburgers in this broader sense were Oswald Külpe (1862–1915), the director of the Würzburg Institute from 1896 until 1909, Narziss Ach (1871–1946), Karl Bühler (1879–1963), Karl Marbe (1869–1953), August Messer (1867–1937), Otto Selz (1881–1943), and Henry Jackson Watt (1879–1925). For the purposes of a rough first overview, we can reduce the Würzburgers' innovations to four ideas.

First, the Würzburgers claimed to have discovered new kinds of mental contents, in addition to the traditional trio of sensations, feelings, and presentations (*Vorstellungen*). The Würzburgers called these new, irreducible mental contents 'situations of consciousness' (*Bewußtseinslagen*), 'awarenesses' (*Bewußtheiten*), and 'thoughts' (*Gedanken*).

Second, the Würzburgers challenged traditional accounts according to which thinking consisted of forming judgements and drawing inferences. Some of the Würzburgers took issue with the specifics of earlier psychological interpretations of judgements, while others abandoned altogether the idea that the psychology of thought should encompass the search for mental representations of judgements. Instead, thought psychologists were to ask more generally, 'What do we experience when we think?'

Third, in their attack on the sufficiency and exhaustiveness of the traditional threesome of sensations, feelings, and presentations, the Würzburgers also rejected associationism. Over and above associations, they argued, one had to assume 'determining tendencies' (*determinierende Tendenzen*) that linked presentations and thoughts together in a goal-oriented fashion, the goal being the solving of a problem or 'task' (*Aufgabe*).

Fourth, and finally, the Würzburgers believed that a retrospective self-observation (*Selbstbeobachtung*) of one's own thought processes was possible and reliable provided it occurred under controlled experimental conditions. The most important of these conditions was the presence of a second person. Successful retrospection demanded an experimental setting with two persons such that the 'retrospectionist' as an experimental subject was assisted in his or her retrospection by an experimenter. The experimenter would first set a task for the subject, a task that would demand thinking to solve it. The subject would try to solve the task and give an answer as soon as he or she felt satisfied with the solution. Immediately afterwards, the subject would furnish a complete account of all that took place in his or her mind between hearing the problem and providing the solution.

The Würzburg Psychological Institute was one of the four main institutes of its kind in Germany during the first decade of the twentieth century. The others were Wilhelm Wundt's (1832–1920) laboratory in Leipzig, Georg Elias Müller's (1850–1934) institute in Göttingen, and Carl Stumpf's (1848–1936) institute in Berlin. Several of the Würzburg psychologists had trained with Wundt or Müller; Külpe had worked under both. The competition and rivalry between these 'schools' was often intense, and it reached its climax during the thought psychology controversy. My main focus will be on the dispute between Würzburg, Leipzig, and Göttingen – and here I shall have more to say on the relationship between the first two than on the relationship between the first and the third. I shall also comment on the involvement of the Berlin Institute and about Külpe's position after he moved from Würzburg to Munich via Bonn.

Both Wundt and Müller published lengthy criticisms of the Würzburgers' thought psychology. Wundt had no quarrel with the Würzburgers' rejection of associationism, but he defended a psychological theory of judgements, denied the existence of 'non-depictive' (*unanschaulich*, imageless) contents of consciousness, and doubted the reliability of introspection with respect to thought processes. Wundt insisted that the process of forming a judgement was open to psychological analysis. It was a process in which a 'dimly conscious, total presentation' unfolded into two fully conscious parts, the subject presentation and the predicate presentations. Prior to this unfolding, total presentations were represented in consciousness as 'intense and clear feelings'. What the Würzburgers claimed to be newly discovered conscious contents were really nothing but these feelings. Thought and other 'higher' mental processes could not be investigated by means of experiment and introspection. Their study demanded a new type of psychology. Using the methods of the humanities, this new 'collective psychology' (*Völkerpsychologie*) analysed thinking by investigating the products of collective thinking, namely language, myth, and custom.

According to Müller and his students at the institute in Göttingen, two kinds of laws governed the train of presentations in consciousness: laws of association and the law of 'perseveration'. Laws of association concerned connection strength between presentations, and the law of perseveration was about the natural tendency of every conscious presentation to remain in consciousness for a while. As Müller pointed out repeatedly against the Würzburgers, there was no need for 'determining tendencies' as a third type of law. All of the Würzburgers' experimental results could be explained by association and perseveration alone. Nor did Müller believe that the Würzburgers had made a convincing case for non-depictive contents of consciousness. Their claims could be shown to be based upon bad observation and sloppy theorising. Finally, although Müller was not sceptical of introspection as such, he felt unhappy about the ways the Würzburgers had employed it in their thought-psychological experiments. They had displayed little awareness of its many possible pitfalls and had failed to train their subjects properly on how to use this complicated method.

In the last few paragraphs I have presented an overview of the technical content of the controversy over the Würzburgers' thought psychology. In Chapters 1 and 2 I will go over this same ground in much greater depth. In Chapters 3 to 6 I shall demonstrate that more was at stake in this controversy than the issues mentioned up to now; that is, that this controversy partly overlapped with a number of other struggles. The fight over the establishment of Würzburg-style thought psychology as an institution within German psychology at large was, at the same time, a debate over the nature, or the relative strength, of a number of already existing institutions. These institutions were the psychological experiment, the research school, psychology as a field of study, the system of university disciplines, the state, and the Catholic and Protestant confessions.

Put differently, at issue in the debate over thought psychology were not only determining tendencies, situations of consciousness, or the pros and cons of introspection. At issue were also the following questions. Under what social arrangement is the human mind best able to think? How should cognitive and social authority be divided between experimenter and experimental subject? How should psychological research schools be organised and run? How should cognitive and social authority be distributed over different subfields of psychology? Where should the discipline of psychology be located institutionally? To what extent should the psychologist seek to contribute directly to social and educational issues? What is the correct way to think about the relationship between the individual and the collective in general, or the citizen and the state in particular? And, which confession – Catholic or Protestant – should define the role of the Christian in early twentieth-century Germany?

In Chapter 3 I link together the facts that the main antagonists of the debate over thought psychology all had radically different theories of the mind; that they worked in psychological research schools with unlike forms of sociability; that they had very distinct visions of the social order in psychological experiments; and that they proposed distinctive ideas on the organisation of psychology as a field of study. I argue that the differences in these four dimensions were systematically connected: Different psychologies were shaped by, or built to justify, different forms of sociability.

Wundt believed that the mind had a strictly hierarchical, two-tier structure and that thinking demanded solitude. The lower processes of the mind (sensations, simple feelings) could be investigated by means of experiments; the higher, thinking processes called for collective psychology. The Leipzig Institute had a social structure befitting these ideas: Wundt was the solitary master psychologist, specialised in the collective-psychological study of thinking, and overseeing the work of the experimentalists from a distance. Moreover, the more demanding the task set for experimental subjects, the greater the need for them to be alone.

As Müller saw it, the struggle amongst presentations over access to consciousness was governed, without exception, by the laws of association and perseveration. The successful presentations were associated strongly with each other and with many others, and they had high degrees of perseveration. Daily life in the Göttingen Institute was likewise prescribed by detailed rules, and experimental subjects were drilled and tightly controlled. Last, but not least, the apex of the Göttingen Institute was occupied by a man who was associated with every single research project carried out in the institute, and who had proved experimentally that he outclassed everyone in perseveration.

The Würzburgers attacked Wundt's and Müller's theories of the mind and replaced them with a much more egalitarian model. They rejected any sharp division between higher and lower processes, and they broke the exclusive rule of association and perseveration. Their egalitarian structure for the mind had its analogues in an egalitarian, two-person experimental setting and might have been informed by the egalitarian social structure of the Würzburg research school.

In sum then, I argue that the antagonists in the controversy over thought psychology censured not only each others' theories of the mind; at the same time – and with the very same words – they also attacked each others' preferred model of sociability in the psychological institute.

In Chapter 4 I seek to make plausible that the controversy over thought psychology was intertwined with three other debates that all concerned the location and position of psychology within the system of academic disciplines. These debates questioned whether psychology was a natural science or else one of the humanities; whether psychology was one, or even the most fundamental, part of philosophy; and whether the time had come to start applying psychology. In these debates, Wundt and the

Würzburgers took very different stances. (Müller largely drops out of the comparisons from here on, because he preferred not to publish his views on political, confessional, and more philosophical issues.)

For Wundt, psychology was the most fundamental of the humanities, and it was conceptually independent of the natural sciences. In particular, Wundt was always most concerned to draw a sharp line of demarcation between psychology and physiology and to insist that psychology was irreducible to physiology. One central ingredient in this demarcation exercise was the idea that higher mental processes were not accessible to experimental study. We already know that the Würzburgers rejected this idea. This rejection was informed by their belief that psychology was a natural science, and that some parts of psychology would ultimately reduce to physiology.

Moreover, for Wundt, psychology needed to remain, first and foremost, a purely theoretical science; the time had not yet come to start thinking seriously about applications to practical problems. The Würzburgers, however, put much emphasis on the need to apply psychology in education, forensic science, and psychiatry. Külpe even suggested that psychology be moved to the medical faculty for just this reason. Thought psychology was an attempt to push psychology into areas in which it could become more directly relevant to practical concerns; after all, thinking was what children needed to be taught to do properly, and thinking was what the psychiatric patient could no longer do clearly. In resisting thought psychology, Wundt also resisted this whole move towards application.

As concerns the relationship between psychology and philosophy, Wundt's position was complex. Although he deemed psychology an independent science, he wished it to remain, at least for the time being, within philosophy departments. This was because psychologists needed to be prevented from becoming mindless experimentalists. At the same time, Wundt also maintained both that central areas of philosophy, like logic, needed to learn from psychology, and that psychology needed to free itself from the influence of logical theories. We already know that Külpe thought differently about the issue of location. To this we can now add that several Würzburgers agreed with those philosophers of the time who deplored the intrusion of psychology into logic and other areas of 'pure' philosophy. And, to make matters worse, some of the Würzburgers relied heavily on one of the leading logicians of the day, Edmund Husserl. And they did so in their thought-psychological writings.

In Chapter 5 I argue that the antagonists of the controversy over thought psychology adhered to different social, ethical, and political philosophies. We should not be surprised that Wundt was so adamant in his refusal to accept a study of thinking based upon 'individualistic' introspection. In his philosophy of psychology, Wundt maintained the primacy of the collective over the individual; in his ethics, he opposed egoism and the individual's right to happiness; and in his political philosophy, he insisted that the state came before the individual citizen. The main whipping boys of Wundt's political philosophy were various forms of political individualism; amongst them Wundt counted Marxism, social democracy, anarchism, and contract theories of the state. Furthermore, Wundt was convinced that political individualism was ultimately based upon psychological individualism. And in his view, Würzburg-style thought psychology was a clear example of psychological individualism.

With this insinuation, Wundt was not far from the truth. Although the Würzburgers did not have a common and worked-out position in social and political philosophy, some of their writings signalled a strong leaning towards methodological individualism, and an opposition to at least *some* aspects of Wundt's collectivistic ethics and politics.

Finally, in Chapter 6, I contend that religion played a significant role in the controversy over thought psychology. To begin with, neo-Thomist philosophers took a strong interest in the Würzburgers' work; this was because the latter studies seemed to confirm some scholastic and neo-Thomistic views of the mind, and of the mind's relation to the world. The neo-Thomists did not detect similar parallels in Wundt's work, and therefore Wundt was a frequent target of their criticism.

Furthermore, Wundt was a Protestant who made no secret of his dislike for Catholicism and neo-Thomism. Wundt's Protestantism was also visible in his psychology and his metaphysics: voluntarism was – by the standards of the time – a Protestant position, and Wundt made it the hinge around which all of his thinking turned. Wundt noted that the Würzburgers supported the alternative position, 'intellectualism', and that they showed affinities with Aristotelian and scholastic doctrines. This gave him an additional impetus for attack.

Of key figures of the Würzburg school, Bühler, Marbe, and

Messer had all, at one stage or another, wanted to become neoscholastic philosophers of sorts. Nevertheless, the most fascinating link between Würzburg thought psychology and Catholicism is the Protestant Külpe. I suggest that Külpe accommodated to the Catholic environment in Würzburg and increasingly took up philosophical positions that – again by the standards of the time – were looked upon as Catholic. His opposition to voluntarism and various other Wundtian metaphysical and epistemological doctrines were all telling in this respect. Thought psychology was inseparable from Külpe's criticism of these doctrines. Because Külpe's attacks on Wundt's metaphysics and epistemology happened at the same time as the Würzburgers' assaults on Wundt's psychology, Wundt could not have missed the link between the two.

Assume that what I argue in Chapters 3 to 6 is true. In that case I have shown two things for one key episode in the history of psychology. I have confirmed the complex interaction between institutions of different kind and size (theories, experiments, university departments, the university itself, the state, the major confessions), and I have demonstrated the intricate ways in which individuals struggle over the establishment, the structure, and the survival of such institutions.

This study has its definite limits and limitations, and I might as well state them myself, before my critics get a chance to do so.

First, the picture I draw of the debate over thought psychology is static rather than dynamic; I do not describe in detail the development of the debate over time, and I do not analyse how the controversy was resolved or faded away. To do so would have meant writing a different book – or doubling the length of this one. Külpe died in 1915, Wundt in 1920, and the last-mentioned year provides the *terminus ad quem* for my study. The debate over thought psychology continued with a partly new set of antagonists. To deal adequately with their exchanges would have meant describing a plethora of further philosophical and psychological positions, and numerous further social conflicts and institutions of Germany and Austria between 1918 and 1933. This is best left for another author.

Second, for the purposes of my argument the differences between Göttingen, Leipzig, and Würzburg are more important than their similarities. But, no doubt, similarities abound. Of course all three institutes had numerous social arrangements in common, worked under similar academic and institutional conditions, and shared many theoretical assumptions. The similarities of the experimental designs of the three schools can be brought into sharp relief by comparison with, say, Charcot's or Galton's schools. The psychologists of all three schools were part of the *Bildungsbürger* elite, and the egalitarianism of the Würzburg school extended only to a tiny group of trained introspectionists. It would not be difficult to build a bridge from this 'egalitarianism' to the elitism of Müller and Wundt.[2]

Third, the analysis presented here could also be pushed further in still other directions. For instance, one might consider more of the research schools involved in the debate over thought psychology or analyse more social/cognitive variables. Thus one could study in detail, for instance, Edward Titchener's laboratory at Cornell, which might enable one to identify features of 'national styles'[3], Bühler's later institutes in Dresden and Vienna, or Stumpf's department in Berlin. Moreover, I have only glancingly addressed the larger institutional, social, and political frameworks in which these research schools were embedded.[4]

Fourth, and finally, I have resisted the temptation to burden this introduction with an extensive methodological chapter. I have written at length elsewhere on questions of historiography and methodological issues in the sociology of knowledge, and the curious reader is invited to turn to these sources (Kusch 1991, 1995a).[5] I also make some methodological remarks in the concluding chapter here (Chapter 7). This text, in any case, is meant to stand on its own feet.

1 The Würzburgers

Introduction

In this chapter I shall give a summary of the thought psychology of the broader Würzburg school between 1900 and 1920. Existing accounts of the Würzburg school (e.g. Humphrey 1963; Mandler and Mandler 1964) mention only the studies by Marbe (1901), Ach (1905), Watt (1905), Messer (1906), Bühler (1907b, 1908c, 1908d), and Selz (1913). I have no quarrel with this selection; these were indeed the studies that aroused the most interest, both within and outside the school. These texts form the backbone of my own overview, too. Nevertheless, I have aimed here for a more complete account of the Würzburgers' work. I thus have also included in my survey many Würzburg studies that did not attract wider attention.

It seems most natural to present the material under three main rubrics: the Würzburgers' new method, that is retrospective self-observation under experimental conditions; the Würzburgers' new conscious mental contents, that is 'situations of consciousness', 'awarenesses', and 'thoughts'; and finally their new 'laws of thought'.

Self-observation

The most important, and most controversial, methodological innovation of the Würzburg school concerned introspection. Already well before the turn of the century, German psychologists and philosophers had frequently debated the proper place of introspection within psychology. Amongst the most contested questions in these debates one must count the following:

- Is it possible to focus one's attention on one's internal mental states and processes *while these occur*, without altering their quality and intensity? In other words, is *simultaneous self-observation* possible and reliable?
- Is it possible to remember or retain one's internal mental states and processes such that these can become an object for reliable observation? That is, is *retrospection* possible and reliable?
- Is there a distinction between (internal) self-observation and self-perception, a distinction that corresponds to the distinction between the observation and perception of external objects and events?
- Can we distinguish between reliable and unreliable (or possible and impossible) forms of self-observation (or retrospection), depending on which types of mental processes or states are observed (e.g. 'lower' processes of sensing or feeling vs. 'higher' processes of thinking)?
- Is the ability to observe reliably one's own mental states and processes while these occur (or the ability to retrospect) (a) common or widespread among members of the human species, (b) dependent upon a special talent (e.g. an artistic talent), (c) dependent upon a special training (e.g. as a scientist, psychologist, experimental subject), (d) dependent upon induced and abnormal states of consciousness (e.g. hypnosis, drugs), (e) dependent upon a special pathological constitution (a split ego) or upon extreme conditions (long-term solitary confinement), (f) dependent upon the talent to empathise with the experiences of others?
- Is self-observation (or retrospection) (a) dangerous, (b) morally dubious, (c) properly conceptualised as an experiment within oneself, (d) properly conceptualised as a splitting of the ego?[1]

The Würzburg psychologists trusted self-observation. More precisely, they believed that a *retrospective self-observation* of one's own thought processes was possible. It was reliable, however, only in the presence of a second person, the experimenter. The experimenter would set a problem for the subject. The subject would try to solve this problem, and then report on what took place in his or her mind during this time.

All of the Würzburg thought psychologists used this method. Different studies employed it in different ways, however, and some of these ways were controversial even amongst the Würzburgers themselves. The three most influential methodological 'models' were those of K. Marbe (1901), N. Ach (1905), and K. Bühler (1907b, 1908c,1908d).

The Marbe model

Most writers at the time agreed that K. Marbe was the first to use and describe the method of retrospection under experimental conditions. In his book *Experimental-Psychological Studies of Judgement: An Introduction to Logic* (*Experimentell-psychologische Untersuchungen über das Urteil: Eine Einleitung in die Logik*, 1901), Marbe followed earlier authors (e.g. Brentano [1874] 1924; Wundt 1888a) in rejecting simultaneous self-observation. Simultaneous self-observation was unreliable because it was, by definition, '*attentive* inner perception'. Focusing one's attention on one's own conscious mental states, while these occurred, could change their quality or nature; for instance, by concentrating one's attention on one's terror, one tended to reduce the latter's intensity. Although Marbe did not put it in so many words, it is clear from his practice that, despite his scepticism *vis-à-vis attentive* inner perception, he trusted *non-attentive* inner perception. Non-attentive perceptions could be recalled after they had been made, and thereby turned into retrospective *observations* (1901: 1–2).

Marbe introduced the idea of a 'division of labour' between experimenter and introspecting experimental subject as a remedy against theoretical bias: 'The psychologist can avoid, or neutralise, this source of error by not relying on his own perceptions and observations. At least some of the needed perceptions should come from people other than the psychologist himself' (1901: 9).

The experiments reported in Marbe's book were meant to answer the following questions. Do all acts of forming (and understanding) judgements have one and the same characteristic phenomenological quality? Is there a specific 'quale' that distinguishes acts of judging from, say, feelings? Various writers at the time had suggested that such quale indeed existed (Brentano [1874] 1924: 266; Sigwart 1889: 25; Wundt 1893: 154; Ziehen 1900: 190). To test these claims, Marbe elicited judgements from his experimental subjects and then asked his subjects to report their conscious mental experiences during the process of forming these judgements. By 'judgement' Marbe here meant 'conscious processes to which the predicates "correct" (*richtig*) or "false" (*falsch*) can be applied in a meaningful way' (1901: 9–10). Marbe's seven 'observers' were professors, university lecturers, and school teachers. Marbe himself acted only as experimenter, never as experimental subject. Here is an example of Marbe's experiments, and of how he reported them:

> *Experimental conditions*: The experimental subject was asked to lift, one after the other, two cylindrical weights. The subject was supposed to judge which one was heavier, and then to turn the heavier one upside down. The two weights were of different weight (25 and 110 grams), but they looked the same.
> *Statement by observer Külpe*: Before turning over the weight, the acoustic word-picture 'turning over' appeared in consciousness. – The experimental subject did not observe any sensations, except sensations of pressure and kinaesthetic sensations.
>
> (1901: 17)

Although this was never made explicit in his 1901 publication, later discussions of his experiments by himself, his students, and his critics indicate that Marbe asked his subjects to report their mental experiences using a specific taxonomy. This taxonomy allowed for sensations, feelings, images, and 'situations of consciousness' (*Bewußtseinslagen*). It is also worth noting that Marbe's model did not include the posing of additional questions to the retrospecting subject (e.g. Marbe 1915: 19).

The Ach model

N. Ach's *On the Activity of the Will and on Thinking* (*Über die Willenstätigkeit und das Denken*, 1905) introduced several new ideas with respect to what he called 'systematic experimental self-observation' (1905: 8–25). First, he introduced the distinction between three periods within psychological experiments: the 'prior period', the 'main period', and the 'after period'. The prior period began from the moment when the subject was prepared – usually with a 'ready' or 'now' shout – for the imminent stimulus and ended with the occurrence of the stimulus. The main period consisted of the time period from the appearance of the stimulus until the subject's reaction, and the after period of the time immediately following that reaction. Outside of the whole experiment, more precisely before the prior period, the subject was given two instructions: the instruction on how to react to the stimulus, and the instruction to report experiences of the prior and main period during the after period. Ach believed that the second of these instructions strengthened the degree to which the experience of the prior and main period survived or 'persevered' during the after period (1905: 9).

The reliance upon 'perseveration' was the second new ingredient of Ach's theory. Here Ach built upon G. E. Müller and E. Pilzecker's work (Müller and Pilzecker 1900). These men had shown that attentively experienced mental contents tended to remain – that is, 'persevere' – in consciousness for some time even after attention had shifted elsewhere. Ach utilised this theory to argue that retrospection did not have to rely on the often unreliable processes of memory. Retrospection was based on perseveration, not on memory. Persevering mental states were of high 'clarity' and 'almost sensual vividness'. They presented the whole experience of the prior and main periods 'at once' and '*in nuce*'. Moreover, they could be studied and observed without attention interfering with the observed experience itself: 'vis-à-vis these persevering presentations, observation proceeds in the very same way in which it takes place with respect to an external event in nature' (1905: 10–12).

Third, because persevering mental states had this 'objective' character, it was possible for retrospectionists both to furnish '*full and complete*' analyses of their experiences and to answer further questions about these experiences put to them by the experimenter (1905: 14). Ach listed a number of such questions:

> The questions concerned the temporal order; thus, for instance, after the experimental subject had given his report, he was asked: what preceded this state of consciousness? What happened between these two events? Did they follow one another immediately? Was there any conscious relationship between them? The simultaneous content was discussed in a similar way; for instance, were the conscious events simultaneous? At which one of these events was attention directed? How was the process given in consciousness? What kinds of characteristics did the event have? Did feelings accompany the event? and so on. Is this event identical with an earlier one? Where do they differ? What were you conscious of at this point?
>
> (1905: 17)

Ach emphasised that in systematic experimental self-observation the experimenters were much more important than they had been in earlier forms of psychological experimentation. In the 'continuous close exchange of ideas' between experimenter and experimental subject, experimenters had to choose their questions carefully, check their subjects' claims, 'fully empathise with the mental state of the observed', avoid suggestion, display 'skill and tact', be ready to 'educate', and be 'devoted to the experience of the experimental subject in an unbiased but critical way'. In turn, the experimental subject had to check and approve the protocol and thus 'make sure that the experimenter has written down a faithful account of the experience'. To be able to control the experimenter properly, the experimental subject needed also to be a trained psychologist. Ach did not always follow his own prescriptions, however. One of his experimental subjects was hypnotised by Ach, and this man at least was 'a psychologically completely uneducated person' (1905: 9, 16–18, 23, 234).

Ach hoped that systematic experimental self-observation would become central to all areas of psychological enquiry:

> As this method will be developed further, the following often-heard objection will disappear; I mean the objection according to which there are areas to which the experimental method cannot be applied. It will become more and more obvious that *psychology* and *experimental psychology* are one and the same thing.
>
> (1905: 21)

In the experiments reported in his book, Ach used twelve subjects, including himself. While he listed his subjects by name early in his book, he did not ascribe introspective reports to his subjects by name. Instead, he assigned a letter to each subject, not telling his readers who was who (1905: VII). In this, as well as in the use of questioning, Ach had been preceded by O. Külpe who had used both procedures in a study on experimental aesthetics two year earlier (Külpe 1903). In another paper, on abstraction, Külpe had not listed the names of his subjects at all (Külpe 1904).

Ach applied the method of systematic experimental self-observation in the context of reaction experiments. Visual stimuli were presented to the subject by means of a 'card changer' (a device that Ach himself had developed and that soon would carry his name). The appearance of a card started the 'Hipp chronoscope', a clock that measured reaction times in milliseconds. The subject reacted either by lifting his or her fingers off a Morse key, or else by shouting a word into a 'voice key' (a tube with a metal plate inside; vibrations of that plate would arrest the clock) (1905: 25).[2] The card changer was controlled by strings extending from the experimental subject to the experimenter's desk. This desk was partially screened off so that the subject could not anticipate the changing of cards. Experimenter and experimental subject were situated in the same room, 2.5 metres apart from one another. Ach mentioned objections to this setup, but went on to suggest that the 'intimate connection' between experimenter and experimental subject made spatial proximity inevitable (1905: 26–27).

Ach reported a considerable number of reaction experiments. These fell into five main groups that can be summarised as in Table 1.1. With respect to each of the series of experiments, Ach listed the central time values for each of the participating subjects (not all of the twelve subjects were used for all of the series) and gave summaries of their self-observing reports for the prior and the main periods.

Table 1.1 The five main groups of reaction experiments in Ach (1905)

	Stimuli	*Required actions*
Series 1 Reactions with simple assignment	White cards	Lift finger off Morse key
Series 2 Reactions with twofold assignment	Cards with the letter 'E' and cards with the letter 'O'	Lift right finger for 'E' and lift left finger for 'O'
Series 3 Reactions with fourfold assignment	Cards with one of the letters 'h', 'd', 'b', or 'k'	Lift right thumb for 'h', lift left thumb for 'd', lift right index finger for 'b', and lift left index finger for 'k'
Series 4 Reactions without assignment of the stimulus	Cards with the letter combination 'rx' and cards with the letter combination 'xr'	Pick one of the following: lift right thumb for 'x', or lift left thumb for 'r'
Series 5 Reactions without assignment of the action	Cards with two numbers, e.g. 2|3, 5|4	Pick one of the following: do nothing, add, subtract, multiply, or divide

The Bühler model

K. Bühler's main thought-psychological work was contained in a three-part paper entitled 'Facts and Problems for a Psychology of Thought Processes' (*Tatsachen und Probleme zu einer Psychologie des Denkenvorgänge,* 1907b, 1908c, 1908d).

Bühler was very eager 'to do everything that will free the experimental subject from the feeling of being one'. Thus Morse key, voice key, Hipp chronoscope, and card changer had to go. Instead, the stimulus was presented verbally by the experimenter, who also operated a stopwatch (accurate to one-fifth of a second) by hand. The subject was asked to listen with closed eyes to avoid being distracted by the experimenter (1907b: 302).

Moreover, Bühler maintained that thought processes could only be observed properly when they occurred slowly and consciously. This in turn meant that the tasks had to be challenging and difficult – subjects would solve easy tasks in a fast, mechanised, and unobservable way. In Bühler's experiments, the experimenter addressed 'thoughts' to his subject to stimulate thoughts – that is, comprehension, comment, or recall – in the subject. Thereafter, the subjects were asked to furnish an 'as-accurate-as-possible account of what [they] had experienced during the time of the experiment'. Openly contradicting Ach, Bühler called the demand for 'complete reports' 'unreasonable' (1907b: 305, 307).

Bühler put much emphasis on the proper selection of tasks for his subjects. Not every subject could be provoked into thinking by the same problem. The experimenter therefore had to react 'lovingly' (*liebevoll*) to differences in tastes (1907b: 313).

Empathy was called for also in other phases of the experimenter's work. An experimental setting was 'natural' only if subjects were permitted to speak about their minds in their own words. To make sure that the subject had understood properly, the experimenter was allowed to resort to the occasional question. Much more important, however, was the experimenter's ability to empathise with subjects: 'He must empathise with their situation…and he must talk to them in their own language. That gives this co-operation a curious, confidential quality'. Empathy was also needed at a later stage, when the experimenter interpreted the protocols (1907b: 308–9).

Bühler also considered the question of whether self-observation under his conditions was reliable and 'objective'. To begin with, there was at present no good reason to distrust self-observation. Only by using it could one discover its strengths and limits. Moreover, there were at least three ways to safeguard or check the results of self-observation. First, one had to use very experienced subjects. Bühler relied almost exclusively on 'professor Külpe and private lecturer (and now professor) Dr Dürr'. Bühler thanked them both for their lectures and conversations and for their role as experimental subjects. He related that some theoretical parts of his paper were no more than edited summaries of their reports, and he praised the 'ideal beautiful co-operation with both men, a co-operation in which each gave what he could, and all were happy when one had advanced a step'. Second, one could check the reports for internal consistency. And third, and most importantly, Bühler believed that self-observational reports of recent thought processes could also be tested against memories of these same thought processes at a later point in time (1907b: 300, 306–7).

Bühler conducted altogether six experimental series. In all six series, the subjects gave self-observational reports after they had completed their task. Suffice it here to mention just two of these series. In the first, subjects were confronted with yes-or-no questions such as: 'Do you know what Eucken means when he speaks of a world-historical apperception?' (1907b: 304). Or, in the third series, the experimenter read out twenty pairs of thoughts (e.g. 'The noble power of thought – The image of Kant'). After each pair, the subjects indicated that they had understood. Later, after a short break, the first elements were read out again, and the subject had to reproduce the corresponding second elements (1908d: 26–29).

Later developments

Other Würzburg studies either followed one of these three models or else combined elements from two or all three. Furthermore, several Würzburgers commented critically on the different models. For instance, Marbe's students, C. O. Taylor (1905/6), M. Beer (1910), G. von Wartensleben (1910), and A. Feuchtwanger (1911), all followed their teacher in providing their experimental subjects with a ready taxonomy in which to describe their experiences, and in not posing additional questions about details of these experiences. Most of the other Würzburg psychologists did not accept either restriction, however. For instance, as just seen, Bühler opposed the prescribed vocabulary on the grounds that it conflicted with the demands of a natural setting. Külpe applauded Bühler's view and noted that – with the help of their experimental subjects – psychologists were relearning to 'speak in the language of real life' (1912c: 1077). O. Selz added that only unconstrained descriptions would be theoretically neutral (1913: 15), and T. Haering added that conceptual 'freedom' alone could lead to the discovery of altogether new phenomena (1913: 317). Finally, J. Lindworsky submitted that working with a fixed set of categories committed one to an atomistic conception of mental life (1916: 7).

When Marbe advocated the fixed vocabulary, and opposed what he called 'free narratives', he too invoked the topoi of theoretical neutrality and of the independence of the experimental subject. If subjects were permitted to use their own language, then it was left to the experimenter to translate their various and divergent descriptions into one common theoretical framework. And over this work of construction the experimental subject had no control. Moreover, Bühler's insistence on the need for 'empathy' was in fact a damning admission. As Marbe saw it, empathy had no place in a proper science (1915: 19).

Marbe was equally hostile towards the 'questioning method'. He demanded that 'all of the experimenter's questions must be couched in the form of new experiments'. Finding the right kind of question might be a form of 'art', as one Würzburger had informally suggested to him, but then again, psychology was a science not an art (1915: 23). Other Würzburgers defended questioning (Schultze 1906; Dürr 1908c; Ach 1910: 8; Selz 1913: 14–15; Lindworsky 1916: 7). For example, F. E. O. Schultze wrote that there was no justification for any '*horror questionis*'; to ask questions was necessary because experimental subjects tended to overlook important features of their experiences, were prone to invent or censor, and were often the victims of fatigue and boredom. All of these factors could be countered with a sophisticated questioning strategy. Moreover, questioning did not influence what subjects experienced; it only affected what they reported about their experiences. Schultze exonerated the method also by pointing out that it had always been used by psychiatrists (1906: 251–2, 255).

Another issue much debated within the Würzburg school was the opposition 'sophisticated instrumentation' versus 'need for a natural setting'. Ach had introduced the voice key and Hipp chronoscope into thought psychology, and H. J. Watt (1905) and A. Messer (1906) had used an almost identical setting. Whether they followed him in this respect, or discovered the same tools independently, was itself a matter of a little priority dispute (Watt 1906a: 83; Messer 1906: 11; Ach 1910: 9). Here it is more important to note that this high-tech machinery could no longer be found in later studies. If these employed a clock at all, it was a stopwatch, operated by hand. Schultze, Bühler, and others aimed for 'natural' experimental settings, settings that were close to life outside of the psychological laboratory. Ach countered by denying that Bühler's interactions with his subjects amounted to real experiments (1910: 17). This charge was taken up in a nasty public exchange between Ach and Selz in the *Zeitschrift für Psychologie* (Ach 1911; Selz 1910, 1911).

Ach was unhappy with Watt's study not only because of the issue of priority; he also rejected the method of 'fractioning the content of consciousness' introduced by Watt upon Külpe's suggestion. In different experiments, Watt asked his subjects to focus, and report, on different phases of the experiment, that is 'the preparation', 'the appearance of the stimulus word', 'the search for the reaction word (if such a search took place at all)', and 'the appearance of the reaction word' (Watt 1905: 316). Ach opposed this with the following argument: 'the parts of a psychological process...are comprehensible only in their relation to the whole. Separated from this relation to the whole, they lose their meaning' (1910: 11). Although Ach did not make it explicit, the same reasoning also applied to Messer's 'interruption' method. To find out how his subjects prepared themselves for the reaction, Messer occasionally aborted the experiment immediately before the appearance of the stimulus and asked what his subjects had experienced up to this point (1906: 7).

Other methodological ideas and ways of reporting were less controversial. Thus there was no discussion over Ach's (1905) and A. Grünbaum's (1908) use of hypnosis; no debate over the sometimes-seen practice of authors acting as both experimenters and experimental subjects; and no objections to Schultze's advocacy of 'occasional systematic self-observation'. Submitting that 'experiences within experiments are rather odd precisely because these experiences are artificially isolated', Schultze called for a recognition of the '*experimentum naturae*'; that is, of the experience of everyday life:

> one gets much closer to the experiences of everyday life if one observes them *during that everyday life*. These experiences should be analysed and described *immediately* – either in a notebook, or in a pocket-size protocol book. Anyone who rejects this method must also reject most of the observations made by psychiatrists in asylums.
>
> (1906: 249–50)

Most Würzburgers picked up Külpe's and Ach's practice of concealing the identity of the authors of introspective reports. This practice was justified by Messer on the grounds that one's reactions in thought experiments allowed others to draw inferences about one's dispositions and character (1906: 33, 43). Later, E. Westphal defended anonymity as a safeguard against false or exaggerated self-attributions coming from vain experimental subjects (1911: 434).

Finally, looking at the whole series of Würzburg studies from 1901 until 1918, one notices that the number of experimental subjects per study increased continuously: The average number of experimental subjects per study between 1901 and 1908 was 5.8, whereas between 1909 and 1918 it was 11.4.[3] At the same time, the population of subjects became more varied: Initially they were predominantly teachers of the Würzburg Institute; later they included university teachers of different scientific disciplines, students, school teachers, army officers, technicians, as well as university and high-school students. As the population became more varied, and less trained in psychological matters, writers within the Würzburg school became increasingly interested in questions of how to train experimental subjects, of how to select good and reliable subjects, and of how to check their reports. Although Westphal (1911) was the first to publish rules on these matters, he acknowledged that his source here was Külpe's lectures. In his major programmatic paper, 'On the Methods of Psychological Research' (*Über die Methoden der psychologischen Forschung*, 1914), Külpe insisted that a battery of checks and tests were needed to control the experimental subjects and their reporting. The experimenter had to check subjects' ability to observe, their goodwill, their moods and dispositions, their knowledge and their level-headedness. He also had to reckon with the possibility that some people were better in focusing on external rather than internal objects, that some had no sense for nuances, and that others were insecure and prone to utter contradictions. Moreover, Külpe advocated four means by which introspective reports were to be checked. First, the experimenter was to consider whether the reaction of the subject was plausible given the task or stimulus; that is, the experimenter was to judge the subject's reaction by the experimenter's expectations. Second, the subject's reports of his or her emotional state had to be checked by measuring the pulse and breathing rate. Third, reports by one and the same subject, given at different times, had to be compared. And fourth, reports by different subjects had to be measured against one another. Furthermore, every unusual or unexpected report was to be treated with the suspicion that the subject might be lying, and the experimenter was to be prepared, at all times, for the need to educate the experimental subject in the importance of 'duty, honesty, sincerity, and objectivity'. Külpe also advocated – openly contradicting Wundt – that the objectivity of the experimental subject could be improved by the subject's ignorance of the overall purpose of the experiment (1914: 1065–70, 1222).

Situation of consciousness, awareness, thought

The Würzburgers did not just perceive themselves as innovators in the realm of psychological experimentation; they also saw themselves as discoverers of new elements and laws of the mind. I shall explain the new elements in this section and leave the new laws until the next.

When they spoke of 'new' elements, the Würzburg psychologists referred to elements not already contained in what they regarded as the received taxonomy of mental experiences. In the Würzburgers' understanding of this received taxonomy, states and events of inner experience fell into three categories. There were sensations (*Empfindungen*), feelings (*Gefühle*), and 'presentations' (*Vorstellungen*). (Sometimes volitions (*Willensakte*) were treated as a fourth category.) By 'presentations' the Würzburgers meant optical, acoustical, and tactile 'depictive' (*anschaulich*) 'images' (*Bilder*), produced by the brain without immediate sensory input. Memory images and fantasy images were the paradigm cases of presentations.

Situation of consciousness

A first new element with respect to this classification was suggested in Marbe (1901), and in a paper by two of his students, A. Mayer and J. Orth (1901). This new category was the '*Bewußtseinslage*', a term that has been rendered into English as 'posture or attitude of consciousness' (Titchener 1909: 100), 'state of consciousness' (Humphrey 1963: 33), or 'disposition of consciousness' (Mandler and Mandler 1964: 142). I prefer the literal translation 'situation of consciousness'.

Mayer and Orth were investigating association. In their experiments, the experimenter called out words to the experimental subject, and the latter was supposed to react with the first word that came to mind. Afterwards the subjects reported on their mental states during the time period between the calling out of the stimulus and their reaction. Under these conditions, the subjects 'often reported experiencing certain conscious events that they clearly felt unable to speak of as either presentations or volitions' (Mayer and Orth 1901: 5). Following a suggestion made to them by Marbe, Mayer and Orth called these 'certain conscious events' 'situations of consciousness'. Sometimes, but not always, subjects could specify a content for these situations of consciousness, and like presentations or volitions, situations of consciousness could be more or less emotionally charged (1901: 6).

Marbe (1901) gave essentially the same account of situations of consciousness:

> Self-observation occasionally encounters certain facts of consciousness the content of which either cannot be characterised at all, or else can be characterised only with great difficulty. These experiences...can appear with or without an emotional intensity. We...shall refer to them as *situations of consciousness*.
>
> (1901: 11–12)

'Situation of consciousness' was one of the categories of the taxonomy used by Marbe's subjects, who all reported instances of this category. Here is an example:

> *Experimental conditions*: The observer was given a thermometer.
> *Sentence shouted out*: What temperature does the thermometer show?
> *Answer of observer Külpe*: The thermometer shows 13 1/4 degrees.
> *Related statement*: The sight of the thermometer immediately triggered the judgement '13 1/4 degrees'. The end of the sentence was only uttered because of a situation of consciousness; the observer characterised this situation of consciousness as a memory. The content of this memory was the instruction to answer in full sentences.
>
> (1901: 37)

Situations of consciousness had various specifiable contents; there were situations of consciousness of coercion, tension, recognition, doubt, ignorance, expectation, contrast, and agreement, amongst others (Marbe 1901: *passim*; cf. Marbe 1915: 31).

Two years after its introduction, the situation of consciousness appeared in a book title, that is in the title of J. Orth's doctoral dissertation *Feeling and Situation of Consciousness* (*Gefühl und Bewußtseinslage*, 1903). As Orth explained halfway through his book, the relationship between feeling and situation of consciousness needed to be investigated because one influential psychologist, Wilhelm Wundt, seemed to subsume situations of consciousness under the category 'feeling' (1903: 75). Accordingly, much of Orth's book was a criticism of Wundt's concept of feeling. Orth's main charge was that Wundt had inflated this concept out of all meaningful proportion:

> Wundt adopts not only the conventional division of sensory versus higher, that is, aesthetic, logical, ethical, and religious feelings; he also speaks of the feelings of activity, of suffering, of knowledge, of recognising, of concept, of doubt, of succeeding and failing, of expecting and resisting, and of much else....One gets here the impression that everything that Wundt is unable to determine and analyse more closely is simply called a 'feeling', and that mere words make up for the task of penetrating carefully into the objects of self-observation.
>
> (1903: 55–6)

As a remedy, Orth recommended restricting feelings to the realm of pleasure and pain. This move made room for 'situation of consciousness' as a distinct category. Orth also sought to relate the situation of consciousness to more established concepts. For instance, he maintained that situations of consciousness were similar to William James's 'fringes of consciousness'; both were phenomena of conscious experience that defied detailed specification (1903: 69; cf. James [1890] 1983: 249, 446–7).

Two groups of situations of consciousness could roughly be distinguished. A smaller group consisted of those situations of consciousness 'that can only be noticed to be occurring but that cannot be characterised in any way'. The larger group consisted of those that have a certain meaning for mental life, a meaning that we are able to state more or less clearly. Drawing on the material of Marbe's (1901) book, Orth showed that such situations of consciousness could be those of doubt, of certainty, of uncertainty, of contrast, or of consent. In principle the meaning and function of any other conscious element could also be carried by a situation of consciousness (1903: 70–2).

The introduction of situations of consciousness into psychology was, in Orth's view, an important step forward. This was the case not least because situations of consciousness included those stages of conscious mental life when no depictive elements (no sensations, feelings, or presentations) were observable. Again Orth made this point in opposition to Wundt. Wundt had claimed that 'feeling is the pioneer of knowledge', meaning that conscious awareness of knowledge was usually preceded by specific feelings. Orth insisted that Wundt's phrase be changed to 'situations of consciousness often are the pioneers of knowledge' (Orth 1903: 73).

Once situations of consciousness had been introduced by Marbe and explained in more detail by Orth, they made a frequent appearance in the introspective reports of subsequent Würzburg studies (e.g. Nanu 1904; Taylor 1905/6; Watt 1905; Wertheimer 1906; Messer 1906; Bühler 1907b, 1908c, 1908d; Beer 1910; Wartensleben 1910; Feuchtwanger 1911; Koffka 1912; Marbe 1913b). From around 1905 onwards, however, situations of consciousness had to face competition from within the Würzburg school. Ach introduced the new category of 'awareness', and a number of authors insisted on a place for 'thoughts'. Only Marbe and his students (Beer, Feuchtwanger, Wartensleben) denied the need for these further categories.

Awareness

Ach (1905) introduced 'awareness' (*Bewußtheit*) as a category broader than Marbe's situation of consciousness (Ach 1905: 236–7). 'Awareness' was needed because all of Ach's experimental subjects reported conscious experiences of the following kind. First, during these experiences subjects had an 'immediate knowledge' of their situation, of their intentions, of some meaning, or of their past or future actions. Second, neither this knowledge nor its contents consisted of sensations, presentations, or feelings. Third, the content of such immediate knowledge could be complex and structured into distinct parts. Fourth, different such immediate knowledges [*sic*!] had different degrees of intensity; even the different parts of one and the same awareness could differ in intensity. Usually the felt intensity of an awareness decreased as this awareness recurred over time. Fifth, the parts of an awareness could initially appear as sensations, presentations, or feelings. But they would quickly lose this vivid or depictive (*anschaulich*) character. Sixth, and finally, awarenesses would usually be accompanied by bodily sensations of muscular tensions, by visual, acoustic, and kinaesthetic sensations, or by memory images (1905: 210–14).

To account for these phenomena, Ach used the theoretical machinery of associative tendencies. His basic idea can perhaps best be explained by using one of his own examples, and by translating his difficult prose into a pictorial representation (1905: 216–23). This example concerned the most important case of awareness, the 'awareness of meaning' (*Bewußtheit der Bedeutung*) (Figure 1.1).

Assume that the subject is confronted with a card with the letters 'b–e–l–l'. Under normal conditions, the sensations caused by these letters will immediately be interpreted or 'apperceived' as the word 'bell'. What happens here, according to Ach, is that the visual image of the written word 'bell' mobilises or excites, below the threshold of consciousness, an associative network of presentations. The visual image of the written word forms something like the core of this network. Other elements of this network are presentations of the acoustic image of the spoken word 'bell', of the acoustic image of the peal of bells, and of the visual image of the typical shape of a bell. These presentations will usually differ with respect to how often they have been conscious before and thus also differ with respect to the strength of their tendency

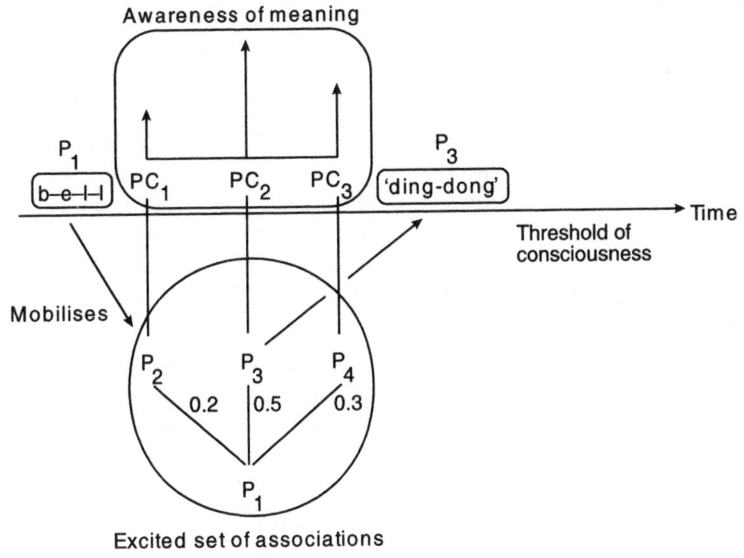

P_1 = visual image of the written word, b–e–l–l
P_2 = acoustic image of the spoken word, 'bell'
P_3 = acoustic image of the peal of bells
P_4 = visual image of the shape of bells

Figure 1.1 A pictorial representation of Ach's theory of awareness

to cross the threshold of consciousness. Roughly, the more often they have been there before, the greater their chance of doing it again. (In Figure 1.1 I have indicated these differences in strength by assigning fictitious numerical values to the three presentations.) Now, for the subject to become *aware* of the meaning of 'b–e–l–l' it is not necessary for any one of these excited presentations to cross the threshold of consciousness. Instead what happens, according to Ach, is that the mere excitation of the associations around the visual image of the word will bring about an awareness of the meaning of the word. Moreover, this awareness is complex and structured, in that it consists of several partial contents, each of which is caused by one of the presentations in the excited set of associations. Furthermore, each of the partial contents has a specific degree of intensity that is a function of the strength of reproductive tendency of the respective presentation.

(I have indicated the degrees of intensity with arrows of uneven length.) Finally, Ach suggested that the presentation with the highest strength of reproductive tendency often subsequently becomes conscious (1905: 216–23).

Ach asserted that an awareness of meaning was almost always accompanied by sensations, or by memories of sensations. Indeed, actual and remembered sensations typically functioned as 'symbols' of, and triggers for, awarenesses of meaning. But Ach stressed that these sensations were not part of the meaning itself and thus merely accidentally related to the given awareness: 'What is essential is the meaning, not the sign'. In the same vein, Ach gave short shrift to the notion that thinking takes place in the form of inner speech. Again, words and phrases were only accidentally connected with awarenesses of meaning (1905: 215, 222–3).

Finally, Ach insisted that the awareness *of meaning* was not the only kind of awareness that could be identified. There were also the 'awareness *of determination*', the 'awareness *of tendency*', and the 'awareness *of relation*'. The awareness of determination was an immediate experience of whether or not a current action happened in accordance with an earlier decision or command. The awareness of tendency was a special case of the awareness of determination. Subjects would couch it in the following terms: 'I know that I still want to do something, or that something is still missing...it's at the tip of my tongue'. And finally, awareness of relation was an awareness of a link between a current and an earlier content. This link could be, for instance, disappointment, doubt, familiarity, or certainty of memory. Ach alleged that Marbe's situation of consciousness was just a special case of the awareness of relation (1905: 231, 233, 236–7).

On the road to thoughts

The third main category of Würzburg thought psychology, 'thought' (*Gedanke*), was introduced and developed gradually from 1902 onwards. First, in a criticism of Mach's sensualistic epistemology, Külpe insisted that a reduction of knowledge to sensations was impossible and that 'thoughts' both had an 'independent meaning' and were governed by 'independent laws'.[4] What exactly he meant by 'thoughts', however, Külpe did not make clear (Külpe 1902: 17–21). Second, in 1905, Taylor noted that his subjects spontaneously 'spoke of "thoughts" when they meant complicated combinations of situations of consciousness and word-presentations, combinations that could not be analysed further' (1905/6: 234–5).

Third, in 1906 Messer went further by submitting that situations of consciousness could best be relabelled as 'thoughts'. As Messer saw it, situations of consciousness included Ach's awarenesses as a special case (1906: 175), and they deserved a proper taxonomy. Such classification had to be based on formal (i.e. logical) features of the content of situations of consciousness because simply *any* content of thought could become conscious in the form of a situation of consciousness. And thus Messer listed situations of consciousness 'of reality', 'of spatial relations', 'of temporal properties and relations', 'of causal relations', 'of teleological relations', 'of logical relations' (e.g. identity, subordination, part and whole), 'of familiarity', 'of unfamiliarity', 'of fit', 'of meaningfulness', 'of correctness', 'of searching', 'of questioning', 'of deliberation', 'of doubt', 'of certainty and uncertainty', 'of difficulty and ease', 'of compulsion', 'of permittedness and forbiddenness', 'of readiness', 'of ability and disability', 'of success and failure', 'of being overwhelmed (by presentations)', 'of emptiness', and 'of helplessness and confusion' (1906: 181–4).

Having first distinguished all these various types of situations of consciousness, Messer then declared the whole term superfluous. In its place he suggested a term used for a different purpose by the logician and psychologist B. Erdmann, that is the term 'unformulated, intuitive thinking', or simply 'thoughts' (*Gedanken*) (cf. Erdmann 1900). The latter term fitted ordinary language much better than did Marbe's neologism. For Messer one of the prime instances of such thoughts was the 'consciousness of sphere' (*Sphärenbewußtsein*); this was the usually non-depictive representation of word meaning (1906: 79, 188).

Fourth, in the same year, 1906, Schultze suggested that all contents of consciousness fell into two categories: 'appearances' (*Erscheinungen*) and 'thoughts'. Because one was dealing here with ultimate elements, neither of the two categories could be defined precisely. Appearances comprised the area of sensations, presentations, and feelings; in short, appearances covered those conscious elements that had a depictive quality, and that could be localised. Thoughts lacked both of these characteristics. One hit upon thoughts 'once one has experienced a situation where one has to say, "I have experienced something" but is unable to find anything that can be classified as an appearance'. Schultze regarded Ach, Binet, Brentano, Husserl, Meinong, and Messer as his forerunners in the study of thoughts (cf. Binet 1903; Husserl [1901] 1984; Meinong 1902, [1904] 1913). Although Schultze insisted that there could be thoughts without appearances, he denied that there could ever be appearances without thought: 'all appearances (presentations, feelings, contents of perceptions) are linked to thoughts, indeed...appearances are besouled by thoughts'. Thoughts provided the meaning and the structure for appearances. Such structuring thoughts could not usually be identified introspectively, however: 'we "think" in perception, too – but not in the sense that we experience "appearances" of thought-activity, but in the sense that the factors of this activity work in us without us actively noticing and feeling it'. Here it was most important not to conflate thoughts with either feelings or situations of consciousness. One could easily convince oneself of the difference between thoughts and feelings by noting that one could have thoughts about the beautiful without having any feeling of pleasure, or that one could have a feeling without a clear meaning. Nor should thoughts be mixed up with 'germ- or cripple-forms of appearances' (*Keim- oder Krüppelformen*). These were pale appearances that occasionally could seem to be free of any image-like quality. Marbe's situations of consciousness belonged in this realm (1906: 258–9, 266, 279, 289, 297, 301, 306, 309).

Fifth, and finally, in a longish review of Ach (1905), Külpe made a number of influential criticisms and proposals (Külpe 1907b). To begin with, Külpe submitted that Marbe's situations of consciousness and Ach's awarenesses were but different names for the same phenomenon, that is for non-depictive (*unanschaulich*) knowledge. Neither concept, however, captured this phenomenon adequately. Situations of consciousness were defined primarily negatively, that is as those mental contents that could not be subsumed under feelings, sensations, or presentations. Ach had gone further by defining awareness in terms of non-depictive knowledge. But, as Külpe pointed out, in more than one place Ach had remained tied to the received view according to which only feelings, sensations, and presentations existed fully and really. For instance, Ach had lamented that awarenesses were difficult to grasp because they lacked depictive or phenomenological features. This showed, in Külpe's view, that Ach continued to regard the traditional threesome as the paradigmatic mental contents. Moreover, Ach had tried to show that awarenesses could have different degrees of intensities – this for Külpe was a faulty assimilation of non-depictive thoughts to sensations. And finally, Ach's theory of awarenesses reduced the latter to mere epiphenomena of the interaction between unconscious presentations (1907b: 602–4).

Progress could only be made, according to Külpe, if thought was acknowledged as a phenomenon in its own right, irreducible to other mental contents. Put more strongly, it had to be acknowledged that thought was not content at all. In addition to studying contents of consciousness, one had to follow Brentano's lead and attend to 'functions' (*Funktionen*) or 'acts', of consciousness.[5] Such functions were not introspectively accessible in the same way as contents were; indeed, they could only be studied through their results. But only an investigation into function would eventually unlock the door to the nature of thinking (1907b: 603).

Bühler's thoughts

'Thoughts' as defined in Bühler's three-part study (1907b, 1908c, 1908d) clearly fulfilled some of the demands that Külpe had formulated. Bühler's starting point was his observation that his subjects' 'experiences of thinking' (*Denkerlebnisse*) consisted of more than just sensations, presentations, feelings, and situations of consciousness (according to Bühler, situations of consciousness were non-depictive and non-intentional states of consciousness). Bühler's subjects also reported other important 'pieces of their [conscious] experience'. These latter pieces had no sensory qualities or intensities, even though they had degrees of clarity, certainty, and liveliness. His subjects referred to these pieces as 'awarenesses', as 'knowledge', and most often, as 'thoughts'. Bühler of course adopted this last-mentioned term. Other than Ach's awarenesses, these thoughts usually occurred without *any* – even without accidental – concomitant depictive contents. And unlike Marbe's situations of consciousness, they were intentional experiences (1907b: 315–16).

Bühler saw these thoughts as the most essential parts of thinking *qua* problem solving. It was obvious that the most essential parts of problem solving could not be presentations: 'Something which is as fragmentary, as sporadic, as chance-like in its appearance in consciousness as are presentations in our thinking experiences, [something like that] cannot possibly be the bearer of the tightly knit and continuing content of thinking'.

Here Bühler chastised Wundt in particular for having postulated that every act of thinking must be represented in consciousness via presentations or feelings. According to Bühler, such opinion was the natural product of bad self-observation, of 'writing table experiments' outside the laboratory, and of 'psychological sensualism'. This influential line of theorising contained the false belief that all contents of consciousness were reducible to sensations. Nor was it correct to assume that thought was carried by optic, acoustic, or motoric word presentations. Word presentations were not essential to thinking. Indeed, the 'deeper' the thinking, the less it relied on words (1907b: 317–18, 320, 323).

Bühler also rejected the notion that thoughts might be 'compressed presentations' of sorts – some pronouncements by Marbe (1901) and Messer (1906) showed that they allegedly had taken this route. And he discarded Ach's theory according to which thoughts, or rather awarenesses, were the conscious reflections of the interactions amongst unconscious presentations. Against the 'compression theory' Bühler argued, for example, that if thoughts really were compressed presentations then they ought to have properties similar to 'normal' presentations. But they did not: it made no sense to speak of the sensory intensity of thoughts. This same argument worked against Ach's reflection theory, too. Additionally, Bühler insisted against Ach that whatever the unconscious causal genesis of thoughts, they should be analysed as conscious phenomena in their own right (1907b: 328–9).

Having shown the insufficiency of alternatives, Bühler proceeded to develop his own theory. To begin with, he defined thoughts as 'the ultimate experienced constituents of our experiences of thinking'. Their further analysis could no longer identify 'pieces' (*Stücke*), but only 'moments' (*Momente*). Pieces here referred to parts that could exist independently of the whole to which they belonged, moments to parts that could not. Bühler borrowed this terminology from the third of Husserl's *Logical Investigations* (*Logische Untersuchungen*, Husserl [1901] 1984). In Bühler's view three such *moments* of thoughts could be identified by focusing on three different *types* of thoughts; in each of these different *types* of thoughts, one of the three moments of thought figured prominently (1907b: 329).

The first of these was the 'consciousness of rules' (*Regelbewusstsein*). Subjects would often report that while trying to solve difficult tasks they would suddenly become aware of 'a knowledge of how to solve such problems'. A paradigmatic case of such 'consciousness of rules' was the 'familiar theorems' that one used in mathematical proofs. The second type of thought was the 'consciousness of relation' (*Beziehungsbewußtsein*); the relation in question could be, for instance, 'opposition' or 'contrast' between two events or objects. And third, there were 'intentions' or, as Husserl had called them, 'purely signitive acts'. These were thoughts *about* some objects or events where those objects or events were 'meant' but not represented in consciousness. For instance, in one of Bühler's experiments, Külpe reported that he 'thought of ancient scepticism...and included in this was a lot; in my consciousness, I somehow had present the whole development of the three periods [of ancient scepticism]' (1907b: 334, 341, 345, 346–7).

Thus, rule, relation, and intention were the three moments in every thought. Every thought had the moment of *intention* because every thought had to be about something or other. Moreover, for the intention to have a content and target, it needed to stand in a *relation* to 'characteristics' (*Wasbestimmtheiten*) of that target. And finally, the relation between a given set of characteristics needed to be rule governed. For instance, Bühler's thought of Külpe was directed, that is it had an intention; what specified this intention were various characteristics by means of which Külpe was identified; and those characteristics were rule bound in that they had to form a coherent whole (1907b: 350).

By 'characteristics' Bühler did not mean sensations and presentations. *What a thought was about* (i.e. its characteristics) could be fully and clearly determined *without* any kind of sensation or presentation figuring in consciousness (1907b: 354). Nor could these characteristics be explained in terms of 'transcendent', real, consciousness-independent objects, or states of the nervous systems. Such explanatory moves were unlicensed because psychology had to remain within the realm of consciousness (1907b: 357). The correct view, Bühler submitted, lay in the direction of Messer's 'consciousness of a sphere': the characterising features of acts of thinking were features that determined the position of the content of the act within a structure of knowledge. To know what a thought was about was to know how the thought's object related to other objects of knowledge, as well as its position in a network of knowledge (1907b: 358).

Later developments

Bühler's thoughts were taken up, and developed further, by students and collaborators of Külpe and Bühler. For instance, F. Hacker relied on Bühler's theory in a study of dreams (1911); B. Schanoff (1911) in an investigation into mathematical calculation; K. Koffka (1912) in a book on presentations; O. Selz (1913) in his work on 'actualisations of knowledge'; L. Rangette (1916) in his enquiry into forms of thinking in mathematics, history, philosophy, and German philology; J. Lindworsky (1916) in his analysis of inferences; and C. Bühler (1918, 1919) in her papers on the genesis of thoughts and on sentence formation. Several of these studies put increasing emphasis on the idea that thoughts and thought moments were essential even within presentations. That is, thoughts were no longer just a fourth, or fifth, category *alongside* others; they were something like the deep structure of all conscious contents (e.g. Koffka 1912: 365).

Külpe himself praised Bühler's work repeatedly (1912b, 1912c, 1914). He also reminded his readers, however, that despite Bühler's important work, there still remained one important lacuna: while Bühler had admirably studied thoughts *qua contents*, no one had as yet tackled the more difficult task of investigating thoughts *qua functions or acts* (1912b: 197, 257).[6]

But Bühler's thoughts also came in for criticism from within the wider community of former collaborators and students of Külpe. E. Dürr (1908b, 1908c) – who had in fact been one of Bühler's two main experimental subjects – raised a number of objections. Of these, the 'reference (*Kundgabe*) error' accusation gained widespread attention, not least because the same point was also made by critics from outside the school. Dürr noted that most reports of thoughts – provided by Bühler's subjects and used by Bühler for his analysis – did not *describe* thoughts psychologically introspectively; that is, they did not describe thoughts as *modes* or *modifications of consciousness*. Instead such reports mentioned and described the *object* or state of affairs that the thought was about. If that was true, however, then Bühler's moments and types of thoughts were not moments and types of psychological entities but moments and types of states of affairs.

Marbe (1915) repudiated all claims according to which situations of consciousness needed to be complemented by, or replaced with, awarenesses or thoughts. As Marbe saw it, whatever was important and noteworthy about awarenesses and thoughts had already been captured by his category. In particular, he rejected Bühler's distinction between intentional thoughts and non-intentional situations of consciousness. After all, doubt was a situation of consciousness even according to Bühler, and doubt was undoubtedly intentional (1915: 38). Moreover, Marbe returned to a point he had already made in his 1901 book. There he had argued that knowledge was a dispositional concept: To have knowledge of something meant to be able to produce true judgements about this something. But knowledge itself could never be given in consciousness (1915: 32; 1901: 92). In Marbe's view this observation pulled the rug out from under both Ach's awarenesses and Bühler's thoughts: both Ach and Bühler claimed that awarenesses and thoughts respectively were states of a non-depictive, immediate knowing. There was no such knowing.

Psychologists close to Bühler were not convinced by either criticism. Against Dürr (and Titchener (1909), who had raised the same objection), Külpe argued that it was of the very essence of thoughts to be a knowledge of objects. In committing the alleged reference error or 'stimulus error' (as Titchener called it), Bühler had only been true to the nature of thought (1912b: 255). And against Marbe's argument concerning the dispositional nature of knowledge, Selz replied that one needed to distinguish between 'dispositional and actual knowledge....Dispositional knowledge can indeed never be given in consciousness. But dispositional knowledge can become actual, and this actual knowledge can sometimes be shown to exist in consciousness'. Ach's awarenesses and Bühler's thoughts were cases of such actualisations of knowledge (Selz 1913: 319).

The laws of thought

The Würzburgers were not just dissatisfied with the received psychological methods and the inherited taxonomy of mental entities. They were also highly critical of traditional associationist interpretations of 'laws of thought'. In their view, associations alone could account for neither one's conscious mental experiences during problem solving, nor for that partly subconscious problem solving itself.

When the Würzburgers attacked associationism, they had roughly the following theory in mind. Association explains why in consciousness one presentation, say *P*, is followed by another presentation, say *Q*. *P* will be followed by *Q*, rather than a third presentation, say *R*, if the degree of associative strength between *P* and *Q* is higher than the degree of associative strength between *P* and *R*. The degree of associative strength between any two presentations is a function of how many times they have jointly been in consciousness before (simultaneously or one right after the other). Finally, presentations are atom-like elements.

Perhaps it is helpful to indicate here the Würzburgers' two basic lines of attack. First, they denied that thinking was a sequence of presentations. Thus even if associationism were true of presentations, it had yet to prove its mettle for thoughts or awarenesses. Second, the Würzburgers emphasised that associations could be overruled. Assume, for instance, that we are dealing with the mind of a person who always takes sugar in his or her coffee. On the associationist theory, in the mind of that person, the presentation 'coffee' will always be followed by the presentation 'sugar'. But now imagine that the person is given the task of baking a cake. The Würzburgers tried to show experimentally that in such a scenario, the presentation 'sugar' is unlikely to be followed by the presentation 'coffee'; it is much more likely to be followed by the presentations 'flour', 'eggs', 'butter', or 'oven'. And thus it seems that we have to go beyond association.

Just as the Würzburgers developed a number of different ideas and approaches with respect to method and mental entities, so also with respect to laws; these need to be distinguished here.

The criticism of classifications of associations

In the early writings of Mayer, Orth, and Marbe, the sufficiency of associationism was not yet at issue. Rather, these authors criticised what they regarded as inappropriate ways of classifying associations. For instance, Orth, in his 'A Criticism of the Classification of Associations' (*Kritik der Assoziationseinteilungen*, 1901), and Thumb and Marbe, in their *Experimental Investigations about the Psychological Foundations of Analogy in Language* (*Experimentelle Untersuchungen über die psychologischen Grundlagen der sprachlichen Analogiebildung*, 1901: esp. 14–16), took issue with a number of such classifications. Wundt was again the chief culprit. Wundt had derived his elaborate taxonomy of associations from experiments by his students. In these experiments, subjects were presented with isolated nouns and asked to respond with the first noun that came to mind, that is the first noun that they associated with the stimulus. Wundt then captured the associative relations between stimulus word and reaction word with concepts such as 'subordination' ('church'–'building'), 'superordination' ('building'–'church'), or 'contrast' ('man'–'woman'). To Marbe's, Orth's, and Thumb's mind, in so doing Wundt had transgressed the realm of psychology: 'Subordination', 'superordination', and 'contrast' were logical, not psychological concepts. A proper *psychological* classification of associations had to attend to the subjects' experience during the search for a reaction word.

On the basis of introspective reports, Mayer and Orth (1901) proposed a new classification. First, two basic kinds of associations could be distinguished: associations where the reaction word came to the subject's mind immediately, without any other events occurring between stimulus and reaction words; and associations where one or several conscious mental events could be observed in the interval between stimulus and reaction words. The latter cases were more frequent than the former, and took longer. Second, the intervening events could be of three kinds: presentations, volitions, and situations of consciousness. Third, more than one intervening event could appear between stimulus and reaction. Fourth, of the three kinds of possible intervening events, presentations were more frequent than the others. Fifth, when the intervening events were emotionally charged, then the association time increased. And sixth, a negative emotional charge ('pain', *Unlust*) led to a longer association time than did a positive emotional charge ('pleasure', *Lust*) (Mayer and Orth 1901).

Judgements

The Würzburgers' assault on the sufficiency of association began with Marbe (1901). Marbe showed to his own satisfaction that none of the best-known psychological theories of judgements fitted with the introspective reports of his subjects. His subjects did not report any of the experiences that these theories had proposed as being characteristic 'qualia' of judgements: that is, acts of accepting or rejecting (Brentano [1874] 1924: 266); the combining of two presentations (Sigwart 1889: 25); the analysis of a presentation into two parts (Wundt 1893: 154); and, especially relevant for our present concerns, *association* (Ziehen 1900: 190). Although Marbe did not emphasise this point, his result was potentially a major blow to associationism. If judgements could not be understood on the basis of associations, and given that thinking consisted of forming judgements, then thinking was due to processes other than association.

Marbe did not draw this conclusion very clearly, however. He did not say that associationism (and related doctrines) needed to be replaced by better *psychological* theories of judgements. Instead, he removed the concept of judgement from psychology altogether. 'Judgement' was not a psychological category because there were no characteristic judgement qualia. 'Judgement' was a logical category only. Put in a nutshell, Marbe suggested that 'judgement' was a relational property that an interpreter ascribed to a conscious process (or its product). The one key condition for such ascription was that the conscious process (or its product, or its meaning) was intended, by its possessor, to stand in a relation of similarity or correspondence to some real or ideal entity. Marbe's theory of judgements was thus both externalist and interpretation based. Figure 1.2 might serve to illustrate the central ingredients of this theory (1901: 52, 92).

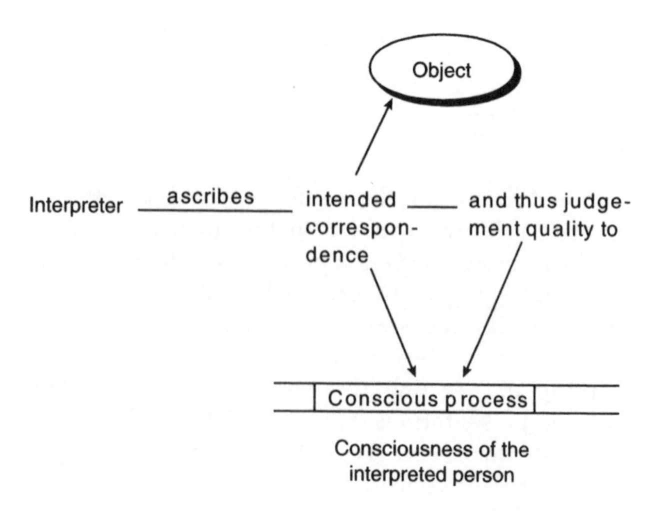

Figure 1.2 A pictorial representation of Marbe's externalist theory of judgement

Külpe and Watt on tasks

Although other Würzburg psychologists accepted Marbe's criticism of earlier theories, they did not accept Marbe's conclusion, that it is his way of removing judgements from the realm of psychology. Other Würzburgers preferred to give a novel *psychological* account of judgements. To understand their alternatives, we need to turn to the concept of 'task'.

This concept appeared first in a study by Külpe on 'abstraction'. By 'abstraction' Külpe meant the process 'by means of which it is possible to have some partial contents of consciousness stand out, and to have some other partial contents of consciousness stand back' (Külpe 1904: 56). Külpe demonstrated that in perception this 'standing out' and 'standing back' could be manipulated by setting varying 'tasks' (*Aufgaben*) for the experimental subjects. In different experimental trials, pictures with a different number of nonsense syllables were briefly (i.e. for 125 milliseconds) projected onto a screen in front of the experimental subject. Varying conditions of the experiment were the number of nonsense syllables in each picture, the number of letters in each syllable, the colour of the letters, the configuration in which the syllables were arranged, and the instruction or task. The four different tasks were to determine the number of visible letters; to ascertain the colours and their rough location in the visual field; to establish the configuration of syllables in a given picture; and to recall as many individual letters as possible together with their rough location. In other words, Külpe first gave the subjects one of the four instructions, and then showed them one picture, that is one configuration of syllables. Subsequently, he asked the subjects to recall as much as they could about each picture. Külpe found that the subjects' memories were most detailed, most often correct, and most often made with certainty when these memories concerned those aspects of a given picture to which the respective task had drawn the subjects' attention (1904: 61).

Külpe's Scottish student H. J. Watt built further upon Külpe's work on tasks (Watt 1905). Watt studied the role of tasks in a series of reaction experiments. Using a card changer, a speaking tube, and a Hipp chronoscope, Watt presented his subjects with written nouns. His subjects' reactions were 'constrained' rather than 'free'; that is, rather than react with the first word that came to mind, his subjects had to call out nouns that stood in specific relationships to the presented word. In different series, these reaction words had to be:

1. a superordinate concept, for example in reaction to 'house', 'artefact';
2. a subordinate concept, for example in reaction to 'house', 'farm house';
3. a part of the entity referred to by the presented word, for example in reaction to 'house', 'door';
4. a whole of which the referred-to entity formed a part, for example in reaction to 'house', 'town';
5. a concept that, like the presented one, was subordinate to the same superordinate concept, for example in reaction to 'house', 'car' (with 'artefact' as the superordinate concept); and
6. a word that referred to an entity that, like the entity referred to by the presented word, was part of the same whole, for example in reaction to 'house', 'street' (with 'town' as the whole).

Subsequent to each experiment, Watt's subjects of course gave retrospective reports.

According to Watt, a number of factors could influence the character and the sequence of the subjects' conscious experiences during the experiments; such factors were the stimulus word, associative tendencies aroused by the stimulus word, moods and feelings before and during the experiment, and the mental attitude adopted in reaction to the task. Watt was mainly concerned with documenting the effect of this last-mentioned factor. He claimed that different tasks created different predispositions for distinct types of associative–reproductive tendencies. These tendencies could be classified in terms of the visual, acoustic, or motoric presentations to which they gave rise (1905: 302–3).

To prove his point, Watt first differentiated between types of solutions. On the one hand, there were 'reproductions with multiple direction', that is cases where the experimental subject sequentially pursued at least two different trains of thought in search of a correct reaction or solution. On the other hand, there also occurred 'reproductions with simple direction'; here the experimental subject reported experiencing a single train of thought leading more or less straight from the stimulus word to the reaction word. Within this latter category, Watt went further and distinguished between (a) cases where the subject did not, or could not, recount any presentations between the stimulus and the response words, (b) cases where the subject experienced a visual presentation, and (c) cases where the subject noted a word presentation, a memory image, or a situation of consciousness (1905: 303–5, 321).

Watt's demonstration of the effect of tasks on reproductive tendencies is most easily understood in the case of reproductions with simple direction. For each experimental subject, and for each of the six different tasks, Watt counted the number of cases of (a) to (c). He then calculated, for each of the six tasks, the percentages of the occurrences of each of the three cases (the total being all reproductions with simple direction). The curves he drew on the basis of these percentages suggested to him that different tasks affected the relative frequency of the three cases (a–c) in a uniform way across all subjects – and this irrespective of whatever their prior (pre-experimental) predispositions towards different types of presentations had been. Watt made this point with a large number of tables and figures; suffice it here to reproduce a part of just one figure. It shows the relative frequency of those cases of reproduction with simple direction in which three subjects reported visual presentations during the solving of the six tasks (Figure 1.3). In Watt's words: 'The form [of the curves] brings out the main thing, to wit, that the change of task has a regular influence on the qualitative content of the train of reaction of each experimental subject' (1905: 315).

Watt claimed that the effect of the task could also be observed with respect to time. For instance, calculating the average times (for the three main subjects) for all reproductions with simple direction (and correct answers) yielded the diagram in Figure 1.4 (1905: 316).

Watt regarded his proof that his time curves could not be explained from an associationist perspective as particularly important. From such an associationist perspective one might argue that Watt's time curves were due not to the tasks but to the strengths of associative links between given stimulus words and correct reaction words. The stronger the link, the faster the reaction.

Watt sought to disprove this proposal by showing that the effect of the task remained visible even when one restricted the data to stimulus–reaction pairs of similar associative strength. His measure for such similarity was the 'familiarity' (*Geläufigkeit*) of such pairs to his subjects. That is to say, the associative strength

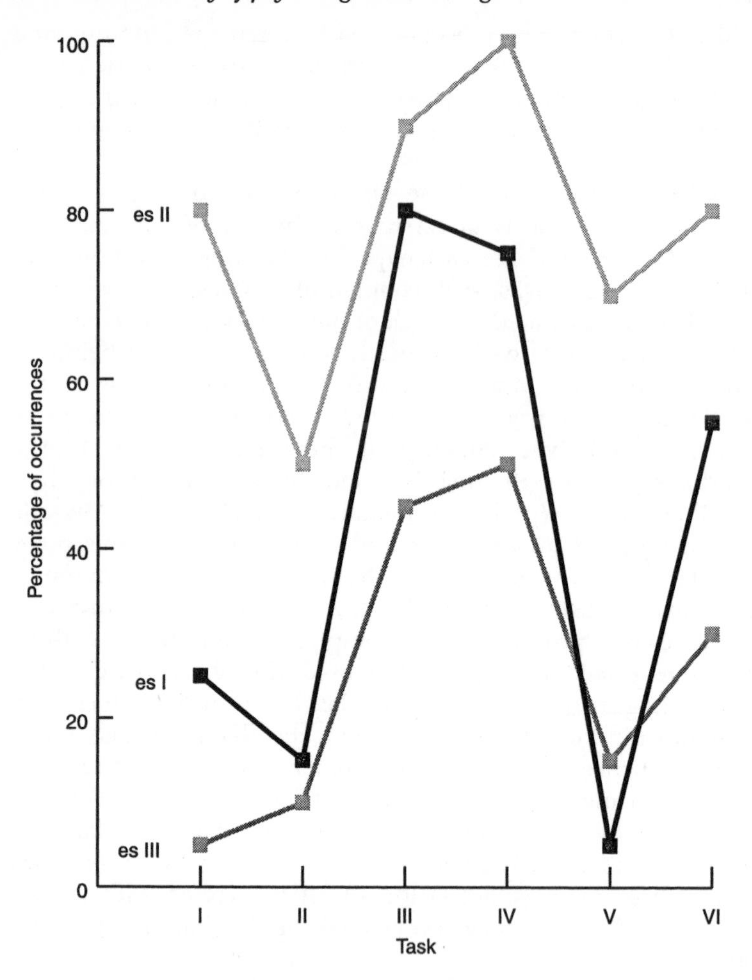

Figure 1.3 The share of visual presentations across Watt's six tasks for three experimental subjects

Source: From Watt (1905: 315)

between two words *A* and *B* was greater than the associative strength between two other words *C* and *D*, if more subjects reacted to *A* with *B* than to *C* with *D*. Watt calculated the number of stimulus–reaction pairs found in all of his subjects (n), all of his subjects except one ($n - 1$), all of his subjects except two ($n - 2$), and all of his subjects except three ($n - 3$). He then drew separate curves for each of these categories and found that the effect of the

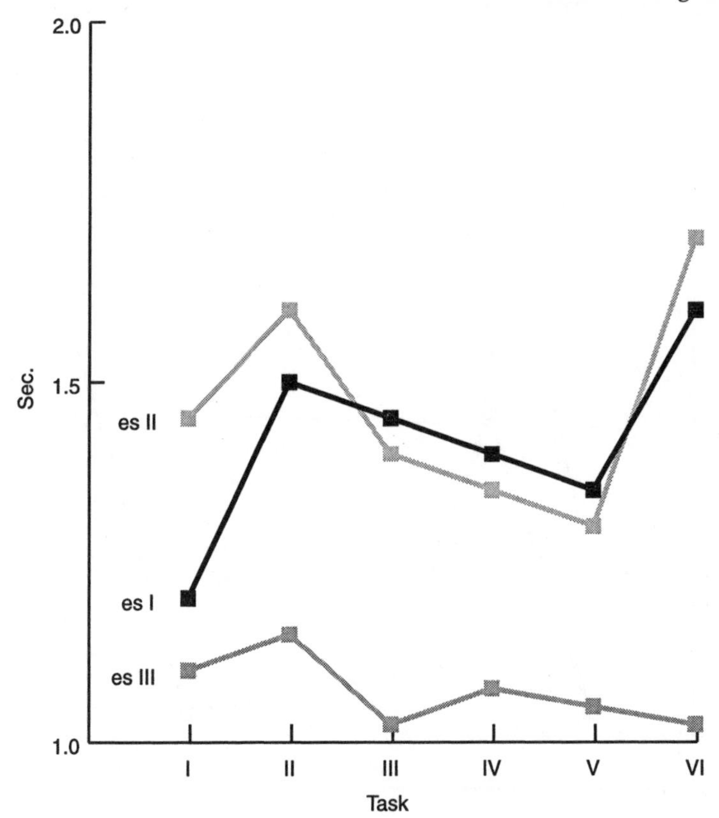

Figure 1.4 Time used by three experimental subjects across Watt's six tasks
Source: From Watt (1905: 316)

task remained visible. Thus the effect of the task was independent of the effects of associative strength (1905: 328, 352) (Figure 1.5).

Once introduced, this analysis of associative strength served Watt also as a means of attacking a rival *non*-associative account of thinking. According to Wundt's psychology, thinking consisted of two kinds of processes. On the one hand, *associative* processes produced many and different presentations. On the other hand, the human will (*apperceptive processes*) chose specific items from the mass of presentations produced by association (e.g. Wundt 1897: 41). Accordingly, Wundt wrote that 'all thinking is an inner choosing' (*Alles Denken ist innere Wahltätigkeit*, quoted in Watt

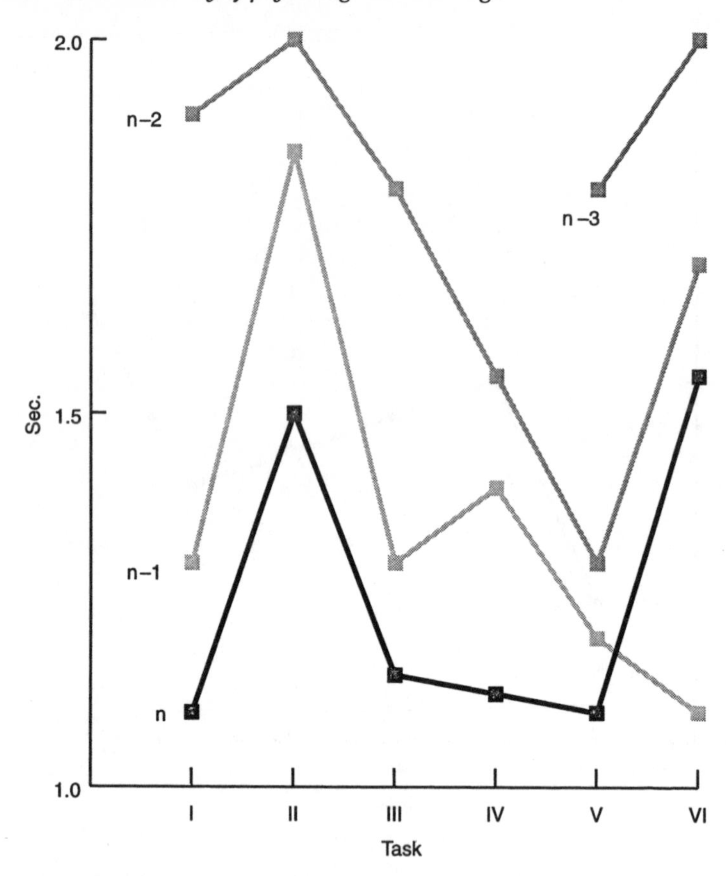

Figure 1.5 Familiarity across the six tasks

Source: From Watt (1905: 328)

1905: 359). At issue was the question of what would happen when different, perhaps contradictory, reproductive tendencies became effective at the same time. Obviously not all of them could 'win' because, at least in reaction experiments, only one reaction word was allowed. According to Wundt, in such cases 'a choice' between these different associations needed to be made. And this choice was the product of the will, or of 'apperception', Wundt's term for the active selecting and structuring of experience by a self-determining, mature person. (The term 'apperception' had earlier been used by Leibniz, Kant, and Herbart. The meaning differed, however, from author to author.)

Much of the second half of Watt's study consisted of an assault on this and related Wundtian doctrines. There was no room for choice here, according to Watt, because – *ceteris paribus* – the more (or most) familiar of two (or more) associations would always win. Watt proved this case to his own satisfaction by focusing on cases in which his subjects had reported two different reproductive tendencies (thus all these were cases of reproduction with multiple directions). From this set of cases, he picked out those that lent themselves to an analysis in terms of familiarity. Particularly telling, in Watt's view, were ten cases where subjects reported experiencing two different, conflicting meanings for given stimulus words. In seven of these cases, the winning meaning was the one that was picked by all subjects (= n); in two cases the winning meaning was the one that was picked by all subjects except one (= $n - 1$). Watt concluded: '*Ceteris paribus* what decides between two tendencies is their respective strength'. The only sense in which one could continue speaking of apperception was as a covering term for such regularities (1905: 358–60, 419, 422).

Finally, Watt used his theory of the task to challenge Marbe's claim that there existed no psychological criterion for judgements. Contrary to what Marbe had alleged, psychology was able to explicate the process of judgement formation. What made a sequence of experiences a judgement was the presence and influence of a task:

> A judgement or an act of thinking is a certain sequence of experiences. The unfolding of this sequence out of its first element, the stimulus, is determined by a psychological factor [i.e. the task]. This factor precedes the sequence....But it continues to exert its influence during the sequence.
>
> (1905: 416)

On Watt's conception of judgements, they did have characteristic qualia, to wit, the qualia related to tasks. At the same time, Watt's theory could also explain why Marbe's subjects had not been able to report such judgement qualia: the task and its qualia preceded the stimulus, and thus it preceded the time period to which Marbe's subjects had attended.

Later work on tasks and judgements

Watt's work became highly influential within the Würzburg school. I shall here mention only a few studies that took Watt's paper as their starting point.

Even the title of Messer's paper 'Experimental-Psychological Investigations on Thinking' (*Experimentell-psychologische Untersuchungen über das Denken*, 1906) indicated a proximity to Watt. After all, Watt's paper had been entitled 'Experimental Contributions to a Theory of Thinking' (*Experimentelle Beiträge zu einer Theorie des Denkens*, 1905). Messer adopted Watt's conception of tasks and claimed to have successfully replicated Watt's experiments (1906: 48–50).

But Messer also suggested some modifications and further developments. In addition to replicating Watt's experiments, Messer did one experiment in which the subjects were free to react with any word that came to mind; that is, in this experiment Messer allowed for free, unconstrained association. He submitted that the retrospective data from this experiment showed that subjects never reacted in a totally arbitrary fashion. Even when they were 'free', as it were, subjects spontaneously set themselves a specific task. They tried to find words that stood in some meaningful relation to the stimulus word. And they were unable to break free of this task, even when asked to react in nonsensical ways. Messer concluded that 'the "association mechanism" that each of us carries within him, is already tuned to "reasonable" thinking; here I mean by "reasonable" a form of thinking that seeks to do justice to the demands of "reality"'. And this was not surprising; after all, associative reproductive tendencies were brought about in and through our contact with the world of physical, psychological, and ideal entities. Messer also noted that in general, subjects did not realise that they were setting themselves the task of reacting in meaningful ways. Here, as often elsewhere, the task worked on an unconscious level (1906: 23–33).

Messer's definition differed from Marbe's. By 'judgement' Messer and his subjects meant 'that process of thinking that finds its complete verbal expression in a ("meaningful") assertion' (1906: 93). Messer addressed the question of whether judgements could be distinguished from mere associative reproductions. Was not the judgement an associative process in which the subject's presentation called forth the predicate association? Messer maintained that the reports of his subjects demanded a negative answer to this question. The processes labelled 'associations' by his subjects differed fundamentally from those that they called 'judgements'. Associations came to mind automatically, whereas judgements had to be 'willed ("intended"), or at least accepted'. To understand this volitional element, one had to refer back to the task: there could be intentions only in response to a task. Messer suggested that tasks were perfectly general phenomena and not an artefact of the laboratory. After all, one spoke of tasks in school and the workplace, too (1906: 105, 107–9).

One of the most taken-for-granted tasks was the task of correctly coming to know, and correctly describing, the world. Because this task was usually taken for granted, it almost always remained unnoticed. No wonder, therefore, that Marbe's subjects had overlooked it, too. And in so far as they had missed it, they had failed to identify a crucial judgement quale (1906: 111).

Messer's study of judgements drew on Husserl ([1901] 1984). For instance, like Husserl, Messer insisted that a proper study of judgements presupposed shifting from a psychology of sensations and presentations to a psychology of perception and thinking. The latter psychology was a psychology of 'everyday conduct': 'In it I am not directed towards sensations but towards things and processes and their properties'. The psychology of sensations, on the other hand, was based upon an 'artificial attitude'. Sensations were a product of abstraction. And for precisely this reason, Messer argued, there was no natural route from the psychology of sensations to a psychology of judgements (1906: 12–13):

> There are people who think that perception and thought can be sufficiently characterised by simply registering the sensations and presentations that accompany perception and thought. These people remind me of someone who tries to understand the nature of money through a study of the raw materials out of which money is made.
>
> (1906: 113)

Messer concluded his investigation into the psychology of judgements by developing a classification. Judgements could be divided, for instance, according to their content (e.g. analytic or synthetic), their relation to other judgements (e.g. new or reproduced), their way of relating to objects (perception judgements versus presentation judgements), and the conduct of the judging subject (e.g. theoretical versus practical) (1906: 115–46).

Two other studies that relied centrally on the task concept were those of A. A. Grünbaum ('On the Abstraction of Identity', *Über die Abstraktion der Gleichheit*, 1908) and E. Westphal ('On Main Tasks and Secondary Tasks in Reaction Experiments', *Über Haupt- und Nebenaufgaben bei Reaktionsversuchen*, 1911). Both of these papers investigated the effects of experimental settings in which subjects had been given both one 'main task' and one 'secondary task'. For instance, the subject had to identify two identical pictures out of two sets of two, three, four, five, or six pictures each (main task), and remember as many of the other pictures as possible (secondary task) (Grünbaum 1908). Or the subject was asked to determine the number of corners of a polygon (main task) and identify the position of the longest side (secondary task) (Westphal 1911). Grünbaum's main result was that an increase of concentration on the main task also improved the performance with respect to the secondary task. This effect could also be observed in a hypnotised subject (1908: 411–12, 477). Westphal introduced a new theory of 'levels of consciousness' and 'mental structure'. Subjects had different degrees of awareness – or different 'levels of consciousness' – of the two tasks during different stages of the process of problem solving. And different subjects solved the two tasks either sequentially or simultaneously. These two dimensions taken together constituted the subjects' 'mental structure' (1911: 305).

Determining tendencies

All of the Würzburg psychologists agreed that thinking – *qua* problem solving – was an intentional activity involving volition. Ach's work provided the Würzburgers' canonical account of how to identify volition within cognition. As we have already seen above, to discuss the relation between volition and cognition brought one face to face with Wundt's theory of thinking as an activity of the will. It therefore is not surprising that Ach critically confronted Wundt's views at many points.

Ach was especially scathing in his criticism of Wundt's idea that one could measure 'choice time' (*Wahlzeit*) (cf. also Watt 1905: 405–7; Külpe 1907b). Wundt thought it possible to measure the time it took a subject to make a decision. All one needed to do was subtract discrimination time from discrimination-plus-decision time. Discrimination time could be measured in discrimination experiments. In such experiments, the subjects were given the instruction to, say, lift their fingers off the Morse key as soon as the stimulus had been recognised. Discrimination-plus-decision time could be obtained if the instruction was changed to this: as soon as you recognise the stimulus, *either* lift the finger off the Morse key, *or* shout 'stop' into the voice key. Here the subjects thus had to choose between two reactions. Ach commented that

> this method must be thrown overboard as unscientific....All one can determine is whether one process takes more, equal, or less time than the other process. But one cannot make any claims about the *absolute duration* of these different processes....the area of reaction experiments provides striking evidence for the claim that this kind of graphic schematism is the most dangerous enemy of psychological research.
>
> (1905: 161)

Another disagreement with Wundt concerned the location of the volitional act within reaction experiments. Wundt had claimed that the volitional act occurred *after* the subject had been presented with the stimulus, and *before* the subject reacted. Ach rejected this claim based on the reports of his subjects. The act of willing occurred in the prior period, that is the period between hearing the instruction and the appearance of the stimulus. Thus, *the execution of the act of willing occurred later than the act of willing itself* (1905: 119). Ach's theory of determining tendencies sought to provide an account of how this could be.

Ach (1905) claimed that there were three kinds of laws, or tendencies, that governed the train of presentations: tendencies of association, tendencies of perseveration, and determining tendencies. The last mentioned 'form the foundation of those mental phenomena which have, since antiquity, been summarised under the concept of activity of the will' (Ach 1905: 187).

The existence of determining tendencies could be demonstrated most dramatically in cases of post-hypnotic suggestion. The hypnotised person acted on the basis of an *external* prior determination that was not introspectively accessible. Subjects in Ach's reaction experiments were *internally* determined, but otherwise the situation was similar. These subjects determined themselves, during the prior period, to act later in a certain way, during the main period. Just like the hypnotised person had no introspective access to external determination, so also experimental subjects, during the main period, were no longer aware of their self-determination. Unlike the hypnotised person, the experimental subject was able to recall the determination later, however. Ach recounted the effect of post-hypnotic suggestion in his book. He had hypnotised one of his male subjects and had given the post-hypnotic command to add up any two numbers that Ach would present to him. The subject complied without hesitation, and without understanding his own action (1905: 190).

Central in Ach's account of the working of determining tendencies were three concepts: 'goal presentation' (*Zielvorstellung*), 'reference presentation' (*Bezugsvorstellung*), and 'realisation' (*Realisierung*). The goal presentation defined what was to be done (in the case of the hypnotised subject, addition). The reference presentation was the apprehended stimulus (the pair of numbers). And the realisation was the result of performing, with respect to the reference presentation, the operation defined by the goal presentation.

The further technical details of Ach's theory can perhaps best be introduced by going through one of Ach's examples, and by relying on a graphical representation (Figure 1.6).

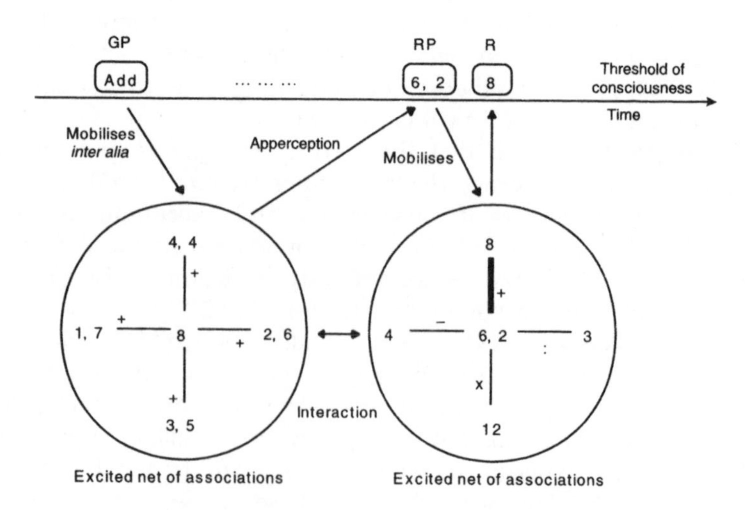

Figure 1.6 A graphical representation of Ach's theory of determining tendencies

In Figure 1.6, the arrow running from left to right represents both the advance of time and the threshold of consciousness: What is above the line is conscious; what is below the line is unconscious. Let 'GP' stand for 'goal presentation', and 'Add' for the content of that presentation. The story in the figure then unfolds as follows. In the light of the instruction, and during the prior period, the subject prepares to add any two numbers that come up in the card changer. The forming of this goal presentation mobilises or excites a number of associations. These are brought into readiness but do not cross the threshold of consciousness. Such associations might, for instance, be those represented in Figure 1.6: various pairs of numbers and their sums.

Another set of associations is mobilised when, at a later point, that is during the main period, the subject is presented with the stimulus, say the numbers 6 and 2. These numbers form the reference presentation (RP). The mobilised or excited associations could well be the ones suggested here: The ordered pair <6, 2> might be associated with the numbers 4, 8, 3, and 12, on the grounds that the latter numbers result from performing various arithmetical operations on <6, 2>.

The reference presentation is not, however, independent of the goal presentation and of the associations mobilised by the goal presentation. The goal presentation and its associations determine how the stimulus is being comprehended – or as Ach says: how the stimulus is being 'apperceived'. Usually subjects did not perceive just two numbers. Rather, these numbers presented themselves to consciousness as already interpreted in the light of the goal presentation. For instance, the numbers might be apprehended with a '+' between them, or as being spatially close to one another, or in the verbalised form '6 plus 2'. Ach spoke of such phenomena as 'apperceptive fusions' and regarded them as being strong evidence for the irreducibility of determining tendencies. Moreover, the associations triggered by the goal presentation, and the associations excited by the reference presentation, interact. As a result, one of the associative ties in the second set, that is the level of excitation of the tie between 8 and the pair <6, 2>, is increased.

Finally, this increased level of excitation between <6, 2> and 8 brings it about that 8 is projected into consciousness as the 'realisation presentation' (R). Occasionally, the goal presentation might reappear prior to the realisation. This possibility is ignored here.

As Ach stressed time and time again, this working of determining tendencies showed that humans were not simple and automatic reaction devices. It was the attitude or goal presentation that determined how the stimulus – here '6, 2' – was processed. If the subject had set a different task, say to subtract rather than to add, then the whole process would have run a different course. In Ach's own words:

> *Thus we see that the ordered and goal-directed course of mental life is determined by the effects of determining tendencies. It is the influence of determining tendencies which secures an independence from accidental external stimuli and from the habitual associative course of presentations.* I cannot here determine the extent of this independence; suffice it here to have pointed out its existence. Note also that this effect of determining tendencies is caused not only by intentions; such tendencies can also be created through *suggestion, command,* and the *setting of a task.*
>
> (1905: 196)

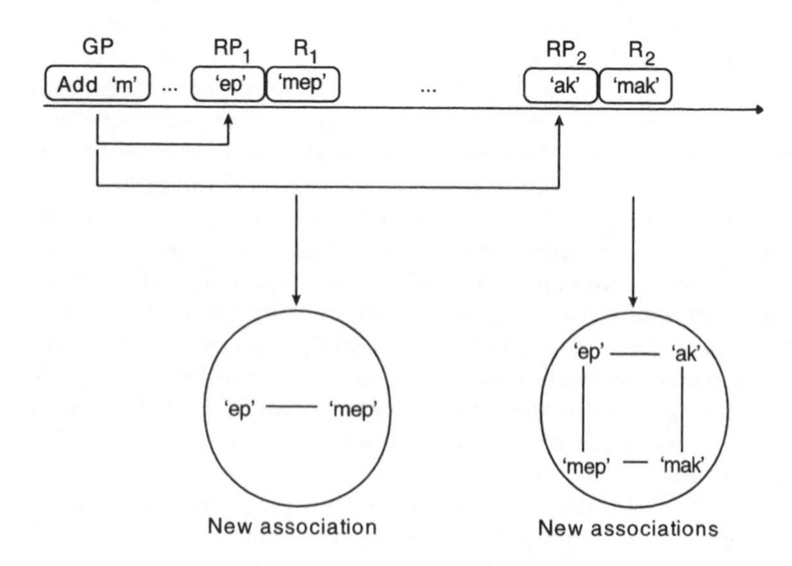

Figure 1.7 The creation of new associations in Ach's theory

In the above example, the independence of determining tendencies from associative tendencies was maintained on the grounds that determining tendencies 'decided', as it were, which associations would come into play. But Ach also had a second way of arriving at the same conclusion: determining tendencies could give rise to new associations. In one variation on the main experiments, Ach instructed his subjects to react to given nonsense syllables with new, *rhyming,* nonsense syllables. Some subjects reported that during the prior period they formulated goal presentations such as 'add an "m" if the nonsense syllable starts with a vowel' (1905: 208). It is easy to see how new associations could now arise (Figure 1.7): there would be new associations between a given nonsense syllable such as 'ep' and the newly created syllable 'mep', as well as new associations between syllables in a given series (1905: 209–10).

Ach's second major book, *On the Act of Willing and Temperament* (*Über den Willensakt und das Temperament,* 1910), tried to marshal still more evidence for the claim that determining tendencies were independent of tendencies of association. Ach devised a new type of experiment. During the first part of the experiment (which lasted for several days) the subjects were repeatedly shown a series of nonsense syllables in a fixed sequential order. As earlier work by Ebbinghaus and Müller had established, this created successive associations between these syllables. For example, after the subject had been exposed to the sequence 'dus–rol–nef–zis' a few times, 'dus' would become associated with 'rol', 'rol' with 'nef', and 'nef' with 'zis'. During the second part of the experiment, the subject was shown one of these syllables at a time and asked to carry out a specific task with respect to it; for example, to invert the two consonants of the syllable. Two things could now happen at this point: either the subject followed the task and reacted to, say, 'dus' with 'sud', or the subject reacted mistakenly by saying 'rol'. Ach interpreted the first reaction as showing that, in the given instance, the determining tendency was stronger than the associative tendency, and the second case as showing the opposite. The fact that the distribution of the two cases could be experimentally manipulated convinced Ach that determining tendencies were different from associative tendencies. Ach concluded:

> The stronger the association, the stronger must be the concentration of the energy of the will, if then it is to overrule the

> effect of the associative reproductive tendency. In other words: we want an intervention of the will to break the artificially created habits. Weak habits call for only a small effort of the will; strong ones, however, call for a very strong role of the impulse of the will. Thus we can indirectly bring about the act of willing at any level and make it accessible to analysis.
>
> (1910: 19)

Tasks and determining tendencies

The similarities between Ach's and Watt's work were not lost on contemporary observers. Not only did Ach and Watt conduct similar experiments with the same high-tech instrumentation, they also arrived at similar results. This raised the issue of priority (Ach 1910: 9; Watt 1906a: 83), an issue that Külpe tried to resolve by recognising Ach's priority and applauding Watt's independence (Külpe 1907b).

Külpe also tried to build a theoretical bridge from Ach to Watt, or from determining tendencies to tasks (Külpe 1907b). In Külpe's view the two conceptions differed only in emphasis: 'The "task" stresses more the instruction...whereas the "determining tendency" highlights more the effect of the task' (1907b: 601). Külpe also underlined the common theme of both studies: Associationism was an insufficient explanation of the will and thinking. And he put Ach's and Watt's work into a wider perspective. The work of both men was not about artificial laboratory phenomena but about 'life': 'One is therefore allowed to say that the simple reactions [studied by Ach and Watt] are elementary cases of those processes that in ordinary life we call "actions"' (1907b: 598).

Of later publications strongly influenced by Ach, at least Karl Koffka's book *On the Analysis of Presentations and Their Laws* (*Zur Analyse der Vorstellungen und ihrer Gesetze,* 1912) must be noted briefly. Koffka distinguished between various kinds of determining tendencies in reaction experiments; for instance between the 'reaction tendency' (*Reaktionstendenz*), which aimed for a speedy completion of the search for a correct reaction word; the 'solution tendency' (*Lösungstendenz*), which brought about a correct reproduction; the 'introspective or description tendency' (*introspektive oder Deskriptionstendenz*), which enabled the subject to report experiences later; or the 'latent attitude' (*latente*

Einstellung). Koffka had most to say about the last-mentioned category. Examples of latent attitudes were the different subjects' tendencies to prioritise speed of response over correctness, or correctness over speed. Latent attitudes were defined as determining tendencies that did not result from an act of will and that usually remained subconscious. Latent attitudes enabled humans to react more quickly because they narrowed down the realm of alternatives. Again, like other determining tendencies, latent attitudes could not be reduced to associative tendencies; latent attitudes could 'superimpose themselves upon associations, develop them further, direct and inhibit them' (1912: 322, 362).

Thoughts and knowledge

Up to this point we have followed one of the Würzburgers' two main lines of attack on associationism mentioned in the introduction to this section; that is, the argument that associative tendencies could be overruled. It remains for me to explain the second line of reasoning. It amounted to insisting that although association might be true for the interaction between presentations, it did not capture the recall and mobilisation of thoughts and knowledge. This idea was introduced – in a very sketchy form – by Bühler and later developed in great detail by Selz (1913). I shall briefly summarise their studies in turn (cf. also Schanoff 1911; Lindworsky 1916).

Bühler argued his case in two publications (1908c, 1908d). These articles dealt with the relations between thoughts.

A first observation concerned the succession of thoughts during the process of problem solving. As Bühler's subjects reported it, thoughts did not follow one another as the associationist would have it; it was not the case that one atom-like thought pulled the next one over the threshold of consciousness. Rather, thoughts were always embedded in a rich texture of knowledge:

> we know e.g. whether we are on the right path, whether we are getting closer to the goal...we know of a later thought how it relates to an earlier one, etc.... All of this knowledge is, at it were, between thoughts.
>
> (1908c: 1)

Moreover, understanding a new thought, say a sentence, was not an associative process in which individual word presentations triggered meaning presentations. The tradition had been right to assume that the understanding of something new involved the mobilisation of something old. But what was mobilised to understand meaningful sentences usually was not presentations but thoughts. Sometimes the mobilised thought would be more general than the encountered one, sometimes more specific. Sometimes understanding meant mobilising a cycle of thoughts within which the new thought could be situated; sometimes the mobilised thought amounted to no more than a label. Most importantly, the relationship between the thought-to-be-understood and the thought-invoked-for-understanding was a relationship between two complete entities, two wholes. It was not a relation between pieces of two aggregates. Again, associationism had no resources for capturing this relation (1908c: 13–18).

Bühler's analysis of thought recall (1908d) further extended his alternative to associationism. His analysis was structured around the following simple model (Figure 1.8). What triggered the reproduction of an earlier thought experience (ee) during a later experience was some 'memory motive' or 'starting point'. This memory motive stood in some sort of 'backwards-looking relation' to the earlier experience. This relation allowed for the

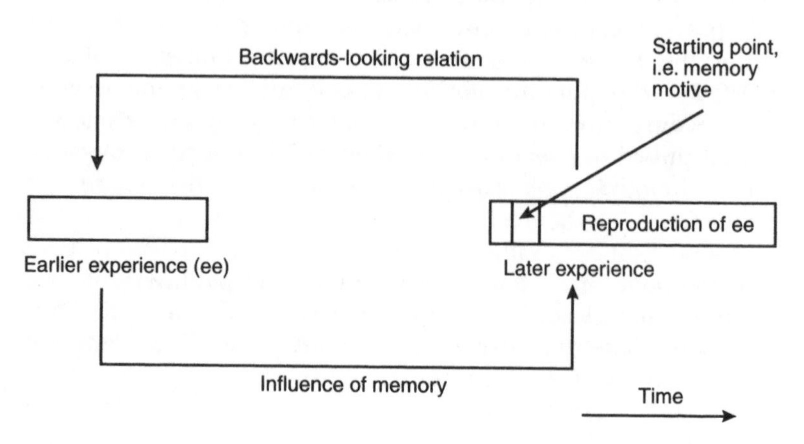

Figure 1.8 A graphical representation of Bühler's analysis of thought recall

retrieval of an earlier thought and thus allowed memory to have an influence on present deliberations. The important point here is that none of the various relationships portrayed in this picture needed to be, and usually were not, relationships of association (1908d).

Selz's 1913 study picked up the Bühlerian theme that to understand thinking and problem solving one must study the relations between wholes and their moments, rather than between aggregates of associated atomistic pieces (1913: 72). Selz contrasted these two conceptions as the 'complex theory' and the 'constellation theory'. Already simple reaction experiments showed that the former was to be preferred over the latter. Asked to react with a 'co-ordinated' concept to the word 'death', one of Selz's subjects answered 'life' and then went on to give the following retrospective report:

> After some reflection, 'life' came to mind as the solution. Before that I had formulated the question in unclear words: 'What is co-ordinated with death?' The solution came to mind as something familiar.
>
> (1913: 28)

Selz interpreted such reports as showing that the reaction 'life' was not reproduced on the basis of an associative link between 'death' and 'life'. Instead it was produced on the basis of the 'familiarity of a relational whole, the awareness (*Bewußtsein*) of the state of affairs (*Sachverhalt*) that death and life...often are put in opposition to one another as parallel concepts'. Selz suggested calling this 'awareness of states of affairs' 'knowledge' (*Wissen*) (1913: 30). Much of Selz's book consisted of developing a theory of 'knowledge actualisation' (*Wissensaktualisierung*) which sought to describe the various ways in which humans managed to mobilise relevant complexes of knowledge.

One key ingredient of this theory was the claim that knowledge actualisation – rather than the association between isolated elements – was at work even when the subject produced the reaction word immediately and without mentioning knowledge in the retrospective report. What made this claim plausible, according to Selz, was the fact that one could see the immediate, seemingly unmediated, reaction as the endpoint of a development that had its starting point in a successive, temporally extended, step-by-step actualisation of knowledge (1913: 84).

Another important point was that knowledge consisted of structured wholes, not of bits and pieces. Even association was primarily a relation between structured wholes. The plausibility of this assumption could be demonstrated even for simple cases such as the association between the German word *sieben* (seven) and the written number '7'. The word *sieben* could not be resolved into associations between the sounds *s*, *ie*, *b*, and *n*. Otherwise the name 'Ibsen' (when spoken with a long German 'i') would trigger the same association to '7'. The same analysis applied to '7' and its two elements ('-' and '/'). The association between *sieben* and '7' was thus between the ordered complex 's–ie–b–e–n' and the ordered complex of '-' and '/' (1913: 89–94).

A similar case could be made for 'complex completion'. A subject who had learnt the complexes '+-+~' and '-++*' would later complete the complex piece '+-+' with '~', and the complex piece '-++' with '*'. Again, it was the order, or structure, of the three elements, '+', '+', and '-' that was decisive. And this structure could not be reduced to associative links between the four isolated elements (1913: 108).

What applied to such simple cases of images also held for the actualisation of knowledge in reaction experiments. To cut a long and highly complicated story short, having heard the instruction, and having been presented with the stimulus, subjects would fuse the two into a first complex, or one meaningful whole, that is 'the whole task' (*Gesamtaufgabe*). This whole task would generate a 'goal awareness' (*Zielbewußtsein*) as a second complex. This second complex was the knowledge of a schematic, abstract state of affairs, a rough knowledge of what a possible solution would have to be like. This would be followed by one or several searches for fitting solutions. These searches would have the structure of a complex completion. For instance, a subject searching for a concept co-ordinated with 'death' would have a 'goal awareness' such as the following: there is a conventional, ordered pair, with death as its second, and the searched-for concept as its first element. The search might be accompanied by a consciousness of ease or difficulty. Generated solution candidates would be measured against the goal awareness and corrected if they failed to fit the whole task (Selz 1913: 181–261).

Judgements revisited

The category of judgement did not figure prominently in Bühler and Selz. We need to note, however, that Bühler was highly critical of Marbe's claim that there were no judgement qualia. In line with his insistence on presenting difficult tasks, Bühler argued that Marbe's subjects had failed to identify such qualia only because the tasks given to them had been too easy. And easy tasks were solved in an automatic and unconscious fashion, by automatic and unconscious judgements (Bühler 1907b: 303).

Marbe did not back down. He replied to Bühler that nothing could be a judgement quale unless it occurred in the case of every single judgement. And to Watt's and Messer's suggestion that the task might provide the psychological criterion for judgements, Marbe countered by doubting that 'the thousands of judgements of everyday conversation ("fine weather today") and many other judgements [could] be regarded as solutions to tasks' (1915: 9–10).

Analytical summary

This chapter has presented an overview of the main methods and results of the broader Würzburg school. Because it is all too easy to get lost in the nitty-gritty of arguments for and against specific claims, I conclude with an analytic summary of the main claims and observations.

The main points regarding the method of *systematic experimental self-observation* (SO) were:

SO.1 Retrospective self-observation is reliable provided it occurs under controlled experimental conditions. The most important of these conditions is the presence of a second person.

SO.2 The 'division of labour' between experimenter and experimental subject is a safeguard against theoretical bias (Marbe 1901).

SO.3 It is best to provide retrospectionists with a fixed taxonomy of conscious mental states (Marbe 1901). This was contested by most other members of the Würzburg school.

SO.4 Within psychological experiments three periods need to be distinguished: the prior, the main, and the after period (Ach 1905).

SO.5 Retrospection is based on perseveration, not on memory processes. Therefore subjects are able to furnish complete accounts of their experiences (Ach 1905). Most other Würzburgers disagreed with this claim.

SO.6 The experimenter should, or is permitted to, pose additional questions (Ach 1905). Marbe rejected the questioning method, while most other members of the Würzburg school accepted it.

SO.7 The relationship between experimenter and experimental subject is a relationship of trust (e.g. Ach 1905; Bühler 1907b).

SO.8 Thought experiments must strive to be natural and close to real-life situations. This rules out the use of instruments such as the voice key or the Hipp chronoscope (Schultze 1906; Bühler 1907b). Ach (1905), Watt (1905), and Messer (1906) disagreed.

SO.9 Thought processes can best be observed when they occur slowly. Therefore the tasks must be difficult (Bühler 1907b).

The central ideas concerning *situation of consciousness, awareness,* and *thought,* that is mental contents (MC), were:

MC.1 Situations of consciousness are non-depictive conscious states that cannot (easily) be described (Marbe 1901; Mayer and Orth 1901).

MC.2 Situations of consciousness must not be confused with feelings. Such confusion is prone to occur if one takes one's starting point from Wundt (Orth 1903).

MC.3 An awareness is a form of non-depictive, immediate knowledge. It has a degree of intensity and can best be explained as the conscious reflection of unconscious associative tendencies. The most important kind of awareness is the awareness of meaning. Awarenesses are usually accompanied by presentations or feelings, but the link between awarenesses and these concomitants is accidental (Ach 1905).

MC.4 Situations of consciousness lack intentionality (Bühler 1907b). Marbe rejected this claim (Marbe 1915).

MC.5 Thoughts are intentional, non-depictive conscious experiences. Usually they occur without any depictive concomitants. The analysis of thoughts is best based on Husserl's terminology. Wundt's theory of thought is the product of sloppy self-observation (Bühler 1907b).

MC.6 Situations of consciousness form a subclass of awareness (Ach 1905). Marbe disagreed (Marbe 1915).

MC.7 Situations of consciousness are not intentional (Bühler 1907b). Marbe rejected this claim (Marbe 1915).

MC.8 Bühler's analysis of thought conflates the thought-act with the object of the act (Dürr 1908a, 1908b; Marbe 1915). Külpe (1912b) defended Bühler against this accusation.

MC.9 There can be no immediate knowledge, because all knowledge is dispositional only (Marbe 1901, 1915). Selz (1913) opposed Marbe on this point.

MC.10 Thoughts as contents need to be distinguished from thoughts as functions, or acts. Brentano was right about the importance of acts (Külpe 1912b).

And the Würzburgers' main claim regarding *laws of thought* (LT) were:

LT.1 Traditional classifications of associations are based on logical criteria and not – as they should be – on self-observation. Wundt's classification is a case in point (Mayer and Orth 1901; Thumb and Marbe 1901).

LT.2 'Judgement' is a logical, not a psychological, category. There are no judgement qualia (Marbe 1901). Watt (1905), Messer (1906), and Bühler (1907b) disagreed.

LT.3 The train of thoughts is determined by tasks. Tasks cannot be reduced to associations (Watt 1905; Messer 1906).

LT.4	Wundt is wrong to claim that 'all thinking is an inner choosing [between presentations]'. The most familiar presentation always wins (Watt 1905).
LT.5	The judgement qualia that Marbe could not find are the qualia of the task (Watt 1905). Marbe (1915) disagreed. Only judgements that emerge slowly and consciously have qualia (Bühler 1907b). Marbe questioned whether this undermined his analysis (Marbe 1915).
LT.6	Choice times cannot be measured. Wundt is wrong to assume otherwise (Ach 1905).
LT.7	Wundt is wrong about the place of volitional acts in reaction experiments. These are situated in the prior period, not in the main (Ach 1905).
LT.8	Determining tendencies are those reproductive tendencies that result from a decision, a suggestion, or a command. They cannot be reduced to other reproductive tendencies, such as tendencies of association or perseveration (Ach 1905, 1910).
LT.9	Thoughts form wholes, not aggregates (Bühler 1908c, 1908d; Selz 1913).
LT.10	Problem solving is a process of knowledge actualisation. The central part of this process is the forming of the whole task and the goal awareness (Selz 1913).

2 Friends and foes

Introduction

In this chapter I shall present a review of positive and negative reactions to the new thought psychology. I shall restrict my review to reactions by German and Austrian writers, and to the time period between 1900 and 1920.

The two most influential critics of the Würzburg work were Wundt and Müller. I shall deal with their criticism at greater length. To understand their lines of attack, it is necessary to have at least a rough understanding of some of the central ideas of their own psychological 'systems'. I shall therefore introduce some of these central ideas here.

A number of critics further developed points first raised by Wundt and Müller. Other critics added new objections to the list. Not all of the reactions to the Würzburgers' thought psychology were negative, however. The neo-Kantians, the neo-Thomists, as well as the director of the Berlin Institute, C. Stumpf, all reacted positively.

Wundt

By far the most influential criticism of the Würzburgers' thought psychology came from the 75-year-old Wilhelm Wundt in 1907. Almost all subsequent commentators on, and proponents of, the new thought psychology quoted and discussed Wundt's critique (Wundt 1907c).

Wundt had little cause to show the Würzburgers much sympathy. As we have seen, several of them attacked key ingredients of his psychology, such as his classification of associations (Mayer and Orth 1901; Thumb and Marbe 1901), his theory of feelings (Orth 1903), his distinction between association and apperception (Watt 1905), his measurement of choice times (Ach 1905), his views on the place of volition in reaction experiments (Ach 1905), and his theory of judgements and thinking (Marbe 1901; Bühler 1907b). And, to make matters worse, usually these attacks were couched in harsh and almost disrespectful terms. Moreover, these assaults marked only the area of *openly pronounced* disagreement. To this area we have to add the set of those rejections and refutations that were not stated explicitly.

First, in 1892 Wundt had published a book-size argument against the use of hypnosis in psychological experiments (Wundt 1892a). Wundt had argued there that hypnosis did not produce reliable data, and that it was immoral to use hypnosis for other than strictly medical reasons. Nevertheless, Ach (1905) and Grünbaum (1908) used hypnosis in some of their thought-psychological experiments.

Second, Wundt had demanded that association experiments should use visual rather than verbal stimuli; that the experimental subject and the experimenter be situated in two different rooms; and that the experimental subject be left to write down retrospective reports on his or her own. Wundt advocated visual stimuli because these were less constraining than verbal stimuli; he called for a spatial separation because the presence of a second person impeded thinking; and he suggested written reports because only they could be done in peace and quiet (Wundt 1903a: 546–9). The Würzburgers, on the other hand, used both verbal and visual stimuli, emphasised the proximity of experimenter and experimental subject, and had the experimenter write down the protocol.

Third, Wundt was also on record as a critic of introspection (1888a). Wundt distinguished sharply between self-observation and self-perception. To perceive one's mental states or processes was to be aware of them, whereas to observe them was to focus on them methodologically and attentively. While allowing for self-perception, Wundt held that self-observation was, strictly speaking, impossible: Focusing one's attention on a mental state or process would destroy or severely alter that state or process. Wundt presumed that only under very special circumstances could inner self-perception be turned into scientific observation: The self-perception had to take place under experimental conditions; it had to concern 'lower' processes; and the so-called introspective reports had to be restricted to judgements of size, intensity, and duration of physical stimuli, supplemented at times by judgements of their simultaneity and succession (Danziger 1990a: 36). Needless to say, the Würzburgers' 'systematic experimental self-observation' went far beyond these limits.

Fourth, precisely because he believed introspection to be highly unreliable when it came to higher mental processes, Wundt had long defended the need for a second, non-experimental kind of psychology. This second kind of psychology was 'collective psychology' (*Völkerpsychologie*). According to Wundt, psychology had two main parts, experimental psychology and collective psychology. Of these, experimental psychology studied the 'lower' psychological processes of sensation and perception, whereas collective psychology investigated the 'higher' processes of thinking. Collective psychology focused on the products of human interaction, namely language, myth, and custom. In each case, the emphasis was on identifying universal structures of the human mind. Collective psychology was needed, Wundt argued, because thinking could not be studied under experimental conditions. Thinking was simply too complicated, and contained far too many variables, for it to be both open to investigation under controlled experimental conditions, and reliably reported after the event (e.g. Wundt 1908b: 163–71; Danziger 1980a: 247; Danziger 1980b: 110, 115). The Würzburgers ignored Wundt's collective psychology – at least in their thought-psychological writings.

Fifth, by the 1890s, Wundt had come to reject the idea of a psychological unconscious (1896: 34). There existed only two realms, the realm of the conscious and the realm of the physical. The first was the topic of psychology, the second the subject matter of the natural sciences. Mental contents could be more or less conscious, that is closer to, or further away from, the focus of attention. But mental contents could not leave consciousness without ceasing to be mental entities. In their thought-psychological studies, the Würzburgers did not explicitly discuss and reject Wundt's views on the unconscious. Most of them took the notion of a psychological unconsciousness for granted, however. For instance, Ach (1905, 1910) clearly assumed that presentations *qua* presentations could rise into, and drop out of, consciousness.

Sixth, the Würzburg psychologists all followed Külpe (1893) and rejected Wundt's distinction between mental and physical causality. In Wundt's view, three characteristics distinguished mental from physical causality. First, whereas for the natural sciences cause and effect were 'separate experiences, *disjecta membra*', in that the causal connection between two events 'comes only from the conceptual connection and treatment of experience', in psychology the connection between psychological elements was not a matter of theory but 'a fact of immediate consciousness'. Second, mental causality could not be reduced to physical causality because the explanation 'of psychological processes is everywhere shot through with value determinations'. And third, psychological causality was distinct from physical causality because 'the formation of mental products which indicate a conscious purposive activity, in which there is a choice between various possible motives, requires a real consideration of purpose' (1894: 43, 98, 108, 117; cf. Mischel 1970).

Seventh, for Wundt the distinction between these two kinds of causality was inseparable from his 'psychophysical parallelism' (1894). This was the idea that the mental and the physical realms could neither be reduced to one another, nor be linked to one another causally. Mental events could only be explained by prior mental events; and physical events only by prior physical events. Wundt had cause to believe that the Würzburgers had followed Külpe (1893) in rejecting his psychophysical parallelism.

Finally, even theoretical disagreements aside, it is not difficult to imagine that Wundt was particularly annoyed by the following aspect of the Würzburgers' work and criticism. Whereas he, the grand old man of German psychology, was singled out for criticism and scorn, other psychologists of the older generation – competitors such as G. E. Müller, T. Lipps, F. Brentano, and C. Stumpf – were adopted and praised. Even philosophers hostile to the advance of experimental psychology, such as E. Husserl, were treated with respect and admiration by the Würzburgers.

Wundt's most important attack on the Würzburg work appeared in 1907, in a paper entitled 'On Interrogation Experiments and the Methods of Thought Psychology' (*Über Ausfrageexperimente und über die Methoden zur Psychologie des Denkens*, 1907c). Wundt concentrated his criticism on Marbe (1901), Ach (1905), and Bühler (1907b).[1]

Wundt began by trying to situate the Würzburg experiments within psychological experimentation more generally. The Würzburg experiments fell within the broad category of 'reaction experiments', that is 'those psychological experiments in which the experimenter brings it about that the observer is affected by a stimulus, and that the observer experiences a mental reaction'. Three features distinguished the Würzburg method from other reaction experiments: the stimulus was a question (e.g. Bühler's 'Do you know what Eucken means when he speaks of a world-historical apperception?'); one did not need any instruments (cf. Marbe 1901);

and the subjects had to give verbal introspective reports after every experimental trial. Wundt proposed 'interrogation experiment' (*Ausfrageexperiment*) as a label for the Würzburg method. This was a natural title not least because Wundt suspected that the setting of school examinations had been the model for systematic experimental self-observation (1907c: 303–5).

Wundt argued that 'interrogation experiments' did not deserve to be called real experiments. He lamented that 'experiment' had come to mean any scenario in which some experimenter produced some effect on some experimental subject. 'Hypnotic experiments' were one example of this inflation. The Würzburg interrogation experiments were another. Indeed, they were 'sham experiments' in so far as they violated all four criteria for good experimental work. First, the observer, that is the experimental subject, was unable to determine alone the occurrence of the process to be observed; this was left to another person, the experimenter. It was the experimenter who 'surprised' the subject with the question as an 'unexpected event'. Under these circumstances the already well-known difficulties of self-observation 'increased infinitely'. Second, the observer could not help disturbing the process to be observed. Wundt suspected that the subjects in the Würzburg-style experiments were impeded in their problem solving by concurrent introspection. The Würzburg experiments presupposed that subjects attended to both the problem and their problem solving. But such 'doubling of the personality' was impossible. Third, the same observation could not be repeated. That is, one could not give the same task to the same subject twice. The second time around the subject would rely on his or her memory of the first time, rather than go through the same reasoning process again. And fourth, the conditions of the occurrence of the observed process could not be varied; a varied task would be a new task altogether (1907c: 302, 310–11, 329–39).

Wundt contrasted these 'sham experiments' with 'perfect' and 'imperfect' experiments. Perfect experiments fulfilled all four criteria, imperfect experiments at least some of them. As examples of perfect experiments, Wundt mentioned E. Meumann's work on temporal presentations, studies on visual illusions, such as the Müller-Lyer, as well as Ebbinghaus's experiments on memory. One positive feature of all of these experiments was that they needed no experimenter as a second person. Experiments on feelings, on the other hand, were bound to be imperfect. This was so mainly because the conditions of repeatability and variability were hard to meet (1907c: 315–6, 322, 328; cf. Ebbinghaus 1885; Meumann 1894).

Wundt was especially adamant about showing that the Würzburgers had not improved the traditional method of introspection. Wundt was critical of introspection in any case. But the Würzburgers had made matters worse. The Würzburg-style introspection was a 'self-observation under the most difficult conditions'. In the Würzburgers' experiments, self-observation was made extremely difficult by the presence of a second person, by the task as an 'unexpected event', by the 'pernicious influence of surprise', by a 'crossfire of questions' – in sum, by a setting that created 'exam pressure' (*Examenspresse*) (1907c: 329, 335, 337, 349).

Not only did Wundt criticise the use of questions as stimuli in reaction experiments, but also he opposed questions by the experimenter to the subject about the retrospective report. Rather than improve recall, the *post factum* questioning led to memory illusions: The experimenter's questioning would make subjects believe that they had to have had certain experiences. They would then report these assumed experiences as facts. Not even careful questioning could avoid being suggestive: 'a question is always a way of influencing someone, never mind how carefully it is couched' (1907c: 338–9). In the Würzburgers' case, Wundt identified two suggestions:

> To examine someone about the content of her earlier self-observations inevitably created two suggestions: a '*hetero*-suggestion' in the observer – caused by the question, and an '*auto*-suggestion' in the experimenter. The experimenter attributes certain conscious experiences to the experimental subject and then goes on to interrogate the subject about just these very assumed experiences. This kind of leading questioning is well known from the practice of many examining magistrates.
>
> (1907c: 340)

Wundt was no less scathing when he discussed 'the main outcome' of the Würzburgers' 'thought experiments' [*sic*!]. The main observation in these experiments amounted to simply 'nothing': 'the observers have observed nothing'. Put differently, the Würzburg subjects had found that 'thoughts' existed without any substrate of sensations, feelings, or presentations. In Wundt's analysis, accepting this result was tantamount to resurrecting the scholastic notion of the *actus purus*. The scholastics believed that because the 'active spirit' was itself independent of all matter, therefore also its products, that is thoughts, had to be 'pure'. According to Wundt, the scholastic doctrine was itself a product of Aristotelian metaphysics and 'naïve' folk psychology: 'still today the naive observer – that is, the observer ignorant of psychological analysis – takes mental processes to be completely immaterial' (1907c: 340, 344–5).

The Würzburgers could have done better, Wundt surmised, if only they had taken on board some key results of his own psychology. Particularly important here was the distinction between consciousness and attention, a distinction the Würzburgers had ignored. But there was a big difference between a mental content being in consciousness ('the mental field of vision') and a mental content being within the focus of attention ('the mental centre of the field of vision'). The Würzburgers had not found depictive (*anschaulich*) substrates for thought within the focus of attention. From this they had drawn the wrong conclusion, to wit, that such depictive substrates did not exist all. But the depictive substrates were there: only they were lurking in the background of consciousness, outside the realm of attention (1907c: 347–50).

Wundt believed that this view of consciousness could even be confirmed introspectively. Allowing for such introspective confirmation did not mean that Wundt retracted from his general pessimism regarding the use of introspection. Introspection concerning thought processes was always 'fragmentary and faint' at best, but occasional and unsystematic retrospection could serve an auxiliary and heuristic function. In any case, unbiased retrospection showed that the process of forming judgements and thoughts never lacked depictive elements. During the first stage of this process, a 'total presentation' (*Gesamtvorstellung*) appeared in the background of consciousness. During the second – sometimes simultaneous – stage, this 'dimly conscious' total presentation became represented in the centre of consciousness, that is in the focus of attention, as an 'intense and clear feeling'. During the third phase the total presentation unfolded into two fully conscious presentations, the subject presentation and the predicate presentation, and this unfolding took place under the control of, and within, the focus of attention. Given this theory Wundt needed neither situations of consciousness nor Bühlerian thoughts: in his scheme both were subsumed under feelings (1907c: 351–56).

Self-observation could get no further than this, however. The laws that governed the unfolding of thoughts in detail could only be found by collective psychology. Its investigations into language were especially important here (1907c: 357).

Wundt concluded by justifying his criticism in general terms. On the one hand, criticism of the Würzburgers' work was timely because their peculiar brand of thought psychology was becoming more and more influential. Many people seemed attracted to it, not least because it was so easy. On the other hand, Wundt suspected that these 'sham experiments' would soon be picked up by applied psychologists. They would apply systematic self-observation in the classroom, or perhaps even sample – with the help of questionnaires – the self-observations of ordinary folks. This would inevitably lead to a further deterioration of scientific standards within psychology (1907c: 359–60).

Müller

Georg Elias Müller's criticism of the (broader) Würzburg school came later than Wundt's, and in three 'instalments'. In 1911 Müller criticised Ach's version of systematic experimental self-observation; in 1913 he attacked the determining tendency, the situation of consciousness and the awareness; and in 1919 he published a detailed criticism of Selz (1913) (Müller 1911, 1913, 1919).

Judged in the light of the following facts, it might seem surprising that Müller directed his criticism primarily against Ach. Ach had been Müller's assistant in 1901, and although Ach had done his doctorate in Würzburg in 1899, he defended his *Habilitation* in Göttingen in 1902. Moreover, in Ach's studies (1905, 1910) Müller was an influence not only as a theorist – Müller's 'perseveration tendency' figured centrally in Ach's work – but also as far as experimental technique was concerned.[2]

Upon closer inspection, however, Müller's discomfort with Ach's work in particular turns out to be understandable and natural. Of the four leading psychological research schools in Germany at the time (Berlin, Göttingen, Leipzig, Würzburg), it was Müller's Göttingen that was most strongly committed to associationism. As Müller undoubtedly perceived the situation, Ach (and later Selz) led the Würzburgers' charge against association psychology. That the attack came from a former student and assistant, and with experimental designs adopted from his own work, can hardly have eased Müller's annoyance.

According to Müller, the mind had three layers. The uppermost layer was consciousness. Consciousness had two key features: it was 'narrow', and it was 'transitory'. In other words, no more than one presentation could be in consciousness at any one time, and most presentations were unable to stay in consciousness for more than a short moment. The narrowness of consciousness created stability; its transitory nature enabled the organism to react quickly to changes in the environment. The other two layers of the mind were below 'the threshold of consciousness': just below that threshold came the level of mobilised or activated presentations, and deeper down, the stratum of presentations without activation (Müller and Pilzecker 1900: 79, 91).

Müller attempted to identify the laws that governed the distribution of presentations over these three layers. Such laws were of two kinds. There were the laws of association and the law of perseveration. Laws of association concerned (the strength of) connections between presentations: that is, how presentations became associated; how repeated triggering of an association would strengthen the connection between the two associated presentations; or how different associations could strengthen or else inhibit one another. The law of perseveration stated that

> having appeared in consciousness, every presentation has the tendency to rise freely into consciousness again. This is the perseveration tendency. It usually weakens quickly. This tendency is stronger the more intensively attention had been directed at the presentation in question. The tendency is increased if the presentation…in question occurs repeatedly.
>
> (1900: 58)

Together with his assistants, F. Schumann and E. Pilzecker, Müller endeavoured to study the workings of these laws experimentally. To cut out the distorting effects of already existing associations (i.e. associations that his subjects brought to the experimental situation), Müller followed Ebbinghaus's famous memory experiments and worked only with nonsense syllables. Müller's memory experiments differed from Ebbinghaus's, however. Ebbinghaus's central measure had been how many rounds of reading were necessary before he (as his own subject) could produce the whole given series of nonsense syllables without mistake. Müller used the number of 'hits' (*Treffer*) instead. In other words, he presented his subjects with elements of an already learned series of nonsense syllables and counted the number of correct answers to the question 'which is the next syllable in line?' Müller's favourite tool for presenting his subjects with series of nonsense syllables was his 'memory drum': this was a horizontal drum with the nonsense syllables attached to its surface, rotating at a controlled speed behind a screen in which a small opening exposed one nonsense syllable at a time (Blumenthal 1985a: 57). The point of all of this was not – as it had been for Ebbinghaus – the discovery of curves of learning and forgetting, but rather the experimental creation and manipulation of associations and perseverations. For instance, nonsense syllables that followed immediately upon one another were more strongly associated than nonsense syllables three or four steps removed from one another.

Most of the data that Müller collected from these experiments were 'objective' rather than 'subjective': They were quantitative and based on observation of others. Müller was not opposed to introspection, however. He regarded the introspective method as an important part of the psychologist's arsenal. But it had to be used carefully and with full awareness of its different forms and potential pitfalls (Müller 1911).

This brings us to Müller's critical discussion of Ach's use of the introspective method. This discussion was part of the first of the three volumes of Müller's *On the Analysis of Memory and the Train of Presentations* (*Zur Analyse der Gedächtnistätigkeit und des Vorstellungsverlaufes*, 1911).

Müller first objected that Ach never gave a full protocol of his interaction with his subjects; he only gave summaries of the retrospective reports and failed to list his questions to his subjects. This made it difficult to understand what exactly had happened during the experiment and during the period of retrospection (1911: 138). It was part and parcel of Ach's insufficient reporting that he left his subjects in anonymity. Müller commented: 'this method does not fulfil the demands of good science. Imagine a historian who mentions the eyewitnesses upon which he relies but who never indicates who of the witnesses made which statement'. Müller also suspected that anonymous subjects would feel less responsible for the quality of their contribution to the experiment (1911: 5–6).

Furthermore, while agreeing that asking questions about the subjects' experiences was sometimes unavoidable or useful, Müller suspected that Ach had queried too often and without proper care. Ach's '*Vielfragerei*' (excessive questioning) had no doubt been suggestive and had led the subjects to reinterpret question-induced fantasies as memories (1911, 91, 93–4, 138).

In addition, Ach showed little awareness of the various ways in which retrospective self-observation could fail. Evidence for this included statements such as the following: in retrospection the subject had her earlier experience '*in nuce*' and 'at once'; or, the retrospectionist was able to furnish – possibly helped by the experimenter's questions – a 'full and complete' report of her experience. Müller was not impressed by Ach's subjects' abilities: 'I would chase away, and treat as a disgrace and as a cheat, any experimental subject who declared himself capable – by focusing attention upon different parts of the persevering content – of answering all or most of [Ach's] questions' (1911: 140–1). Müller also found fault that Ach had given no experimental data in support of his claim that the pre-experimental intention to later report one's experiences improved the quality of the retrospection (1911: 141).

Moreover, according to Müller, Ach and the other Würzburgers had not properly trained their subjects. In particular, the experimenters had not taught their subjects a healthy scepticism towards their (i.e. the subjects') own presentations during the phase of retrospective reporting. The Würzburgers also ignored the obvious principle that 'not remembered' did not always mean 'not occurred'. One's inability to recall a state of mind was not conclusive proof that such state of mind had not occurred (1911: 109, 142).

Finally, the Würzburgers had also overlooked the problems that resulted from demanding retrospective reports after every single experimental trial. The main danger here was that such repeated reporting destroyed the naturalness of the reaction (1911: 128, 135).

Müller was no less dismissive when, two years later, he turned to a critical discussion of determining tendencies and non-depictive conscious experiences. As far as determining tendencies were concerned, Müller was convinced that there was no evidence for the existence of such 'mysterious…life-forces'. The phenomena that were adduced as proof of their existence could all be explained on the basis of association and perseveration alone (1913: 478).

Put in a nutshell, the following was an account of what 'really' happened when the subject solved a task such as 'add the two numbers that I shall show you shortly'. During the first time interval, say $t - 1$, the subject heard the task. This interval was followed by the 'preparatory phase', say $t - 2$. During $t - 2$ the task activated a problem-solving method from memory. This happened by means of an existing association between the (type of) task and the method. The method could now either enter consciousness – thereby further strengthening its association with the task – or remain on the level of unconscious but mobilised presentations. Moreover, the activation of the method brought with it a mobilisation of possible solutions (say, equations such as 2 + 3 = 5, 7 + 8 = 15, etc.). During the third interval, *t* – 3, the subject was presented with two numbers. These numbers entered a consciousness that by now had a specific 'constellation'. This constellation was the links that had been created, strengthened, or mobilised during *t* – 2. In the next phase, *t* – 4, a number of different things could happen. Two of these possibilities were particularly important.

On the one hand, the task could now reappear in consciousness. This was possible because there existed an associative link between the task and the appearance of two numbers. When the task did appear in consciousness, however, it could also pull the problem-solving method over the threshold, and increase the activation of the set of possible solutions. Problem solving then happened fully on the basis of association. The associative compound of the stimulus numbers, the task, and the mobilised solution candidates pulled the correct sum into consciousness.

On the other hand, the task did not need to reappear in consciousness. Sometimes the solution would surface in consciousness straight away, that is right after the subject had seen the stimulus pair of numbers. In this case, the task as well as the solution candidates remained 'latent'; they were mobilised, effective as far as associations were concerned, but not conscious. This case did not fall outside the phenomena that could be explained by means of associations (1913: 427–73).

Müller was equally unimpressed with the argument of Ach (1910). Recall that Ach first created an association between two syllables, for instance 'dus–rol', and then set a task such as 'find a rhyme' with respect to the first element. If the subject reacted with 'rol', then association was assumed to be stronger than determination. If the subject reacted with something like 'bus', then determination was alleged to be stronger than association. Müller did not accept that this was an opposition between associative and determining tendencies. For him the competition in this case was between two associative tendencies: the first linked the syllables, the second the task of rhyming, a problem-solution strategy, mobilised possible solutions, and the shown syllable (1913: 480).

Having dismissed Ach's work in detail, Müller went on to make briefer remarks on criticisms of associationism in Dürr (1907), Messer (1908), Koffka (1912), and Külpe (1912c). By and large, Müller confined himself to reminding these authors that Ach had not – as they assumed – refuted associationism. One important new point came up in Müller's discussion of Koffka, however. Koffka (and Bühler) had used the alleged existence of non-depictive elements of consciousness as an argument against associationism. Müller's reply centred around the claim that Koffka and others were wrong to restrict the use of 'presentation' to depictive elements. Nothing was gained by this restriction. Dropping it, however, meant a return to the traditional claim that there were just three kinds of conscious mental contents: presentations, feelings, and volitions. Moreover, Müller insisted that *even if* non-depictive mental contents existed, this did not amount to a refutation of associationism. The laws of associations were formal and general. They contained no restrictions concerning the elements to which they could be applied (1913: 495).

In fact, Müller did not believe that the Würzburgers in general, and Ach in particular, had made a convincing case for non-depictive contents of consciousness. Their claims were based upon poor observation and sloppy theorising. What Marbe and Orth had called 'situations of consciousness' were actually unclear, hazy presentation-images. And what Ach had dubbed 'awareness of tendency', 'awareness of meaning', or 'awareness of determination' turned out, upon closer in(tro)spection, to be hazy visual or kinaesthetic presentations. But why then had the Würzburg experimental subjects reported such presentations as situations of consciousness and as awarenesses? Müller suspected that subjects came to speak of such entities only because the terminology had been suggested to them. If only the experimental subjects had been given a chance to speak their own minds, and if only they had been fully honest, they would indeed have spoken of fleeting, hard-to-remember, and hazy presentations: 'But if one has equipped them with the expressions "awareness" or "situation of consciousness" then they will of course tend to say that such awareness or situation of consciousness has been present' (1913: 528–33).

Müller presented a separate argument against Achian awareness as an immediate, non-depictive knowledge. Knowledge was a disposition, not a mental occurrence. Thus Müller's knowledge of French was nothing but the mental disposition that enabled him to translate French words into German words and to utter well-formed French sentences. The existence of a disposition could not be directly experienced; that oneself, or some other person, had a disposition was something that could only be inferred. Müller realised that his analysis of knowledge coincided with Marbe's. This led him to suspect that Ach had picked his Würzburg subjects in a biased way: 'It is noteworthy that amongst the Würzburg experimental subjects upon whose reports Ach bases his doctrine of awarenesses, one cannot find Marbe – and this despite the fact that Marbe was still in Würzburg at the time.' Concerning those subjects that Ach ended up using, Müller submitted that either they had made mistakes or else they had been misinterpreted by Ach. Some of the blame, however, had to rest with these subjects, too:

> my criticism of the Achian method is directed not only at Ach but also at least against those of his experimental subjects who are academic representatives of psychology, that is, Bühler, Dürr, and Külpe. If only these experimental subjects had behaved in a critical fashion, and if only they had protested against Ach's method, then much damage could have been prevented.
>
> (1913: 535, cf. 542, 544)

Müller's review of Selz (1913) repeated many of the themes already mentioned (Müller 1919). Again, determining tendencies were likened to teleological life-forces, and again Müller marvelled at the fact that 'as soon as an experimenter belongs to the Külpe school, fate sends him experimental subjects who are graced by God and possess non-depictive memories; a type of memory that I and most other mortals lack' (1919: 119). In great detail Müller turned on Selz's idea according to which the existence of complexes was proof for the insufficiency of associationism. As Müller saw it, such insinuation was based upon an inadequate knowledge of associationism:

> Already during the time when Külpe was still a student here in Göttingen, I presented in my lectures the following doctrines: presentations can form unified complexes; such complexes can be associated with other complexes or with single presentations; and any part of such a complex has the tendency to reproduce the whole complex, and in the right order, and so forth.
>
> (1919: 109)

Other critical reactions

Wundt and Müller did not stand alone in reacting critically to the Würzburgers' new thought psychology. Views that were close, and sympathetic, to Wundt's criticism were expressed by G. F. Lipps

(1906), E. Meumann (1907a, 1907b), G. Deuchler (1909), G. Reichwein (1910), and G. Anschütz (1911, 1912a, 1912b, 1913a, 1913b, 1913c).

In a review of Ach (1905), G. F. Lipps questioned whether systematic experimental self-observation was really any different from old-fashioned armchair introspection. The only way to improve upon the latter was to complement the subjective, self-observational data with objective measurements of time and the number of correct or incorrect reactions. Ach had done too little in this respect, and his corpus of data was much too small (1906: 678).

In 1907, E. Meumann published both a criticism of Watt's and Messer's time measurements (1907b), and a detailed summary and review of Wundt's criticism (1907a). In the latter publication, Meumann agreed that all questions were potentially suggestive. The only exception to this rule was this query: '*What* have you observed about the train of events?' And Meumann continued:

> But even here suggestion – and thus the breeding of school-internal dogmas – can only be avoided if the observers are not talking to one another or to the experimenter about the interpretation of the experiments. It is also imperative that the observers have neither read nor thought about the problems in question...in the Würzburg experiments, these conditions have not been met.
>
> (1907a: 129)

According to Meumann, Wundt was also right to claim that the Würzburg subjects had 'observed nothing'. This verdict was passed on Bühler's, Marbe's, and Messer's subjects in particular (1907a: 131). Curiously enough, one year later, in Meumann's book *Intelligence and the Will* (*Intelligenz und Wille*, 1908), the assessment of the Würzburg work was more positive. Now Meumann regarded the discovery of non-depictive thoughts and of tasks as a major achievement (1908: 156).[3]

G. Deuchler (1909) concurred with Wundt that the presence of the experimenter distracted the experimental subject; that all questions were suggestive; and that the Würzburgers had neglected the collection of objective data. Introspective reports had to be checked against such data. Deuchler was particularly unhappy about the Würzburgers' suggestion according to which a given introspective report was reliable as long as it agreed with other reports: 'it amounts to determining the reliability of observations on the basis of a vote'. Deuchler called on experimenters to formulate strict rules for their subjects to follow. Nevertheless, he granted that ultimately experimenters had to trust the self-control of their subjects (1909: 381, 386, 388, 390–1).

G. Reichwein's dissertation, *Recent Studies about the Psychology of Thought* (*Die neueren Untersuchungen über die Psychologie des Denkens*, 1910), was a detailed, book-size criticism of the studies of Marbe, Watt, Messer, Ach, and Bühler. The criticism was strongly informed by Wundt. With respect to Marbe's work on judgements, Reichwein objected, *à la* Bühler, that Marbe had overlooked the possibility that judgements might become mechanised, and that judgement qualia might be observable only in the case of unmechanised judgements. The looked-for quale was the experience, first described by Wundt, in which the subject presentation became separate from the predicate presentation (1910: 12, 20). To Ach, Watt, and Messer, Reichwein suggested that their 'determining tendencies', 'tasks', or 'consciousnesses [*sic*!] of sphere' were in fact nothing but Wundtian 'total presentations' residing in the background of consciousness (1910: 31, 52). Ach's awarenesses and Bühler's thoughts fared no better. The only reason why Bühler's subjects were unable to report sensations, presentations, and feelings – as running alongside their thoughts – was that Bühler's tasks had been so difficult. The subjects' attention had been so completely captured by the task that they failed to notice the depictive elements (1910: 115). In two respects Reichwein was willing to side with the Würzburgers against Wundt, however: he had no objections to the use of questions, and he valued the 'division of labour' between experimenter and experimental subject. Wundt exaggerated the experimental subject's need for solitude (1910: 77, 151).

Finally, G. Anschütz criticised the Würzburg work in a number of publications. He criticised the Würzburgers' use of philosophers such as Husserl and lamented the alleged vagueness of their key concepts (1912b: 204, 206). He also contradicted Külpe arguing that situations of consciousness and awarenesses were not really different from Wundt's feelings of meaning, and he attacked what he perceived as Külpe's thought-psychological imperialism (1912b: 198). This attack on Külpe (1912b) led to a series of increasingly hostile exchanges between Anschütz and W. Köhler in the *Zeitschrift für Psychologie* (Anschütz 1912b, 1913b, 1913c; Köhler 1913a, 1913b, 1913c).

Anschütz's most detailed critical evaluation of the Würzburg methodology was published in his book *Intelligence* (*Die Intelligenz*,

1913a). The Würzburg experimenters created unnatural situations, failed to recognise the subjects' need for solitude, and did not properly select appropriate subjects (1913a: 190, 208). If proof was needed for the last-mentioned claim, one only needed to note the negative result of Marbe's study: the subjects' inability to identify judgement qualia clearly attested to their incompetence (1913a: 197). Moreover, Anschütz was willing to accept the results of systematic experimental self-observation only if a variety of different subjects had been used, if their reports had been subjected to statistical analysis, and if the experimenter had obtained information about his subjects' character (1911: 487).

Other critics of the Würzburg school shared Müller's agenda of defending associationist psychology. One particularly influential paper was that by G. Moskiewicz, 'On the Psychology of Thought' (*Zur Psychologie des Denkens*, 1910b; cf. also Moskiewicz 1910a). Moskiewicz regretted that so many psychologists of thought preoccupied themselves with the study of judgements:

> It is true that we think in judgements; but individual judgements are only the elements, the building blocks of thinking, not thinking itself. Thinking itself is the quite unique sequence of judgements and presentations, and we do not understand this sequence any better by developing psychological analyses of the judgement.
>
> (1910b: 311)

Furthermore, Moskiewicz believed that a slightly modified associationism could capture much of what was correct and important about Ach's determining tendencies. The needed modification was the assumption that presentations could form constellations. Under one constellation, the presentation 'clouds' would reproduce the presentation 'rain'; under a different constellation, the presentation 'clouds' could reproduce the presentation 'snow'. In the first case, the constellation was that of 'cloud-and-summer'; in the second case, the constellation was that of 'cloud-and-winter' (1910b: 330–6).

For committed followers of Müller, the Würzburgers had completely failed in their attempt to disprove the all-pervasiveness of associations. Writing after Müller's criticism, H. Henning (1919) insisted that there existed, for the moment, no further need even to discuss the Würzburg work 'critically and polemically'; Müller's refutation had allegedly been so effective that 'no researcher seems to be willing to repeat the old mistakes in new publications' (1919: 2). Nevertheless, Henning felt that at least one more stab at Bühler's method could not hurt. Bühler's method of presenting his subjects with difficult tasks had been a fatal mistake. It was as if Hering and Helmholtz had tried to study perception by asking subjects to give a detailed description of a landscape (1919: 7).

Another student of Müller, W. Baade, was slightly more constructive in his attitude towards the Würzburgers. He too deemed their experimental work insufficient, but he made a number of suggestions on how it needed to be improved. In this spirit he suggested, amongst other things, the use of an 'introvocation' (i.e. interruption) method, and the employment of recording devices (1913a, 1913b, 1918).

Not all of the critical reactions to the Würzburg work came from within Wundt or Müller's orbits. For instance, E. Wentscher (1910) objected that Ach had contributed little to a proper understanding of the will. Reaction experiments were a far cry from human decision making in everyday contexts. Moreover, determining tendencies were only a necessary, but not a sufficient, condition for willing. Phenomena of determination occurred all throughout mental life, but not all of such phenomena were phenomena of willing (1910: 178–83). E. Schrader (1905: 57–60) insisted that Marbe had rushed into experiments before he was clear about the conceptual issues; thus it was no surprise that his subjects were unable to find judgement qualia, and that his concept 'situation of consciousness' remained vague. T. Elsenhans (1912) complained that the Würzburgers' systematic experimental self-observations were no real improvement on traditional introspection and that Bühlerian thoughts and Achian awarenesses were little more than artefacts of the laboratory. They were the artefacts of an experimental design in which subjects – under time pressure – were more or less forced to disregard depictive contents (1912: 205). H. Maier distinguished between primary and secondary thinking and argued that the Würzburgers had studied only the latter, not the former. Primary thinking was tied to depictive presentations, whereas secondary thinking was routinised and mechanised and happened on the basis of presentations that had largely lost their depictive character. As Maier saw it, the new thought psychology was mistaken in equating mechanical and routinised forms of thinking with thinking itself (1914a: 352–58; 1914b). Finally, M. Frischeisen-Köhler regretted that thought psychology had failed to make any contribution to understanding issues such as historical life, the development of languages, myths, or religion, or the economy (1918: 12).

Positive reactions

Many reactions to the Würzburgers' experiments were far from critical, however. A considerable number of philosophers praised their results and heralded the Würzburg research programme as 'a truly important event in the history of psychology' (Popp 1913: 58).

Particularly friendly was the reaction amongst neo-Kantians. H. Rickert applauded Marbe's book (1901). Marbe's main thesis – that 'judgement' was a logical and not a psychological category – was grist for the neo-Kantian anti-psychologistic mill (Rickert 1904: 94). Ach's work (1905) was reviewed by J. Cohn and praised for its attempts to bring 'ever higher and ever more complex problems into the realm of experimental work' (1906: 1684). E. Cassirer's massive study, *The Concepts of Substance and Functions* (*Substanzbegriff und Funktionsbegriff,* 1910), congratulated thought psychology on its discovery of non-depictive elements. Nevertheless, Cassirer hoped that thought psychologists would now go further and investigate how the mind created and constructed meaningful sentences (1910: 458–9). And R. Hönigswald published a forty-page article on 'Questions of Principle Regarding Thought Psychology' (*Prinzipienfragen der Denkpsychologie,* 1913) in which he credited Würzburg thought psychology not only with having refuted associationism, but also with having recaptured key Kantian insights. Such insights included that thinking was reflexive – the thought 'I think' must be able to accompany all of my thinking processes – and that non-depictive concepts or meanings were involved in, and intermingled with, all mental contents, including depictive presentations (Hönigswald 1913: 210, 212, 217, 226).

Also largely supportive of the Würzburg work were four important neoscholastic philosophers, C. Baeumker (1921 [1913]), J. Geyser (1908a, 1908b, 1909a, 1910, 1914b), M. Grabmann (1916), and C. Gutberlet (1908, 1915, 1923). The recurring main theme in the writings of these authors was the relationship between Ach's awarenesses and Bühler's thoughts on the one hand, and the scholastics' *species intelligibiles* and *conceptus mentis,* on the other. To begin with, the neoscholastics criticised Wundt's suggestion according to which the Würzburgers' thoughts and awarenesses were identical with the medieval *actus purus;* as the neoscholastics reminded their readers, *actus purus* did not refer to mental acts, but to God himself (Geyser 1908a: 101; Gutberlet 1915: 74–5). Although the Catholic philosophers felt that Ach and Bühler in particular had come close to Aristotelian and scholastic views of the mind, they also stressed important differences. Ach's awarenesses were not sufficiently 'pure', and they could be construed as having no further function than to announce, to consciousness, the arrival of presentations. Bühler exaggerated the extent to which thinking was independent of depictive elements. Bühler's belief that thoughts could occur without any concomitant depictive elements violated the Thomistic (and Aristotelian) principle that thinking was tied to '*phantasmata*' (*impossibile est intellectum nostrum...aliquid intelligere actu nisi convertendo se ad phantasmata* Geyser 1909a: 63; Baeumker 1921 [1913]: 114; Gutberlet 1915: 84; Grabmann 1916: 359). Nevertheless, there was no doubt in the mind of these neo-Thomists that the Würzburg work was a decisive advance over both associationism and Wundtian psychology. Moreover, Grabmann saw systematic experimental self-observation as a continuation of Augustine's and St Thomas's 'look inside' (*Innenschau*) (1916: 358). And Gutberlet made clear that for neoscholastic philosophers the Würzburg work was also of metaphysical significance: the emphasis on the existence of 'pure mental acts' strengthened 'the case for the immateriality of the soul against materialism' (1923: 26).

Other writers who commented positively are difficult to group. The leader of the fourth main psychological institute in Germany at the time, C. Stumpf of Berlin, listed the Würzburg studies amongst those that had shown that 'conceptual thinking, in all its operations...is much more independent of appearances (images) than association psychology has believed and taught for a long time' (Stumpf 1906: 25). Stumpf's endorsement was not surprising given that the Würzburgers often referred to Stumpf's work as an important independent confirmation for their own claims and approach (e.g. Bühler 1908b; Külpe 1907b). R. Müller-Freienfels (1912) claimed to have independently arrived at the same position as the Würzburgers, including their rejection of associationism (cf. also A. Brunswig 1910: 58, 72; W. Popp 1913: 57–8). W. Specht (1909) regarded the Würzburg work as *the* key to understanding pathological phenomena of attention. The importance of thought psychology for pathological phenomena was also underlined by K. T. Oesterreich (1910). And the neovitalist H. Driesch adopted both the Würzburgers' non-depictive thought and their concept of 'task'.

In his book *Logic as Task* (*Die Logik als Aufgabe*, 1913), Driesch undertook to conceptualise systematically humans' relationship to the world as a task-solving exercise. The ultimate task of all knowledge was, for Driesch, 'the creation of order'.

Mixed reactions

To conclude this review of reactions to the Würzburg work, I need to mention a few authors who mixed approval with criticism.

P. Petersen (1913: 194) listed the Würzburgers amongst those philosophers and psychologists who had succeeded in separating logic from psychology. One year later, the same author noted that the Würzburg studies had one-sidedly focused their attention on 'scientific thinking' and argued that to capture other modes of thinking, the Würzburg methodology needed to be supplemented by 'phenomenological intuition' (1914: 182). A similar mixed approach was advocated by W. Betz (1910). Although he agreed with Bühler's results, Betz believed that the sceptics could only be won over once the thought psychologist went beyond the laboratory and collected his evidence 'on the street and in everyday life' (1910: 267).

The most influential writer within the category of 'mixed reactions' was E. von Aster (1908). Although Aster criticised Bühler's work in detail, he was not opposed to the Würzburg programme in general. Instead, he advocated a return from thoughts to situations of consciousness. Aster's starting point was Bühler's distinction between intentional thoughts and non-intentional situations of consciousness. What did it mean to characterise thoughts and their moments? What did it mean to describe experiences such as 'consciousness of…', 'knowledge of…', 'consciousness that…' or 'the thought of…'? According to Aster it meant describing mental states in their non-intentional features; it meant characterising mental states in isolation from the objects towards which they were directed. And this described non-intentional situations of consciousness. Aster claimed that Bühler had confused the object of given conscious experiences with those conscious experiences themselves. Additionally, Bühler had also failed to recognise that his subjects often *expressed* rather than described their experiences: an utterance such as 'Oh well, this is a difficult problem', as part of an experimental report, did not describe the structure of an experience of thinking; it expressed such an experience (1908: 66–7).

Analytical summary

Because in later parts of this study the reactions of Wundt, Müller, and the neo-Thomists will be particularly important, I shall confine my summary to them, listing here the main objections and comments.

Wundt's (W) objections concerned all three main areas of the Würzburg work. Indeed, Wundt did not accept a single one of the twenty-nine main methodological and substantial theses of the Würzburgers that I listed in the summary of the last chapter. The only issue that Wundt and the Würzburgers agreed on was their rejection of associationism. This agreement did not mean much, however, because Wundt did not adopt the Würzburgers' alternative. They in turn rejected his.

W.1 The Würzburgers' experiments are 'interrogation experiments'. They are modelled on the school examination.

W.2 The Würzburgers' experiments are 'sham experiments', that is they are not experiments at all. They do not fulfil any one of the four conditions that a perfect experiment satisfied.

W.3 Perfect experiments are experiments that can be carried out on one's own.

W.4 Würzburg-type retrospection is self-observation under the most difficult conditions. The presence of a second person is a disturbance.

W.5 The questioning method introduces the danger of suggestion.

W.6 The Würzburgers' non-depictive contents are a resurrection of the scholastics' *actus purus*.

W.7 The Würzburgers conflate consciousness with attention.

W.8 Only collective psychology can ultimately clarify the nature of thought.

W.9 Thought psychology will have disastrous effects once it is combined with applied psychology.

Müller (M) was marginally closer to the Würzburgers than was Wundt. As opposed to Wundt, Müller accepted systematic experimental self-observation, at least in its broad outlines. And several of the Würzburgers made use of Müller's perseveration tendencies. Müller criticised non-depictive mental contents, but he reserved his greatest critical effort for his attack on determining tendencies. Müller's central goal was of course the defence of associationism.

M.1 Ach's way of reporting on his experiments is unsatisfactory. For instance, he should mention the authors of retrospective reports.

M.2 Asking questions is permissible, but Ach asks too many questions.

M.3 The Würzburgers disregard the many ways in which self-observation can fail.

M.4 The Würzburgers fail to train and control their subjects.

M.5 A combination of tendencies of association and perseveration can explain all aspects of the train of thought. Determining tendencies do not exist.

M.6 The traditional taxonomy of conscious elements (presentations, feelings, volitions) is sufficient. There is no reason to confine presentations to being depictive.

M.7 The laws of association do not just cover presentations. They cover all mental contents, whatever they may be.

M.8 The Würzburgers' allegedly non-depictive contents (situations of consciousness, awareness) are actually unclear, or hazy presentations. Subjects find such contents only because they have been suggested to them by the experimenters and their taxonomy.

M.9 There can be no immediate knowledge, because all knowledge is dispositional only.

M.10 The existence of complexes does not prove anything against associationism.

The main thing to bear in mind about the neo-Kantians is that their attitude towards the Würzburgers' thought psychology was a very positive one. Rickert applauded Marbe's anti-psychologism, and Cassirer and Hönigswald congratulated the Würzburgers on their refutation of associationism and on their discovery of non-depictive elements.

The same is also worth remembering about the neo-Thomists.

They sided with the Würzburgers against Wundt as far as the *actus purus* issue was concerned, and they applauded the proximity of Würzburg doctrines to the views of Aristotle, Augustin, and St Thomas.

3 Recluse or drillmaster versus interlocutor and interrogator

Introduction

In the last two chapters, I have summarised the technical content of the controversy over the Würzburgers' new thought psychology. In the next four chapters, I shall attempt a sociological analysis of this controversy.

In this chapter, I shall take as my starting point the differences in the social order of the respective research schools in general, and the three directors' (Külpe, Müller, Wundt) styles of leadership in particular. To have handy labels for these three styles, I shall speak of Wundt as the 'recluse', Müller as the 'drillmaster', and Külpe as the 'interlocutor'. I shall show that these three styles of leadership varied systematically both with substantially different theories of the human mind, and with different notions of how cognitive and social authority ought to be divided between experimenter and experimental subject. Lest my interpretative model be suspected of being static and deterministic, I shall conclude by making some observations on the development of the controversy. I shall explain that, although Külpe did not change his style of leadership, several Würzburgers – including Külpe himself – were ready to redefine the role of the experimenter. The experimenter changed from being an interlocutor to being an 'interrogator'.

The recluse

The following idea permeated much of Wundt's reasoning about the psychological and the social realms: the notion that thinking in general, and philosophical theorising in particular, demanded solitude. Wundt did not invent this idea, of course. Indeed, the link between thinking and solitude has been a pervasive topos of

Western culture; according to this topos, genuine, authentic, and successful knowledge acquisition is possible only for an individual who is able to escape the disturbing and corrupting influence and interference of others. Thinkers must be directed exclusively at the truth, not at their fellows. Only after the thinkers have obtained their knowledge will they emerge from their solitude and inform the community of their insights (Shapin 1990).

In making use of this topos, Wundt did not stand alone. In German university philosophy of the time, the topos of the philosophical recluse was widespread and popular. Undoubtedly the ideal of inward self-cultivation, central to the new elite of the academically educated, the *Bildungsbürgertum*, was an important source here (Nipperdey 1990: 382–9). The recluse topos also supplied a convenient way of characterising the philosopher as different from the natural scientist and the technician. After all, the rise of scientific research schools had brought home to everyone that science was a collective and communal activity. At the same time, the use of the topos in academic philosophy also reflected a real and intended isolation. For much of the second half of the nineteenth century academic philosophy was pushed into isolation, first by attacks from natural scientists, second by insinuations from non-academic philosophers, and third by a political climate – the *Nachmärz* period – in which any public philosophical–political engagement carried severe risks for one's professional career (Köhnke 1986). It is not surprising then that we find solitude treated as a philosophical virtue.[1]

The fact that German university philosophy of the period was predominantly clustered around so-called schools does not contradict my claim that the recluse theme was all pervasive. Historians of science have shown that research schools can have very different forms of sociability. As J. S. Fruton notes, at 'one extreme, the leader has been a quasi-military director of the work of subordinates, and at the other, a senior counsellor in the independent efforts of his junior associates' (1990: 2). Or, to put it another way, the social order of a research school can range from a circle of equals who seek knowledge by engaging in constant debate to a strict hierarchy in which a single prophet-like recluse oversees and directs disciples from a carefully orchestrated distance. The latter model was familiar to Wundt from his time as assistant in Helmholtz's physiological laboratory in Heidelberg between 1858 and 1863 (Turner 1993: 89).

It is intriguing to see how strongly the topos of the recluse structured Wundt's interaction with his students. According to contemporary sources, in his institute Wundt played the role of a remote 'sage'. Wundt's students complained bitterly that Wundt did nothing for them and that he reacted to their attempts at independent work 'somewhat arrogantly and without any attempt to understand them' (Hellpach 1948: 175). Students felt exploited by this 'old feudal overlord' (Hall 1921: 155), and they were put off by Wundt's 'cold intellectuality' (Pintner 1921: 187). They also disliked his dictatorial ways. As one student described Wundt's practice of assigning research topics:

> he had in his hand a memorandum containing a list of subjects for research, and taking us in the order in which we stood – there was no question of our being seated – assigned the topics and hours to us by a one-to-one correspondence.
>
> (McKeen Cattell 1921: 156)

Students who wanted to talk to the master were advised to 'wait...for His Excellency to pass from the laboratory down the corridor to his lecture room. Those who had missed him were directed to take up a position at a certain place on Thomas Ring [Street] that he was known to pass daily with clock-like regularity', clearly imitating Kant's walks around the streets of Königsberg a hundred years earlier (Arps 1921: 186; cf. Hellpach 1948: 174).[2] Wundt also let it be known that 'no one should speak to...[him] about any of his forthcoming books' (Judd 1921: 176) and that 'he could be very hard towards anyone who took issue with him concerning any of his published doctrines'. As his students saw it, Wundt did not want to stimulate but to pre-empt their thinking: 'Wundt thought for us: there were no problems left over for us to try our teeth on' (Tawney 1921: 181).

A less straightforward ingredient of the German model of the philosophical recluse was lectures given to large audiences. The sage did not talk to his students and he did not test his ideas in debates; instead he preached in a social setting that made discussion practically impossible. Moreover, the packed lecture hall attested to the authenticity of the knowledge obtained in solitude, and it underlined the belief that the solitary sage spoke to all and none. Lacking the rhetorical talent of some other German philosophers, Wundt had to work hard to fill the bill. And he succeeded: his carefully prepared lectures were attended by more than 600 people. Some listeners had to stand in the corridor, others around the podium. Conveniently timetabled from 5:15 p.m. until 6:00 p.m. on Wednesdays and Saturdays so that townsfolk could be present, Wundt's lectures were social events in Leipzig and were attended by 'many visitors from town, amongst them tradesmen and *Reichsgerichtsräte,* members of the conservatory, as well as high-school teachers' (Hellpach 1948: 170–1).

It is also worth pointing out that a German professor was expected to cover, in his lectures and books, the entire field of knowledge in his discipline. The enormity of the task was a powerful incentive to use the students as subordinates who could provide research materials, rather than to treat them as equals, as independent researchers in their own right. Wundt behaved accordingly. Some students felt that Wundt needed them 'chiefly to read for him' and thus supply him with notes and material for his 'amazingly clear and popular lectures' (Hall 1921: 155).

Although Wundt included demonstrations as part of his lectures (Judd 1921: 174), he did not in fact participate in the experimental work of his famous laboratory. In the early 1880s he had occasionally acted as experimental subject but later this practice stopped. The running of the laboratory was left to *Institutsassistenten* whom Wundt selected almost at random (Hellpach 1948: 176). Wundt 'merely moved over the waters of experimental work like a holy spirit' (Hellpach 1948: 174), and only 'once in a great while...would [he] make a tour of the work rooms' (Judd 1921: 174).[3]

On its own, the fact that Wundt distanced himself from his students and invested most of his time in his lectures is no more than a curiosity. It becomes more significant, however, once we note that Wundt's preferred model of social order – hierarchical structure with a recluse at the top – was reflected in his psychological theory of the human mind and in his conception of the internal structure of psychology as a field of knowledge. Psychological order repeated and justified social order.

In his theory of the human mind, Wundt made a sharp division between the 'lower' processes of, for instance, perceiving, feeling, willing, or forming presentations, and the 'higher' processes of, first and foremost, thinking. Although the higher processes presupposed the lower, the two realms were governed by different laws: the lower by the laws of association, the higher by the laws of 'apperception'. Apperception could be either active or passive. Passive apperception was the process in which a presentation or feeling became the focus of attention. Active apperception was the same as a choice between presentations or feelings. Apperception, and especially active apperception, was thus an activity of the will (Wundt 1903a: 341–5). Active apperception was Wundt's term for the selecting and structuring of experience by a self-determining, mature person (Wundt 1908b: 202; cf. Blumenthal 1970: 14–15; Weimer 1974: 245).

The products of the higher mental processes could not be explained from below. The reason for this was that active apperception was always 'creative': in modern jargon, the output of active apperception was always more than the sum of the inputs.[4] In Wundt's own words:

> The mechanism of associations is…the preparatory workshop for thinking:…that is, this mechanism constantly seeks to produce numerous links, and amongst those links there regularly are some that fit the purposes of thinking. Different associations end up fighting with one another, but then the purpose of the train of thought – a purpose fixed by the will [i.e., by active apperception] – prefers one link over others.
>
> (Wundt 1897: 41)

The attribute 'higher' in 'higher processes' was not just a convenient spatial metaphor; it was also meant to signal a difference in value. Wundt made this point explicitly in arguing that psychological theorising about the mind should follow the natural tendency of mental life to use evaluative terms (1908b: 209).

The term 'higher' also expressed something like 'above experimentation'. Because of the many parameters of apperception, the higher – thinking – processes could not be controlled and manipulated under experimental conditions. It was for this reason that their study had to be left to 'collective psychology' (*Völkerpsychologie*), that is the non-experimental psychological study of language, myth, and custom.[5]

Moreover, because collective psychology studied the 'higher' processes, it was itself 'higher' with respect to experimental psychology. At one point, Wundt went even further by equating collective psychology with psychology proper. He wrote of experimental psychology as 'the *Vorschule* [nursery school] for the [!!] psychologist'. In the same context, Wundt added that 'one does not need any special prophetic ability to predict that not too far in the future the experimental areas of psychology will move into the background' (1908b: VIII). Calling experimental psychology the nursery school for the psychologist in the full sense of the word was of course in line with Wundt's speaking (above) of the lower processes as a mere 'preparatory workshop' for thinking proper.

Given Wundt's view of experimental psychology, it is not surprising that his students were expected to learn the skills of experimental work. What *is* surprising is the fact that Wundt made little effort to train any of them in the 'higher' area of psychology. True, he did lecture on collective psychology to more than 600 people, but that was about as far as he went, although his dominant position in Leipzig would undoubtedly have enabled him to do much more. For instance, he could easily have handed out dozens of dissertation topics on collective psychology – Wundt supervised 186 dissertations during his Leipzig period. Instead, he reserved that area of psychology which in his own interpretation was the higher and more important one, for the exclusive attention of a single individual: himself. His students were not allowed to leave the 'preparatory workshop'.

Wundt's active creation of a parallel between the social order in the Leipzig Institute, his theory of the mind, and his view of psychology could hardly be more striking. In each case we find a sharp division between higher and lower, and in each case the higher was not just quantitatively but qualitatively set apart from the lower. Thinking had its own laws, collective psychology had its own data and its own subject matter, and Wundt not only commanded the experimental work, but also studied an area of psychology that he kept for himself. Obviously, the two-tier hierarchy of the mind could be used as a justification for the two-tier hierarchy within psychology as a field of knowledge, and both in turn could be used to lend plausibility to the hierarchical 'division of labour' in Wundt's institute.

To show that Wundt believed thinking and other 'higher' processes formed their own unique realm, on the one hand, and that he monopolised their study, on the other hand, is not yet enough to prove that for Wundt, thinking and observing themselves demanded solitude. We can establish this last point, however, by returning to some of Wundt's pronouncements on psychological experiments. Four of these pronouncements are of special importance for my argument.

1 *Perfect experiments can be carried out on one's own.* We have already encountered this claim in Wundt's criticism of the Würzburg school. To repeat the main point, according to Wundt, in perfect psychological experiments the presence of a

second person is not necessary, that is the experimenter is either identical with the experimental subject or else replaceable by a machine. The experimental work that Wundt singled out for special praise either could have been done by a single person or else had in fact been done by a solitary individual. This emphasis on the perfect experiment as one in which others are dispensable not only fitted well with the solitude topos, but also marked an important difference from the Würzburgers' emphasis on community.

2 *The presence of others impairs one's ability to think and observe.* The topos of the corrupting influence of others upon one's thinking surfaced clearly in another passage of the same critique. Wundt insisted that thinking and observing in the presence of other persons would always be very difficult:

> Let me mention another factor…the need to isolate the observer as much as possible, and in particular to keep away the disruption that is inevitably caused by the presence of other persons.…I doubt that amongst the participants in the interrogation experiments there are many who are indifferent to whether they are alone in their study or whether they are in the company of others, or even whether they are being observed by those in their presence. I admit that of all the incomprehensible features of the [Würzburg] interrogation experiments the most incomprehensible is that their advocates not only overlook this disturbing factor, they even try to turn it into a virtue by using the completely misplaced notion of a 'division of labour'.
>
> (1907c: 336–7)

3 *The cognitive authority of the experimental subject must be preserved.* Although the ideal experiment was one that was carried out in solitude, Wundt did allow for psychological experiments in which the roles of experimenter and experimental subject were distributed between two different persons. But given such experimental design, it was of great concern to Wundt that the experimental, observing subject retained his or her autonomy, free will, and cognitive authority. Wundt already had underlined this point in 1892, when he criticised hypnotic experiments (Wundt 1892a). There he advanced the moral argument that

> hypnosis is a state in which the ability to exercise one's free will is removed....With the exception of those cases that are explicitly sanctioned by law...no one is master over his own person to such an extent that he has the right to make another person absolute master over himself....The least moral of all possible relationships between humans is that in which one human being becomes a machine in the hands of another. And this holds not only for cases when the human-turned-machine is abused for immoral purposes; the relationship as such, independent of its use, is immoral.
>
> (1892a: 40–1)

In line with this consideration, Wundt held that there was only one morally acceptable form of hypnotic experiment: a self-hypnosis carried out by a trained psychologist (1892a: 102–3).

As already mentioned, Wundt saw the Würzburg experiments as continuous with hypnotic experiments. In both cases the self-observation was induced and influenced by suggestion, and in both cases it was assumed that 'the only essential property of the experiment is that some person *A* exercises an influence on some other person *B*'. In other words, advocates of hypnotic experiments as well as Würzburgers believed that a psychological experiment needed two people, and both schools of thought placed the initiative and authority with the experimenter (1907c: 302, 310–11, 339–40).

Wundt tried to make explicit what he perceived as the loss of autonomy of the experimental subject in the Würzburg experiments by labelling the latter 'interrogation experiments'. 'Interrogation' here referred to tests but also to cross-examinations in court. Wundt suspected that the school exam had in fact been the model for the Würzburg experimental setting. He painted a dramatic picture of the 'despair' (*Not*) of the interrogated person, and contrasted Würzburger interrogation with the autonomous 'older form of self-observation in which one and the same person posed questions for himself in order to answer them' (1907c: 305).

4 *The ideal psychological observer, that is the experimental subject, is a trained psychologist.* Wundt believed that the experimental subject rather than the experimenter was the true observer in psychological experiments and that psychological – just like any other scientific – observation demanded proper training.

Thus the primary experimental subject had to be a psychologist. Wundt insisted time and again that the heart of experimental psychology was a trained psychologist's self-perception under experimental conditions. Interestingly enough, one of Wundt's arguments for using trained psychologists was that only they would fully understand the overall purpose of the experiment. Again we encounter here an insistence on the autonomy and self-determination of the experimental subject (1892a: 40; 1907c: 331). At the same time, Wundt's stress on trained psychologists as experimental subjects also had the implication that proper experimental psychological research did not have to extend beyond university walls: the subjects needed for one's experiments could and should be one's psychologist colleagues.

Wundt's hostility towards both applied psychology and questionnaire methods was undoubtedly motivated by his twin concerns of keeping experimental psychology inside laboratory walls, and defending the solitary autonomy of the experimental subject–psychologist. For instance, Wundt suspected that the Würzburg experiments would lead to the use of questionnaire methods – asking ordinary folks about their self-observations (1907c: 359–60) – and he argued elsewhere that the transferring of the methods and theories of applied psychology back into pure psychology would have disastrous repercussions for the latter (1910c).

We can now summarise Wundt's hierarchical recluse model for the social-cum-cognitive order of psychology. The proper place for the study of psychology was the psychological institute. The internal structure of such an institute consisted of a lower stratum of experimental psychologists – set apart from non-psychologists by their ability to retain autonomy under experimental conditions – and a higher stratum consisting of the solitary master psychologist. The latter owed his dominance and solitude to his subject matter: the thinking mind. Because the study of the 'higher' mental processes demanded a 'higher' form of psychology, the master psychologist rightly controlled the 'lower' experimental study directed at the 'lower' mental processes and carried out by the 'lower' ranks of students and assistants. And because thinking itself demands solitude – this proposition being supported by experimental work and by common sense – the master psychologist had better safeguard such solitude for himself as well.

Wundt saw all of this threatened by the Würzburger theories and practices. The Würzburg school challenged the hierarchical model of the mind, made the experimenter an essential figure in psychological experiments, regarded thinking as a proper subject matter for experimental psychology, denied the virtues of solitude, lessened the emphasis upon autonomy and competence of the experimental subject, opened pure psychology to influences from applied psychology, and thereby totally undermined the social order in Wundt's institute.

The drillmaster

G. E. Müller's psychological institute in Göttingen was a very different place from Wundt's research school in Leipzig.[6] The Göttingen Institute too had a hierarchical social structure, but Müller did not play the role of the aloof sage. In constructing his scientific identity, Müller drew on different cultural resources: on the popular connection between madness and genius; the association between the search for truth and a military engagement; the idea of the pure, restless, and incorruptible searcher for truth who has no time for either politicking or politeness; and the image of the Prussian drillmaster who is harsh towards his recruits only because toughness increases their chance of survival in later battles.

The social interaction between Müller and his students was structured by a tight regime of rules, drills, and checks. As one student enthusiastically described the arrangements:

> [Müller] strictly kept watch over everybody's adherence to the rules that concerned the possessions and tools of the institute. Only those who have experienced how harshly Müller judged users of the institute who committed the slightest offence (and I mean offences that were treated as bagatelles elsewhere); only those who have read his rules of conduct for the institute, rules that advised 'undisciplined personalities' to seek 'their area of activity' elsewhere; only those who have witnessed how a little piece of wood that at some point had become a possession of the institute could become the object of a long search, and the occasion for a principled speech about order and discipline – only those are able to appreciate that the purest form of the Prussian Spirit ruled at the institute in Göttingen.
>
> (Kroh 1935: 167)

The students' programme of studies was strictly prescribed. In addition to attending Müller's own lectures, students were expected to take courses in physiology, physics, and precision engineering (so that they could construct their own instruments). Müller intervened immediately when students started to take an interest in other subjects, such as philosophy, that he regarded as irrelevant (Düker 1972: 43–4; Fröbes 1961: 128; Katz 1936: 237; Katz 1972: 107, 109).

Moreover, Müller was directly involved in every study done in his laboratory. First, 'the choosing of a topic happened almost without exception upon Müller's personal suggestion' (Katz 1936: 237, but cf. Kroh 1935: 177). Second, one of the rules of the institute was that all staff and students had to be present in their offices every morning; during the mornings Müller would visit everyone and check up on the latest progress. Third, Müller had to be one of the experimental subjects of every study done in his laboratory (Katz 1972: 106). And fourth, Müller would go through every line of his students' papers and dissertations, criticising them in great detail. This is how one student described his work under Müller:

> Directly upon my entry into his classes, Müller outlined a plan for some original research in which I was to supervise the whole experimental work.…I was to study the literature according to his direction. He planned the experiments and ran through them, using himself as a subject. Later, in the next semester, I was called upon to develop these experiments, using others as well as Müller as subjects. This work, for which of course credit must be given entirely to the master, has this advantage: The student learns the procedure in detail. Müller…criticised my presentation heartlessly in all its details.…The final revision, for the sake of safety, he took upon himself.
>
> (Fröbes 1961: 128)

Müller's line-by-line criticism of students' work was much feared by them. It happened in almost ritualistic fashion. The student was called to Müller's private apartment in the evening. Mrs Müller opened the door and tried to console the student by saying '*Schwere Zeit*' (hard times). Then came the 'purgatory', as the students called it. Müller was critical, coarse, and insulting. Some students were so distressed by the treatment that they became suicidal (Katz 1972: 106). Only after every line and every graph had been checked and corrected did Müller allow the student to publish (Katz 1972: 179).

We find the same mixture of drill and control when we turn from the social organisation of the Göttingen Institute to Müller's memory experiments (Müller and Schumann 1894; Müller and Pilzecker 1900). I have already introduced the basic scenario of these experiments: The experimental subjects were placed in front of the memory drum and asked to learn long lists of nonsense syllables. The created associations between successive nonsense syllables could then be studied with the 'method of hits' (*Treffermethode*): the subjects were presented with single elements of the series and had to react with what they recalled as the following element. For each experimental series, the subject had to spend one hour per day, on up to fifty successive days, in front of the memory drum. Staring at the revolving drum made some subjects experience dizziness; the sheer tedium of the task must have been a contributing cause of the discomfort. No wonder that according to Müller and Schumann, working as their experimental subject demanded a good deal of 'sacrifice' (1894: 264).

The term 'drill' seems like an entirely adequate characterisation of this experimental setting. Just like in parade-ground drills, so also in the memory drum experiments, the actions of the subjects were highly constrained, repetitious, and somewhat mindless. Usually, the subject learned nonsense syllables without actually knowing what was the purpose of the experiment. The parallel between the parade ground and the memory experiment is strengthened further by the vocabulary used in the context of the latter: 'full hits', 'partial hits', 'drum', and 'sacrifice'.

Unlike Wundt, Müller did not insist that the experimental subject be a trained psychologist. Thus of the thirteen experimental subjects used in Müller and Pilzecker (1900), only two – Müller and Pilzecker – were fully trained psychologists. Of the others, one was a psychology student, and the others members of other disciplines, or people without a position in academia. Müller's and Pilzecker's wives figured particularly prominently. Of the forty-one experimental series, nineteen were done with Frau Müller and Pilzecker. Particularly willing to 'sacrifice' herself was Müller's wife; she spent a total of 261 hours (twelve series) in front of the memory drum. Müller and Pilzecker themselves each acted only once as experimental subject.

In light of the above, it will not come as a surprise that Müller found the Würzburg experiments lacking in rigour and control. In the context of his criticism of the Würzburgers' methods, Müller stressed the need for proper training of inexperienced subjects; the importance of educating them in honesty and accuracy; and the necessity of checking their accounts through questioning, through interruption, as well as through observation of their outward behaviour. He also underlined the desirability of using the same subjects for long periods of time (preferably of up to 3–5 years) (1911: 109, 112–13, 117, 170).

To start connecting Müller's preferred model of social order – the parade ground with himself as the drillmaster – to his theory of the mind, I first need to say a bit more about this theory. We have already seen that Müller's theory of thinking was associationistic (cf. Katz 1936: 235: 'Within psychology Müller was wholly under the influence of classical association psychology'). However, Müller's specific brand of associationism owed as much to Herbart as it owed to the British tradition from Bain to Mill. Particularly Herbartian were Müller's notion of 'threshold of consciousness' and the idea that presentations fought for entry into narrow and transitory consciousness. Presentations that made it into consciousness were called *überwertig* ('above value', or, 'of high value'), those that did not were labelled *unterwertig* ('below value', or, 'of low value'). Like Herbart and several of the British writers, Müller saw associationistic psychology as a 'mechanics of presentations' (Müller and Pilzecker 1900: 78); that is, he conceived of psychology as trying to do for presentations what physical mechanics did for physical bodies. Where the physicist spoke of attracting and repelling forces, the psychologist spoke of reproductive tendencies and forces (1900: 269).

Together with his co-authors Pilzecker and Schumann, Müller identified, amongst others, the following laws of associative reproduction:

> the relative number of hits increases with the number of repetitions (1900: 30).

> Of any given group of equally old and high-value associations the following holds: Those associations that become effective more quickly are also those that retain their high value for a longer time period (1900: 44).

> Within certain limits, and given an equal number of hits, older associations result in longer reproduction times than younger associations (1900: 47).

> Assume that syllable *b* has already been associated with syllable *a*; if subsequently *a* is read in the new syllable combination *a c*, then the association *a-b* will in general also be excited ('co-excited') (1900: 134).

> In associative co-excitation...stronger associations profit more than weak associations (1900: 137).

In formulating such laws on the basis of laboratory experiments, Müller and his co-workers went beyond earlier associationism.

In light of the contemporaneous debate over the sufficiency of associationism, it was even more important, however, that Müller complemented association by perseveration; that is, that he showed that 'every presentation has, after its appearance in consciousness...[the] tendency to rise freely into consciousness again'. In the memory drum experiments, the perseveration tendency surfaced in the following phenomena: after a training session, some syllables would suddenly reappear in consciousness; or, in the test session, subjects would recall a recently seen or uttered syllable rather than the correct one (1900: 58, 62).

Müller emphasised that only association and perseveration tendencies together could fully account for the nature of human thinking. Both tendencies could be studied experimentally – Müller showed no interest in a Wundtian non-experimental collective psychology. Perseveration tendencies had several important functions. First, they allowed thinking to cope with interruptions. The last presentation before an interruption was retained and could freely rise again after the disturbance. Furthermore, perseveration enabled younger and weaker associations to catch up with older ones. The strength of an association was a function both of how often it had been in consciousness before, and of the number of other associations to which it was linked. In the absence of perseveration, younger associations would be unable to catch up, and this could be dangerous to the organism: 'Things change, and the effectiveness of our associations must adapt to the changes in things' (1900: 75). Finally, perseveration helped the individual to become independent of 'sensual needs, external situations, and talk of other people'. Nothing over and above association mechanisms was needed as long as the individual was directed by ever-changing needs, situations, and talk. 'Yet', Müller insisted, 'the human being is a being that can also pursue goals that reach further.…We owe it in great part to perseveration that we can hold on to such goals, and that we don't loose them in new situations'. In addition to fixing goals more permanently, perseveration also made possible 'the pursuit of certain reflections. In so far as thinking consists only of…the processing of the immediately given, associations suffice. But when we rise in thinking above this standpoint…then perseveration is involved'. And in an aside against Wundt, Müller continued: 'There is no need to explain this kind of thinking and doing through a "will that guides presentations" or some such device' (1900: 77). Finally, because of the importance of perseveration for thinking, Müller suggested that educationists try to find ways to further and strengthen the development of perseveration in children (1900: 78).

The above rough sketch of Müller's theory suffices for my present purpose of establishing a link between Müller's ideal of social order and his conception of the psychological order of the human mind. To begin with, it is of some interest to point out that before Müller, there already existed a direct link between Herbartian psychology and political science. Herbart himself had provided a political interpretation of what we might call his 'threshold psychology'. In the second part of his *Psychology as a Science* (*Psychologie als Wissenschaft*, 1850), Herbart had followed Plato in proposing that one might understand human individual psychology better by 'studying its enlarged image in the state' (1850: 22). The conflict of individual presentations within the individual human soul had its parallel in the struggle between various groups in society. And the threshold of consciousness had its analogue in the 'threshold of influence in society'. Just as a few strong presentations could inhibit the rising of numerous other, weaker presentations, so also a few powerful individuals could keep the masses in check (1850: 32–3). In fact, Herbart sought to formulate various principles of politics in direct parallel to his psychology of the individual. Thus he warned, for instance, against the 'persistent mistake of bad politics that seeks to suppress the powers with which it would do much better to align itself' (1850: 41). These Herbartian ideas on politics as a model for, and application of, threshold psychology were not forgotten by the turn of the century. For example, Wundt referred to them in more than one place (e.g. Wundt 1920a: 214).

Müller did not follow Herbart's lead. Rather than link psychology to political science, he preferred the more naturalistic turn of connecting psychology to evolutionary biology. For Müller the struggle between presentations *within* one individual was directly linked to the struggle for survival *between* individuals. The individual's success in the latter fight depended on having the right kinds of rules and tendencies for the former (1900: VII, 75). The mechanics of presentations was thus itself part and parcel of an evolutionary, Darwinian, conception of the human being. Müller did not make any explicit pronouncements, however, on the question of whether there were any analogies between the intra-subjective and inter-subjective rules and laws.

Even in the absence of such explicit pronouncements, it is striking to note that Müller himself behaved *as if* such analogies existed. In other words, Müller behaved – both within his institute and within the wider psychological community – as if the social world was the mind writ large, and as if there was a direct parallel between the threshold of consciousness and the threshold of power and influence.

For instance, recall that the associative strength of a presentation was a function of its age, of how often it had been in consciousness before, and of the number of other presentations to which it was linked. That fits well with Müller's insistence that he had to be involved – as experimental subject, supervisor, or co-author – with every research project in his institute. The case of co-authorship is especially striking. Whereas Külpe, Stumpf, or Wundt never co-authored papers or books with others, Müller did so repeatedly but only with junior researchers of his own institute. The 'Matthew Effect' (Merton 1968), which no doubt came into play here, had its analogue in the co-excitation principle, according to which older and already stronger associations profited more from co-excitation than younger and weaker ones (1900: 136–7).

Particularly striking is the parallel between the threshold of consciousness (on the side of psychological order) and the threshold of publishing, the 'purgatory' (on the side of social order in Müller's institute). The students could get over the latter threshold only by undergoing Müller's regime of drill and humiliation and by co-operating closely with him. Ultimately, it was the strength of their attachment to Müller – the master presentation, as it were – in experimental technique as well as theoretical matters, that allowed them to make it from the purgatory of rejection into the heaven of acceptance.

Müller was also a master of networking outside of his institute. This man, who once said of himself that he would have loved to become a politician (Misch 1935: 52), turned out to be a very successful academic strategist indeed. In 1890 he became one of the contributing editors of the new major journal, the *Zeitschrift für Psychologie und Physiologie der Sinnesorgane*. When the journal later split into two sections, for pure psychology and sense physiology, he was the only person sitting on both editorial boards (Kroh 1935: 150). From 1908 onwards, the editorship of the section for pure psychology was firmly in the hands of Müller's students, and the *Zeitschrift* became referred to as 'Müller's mouthpiece'. Moreover, in 1904 Müller became the chairman of the newly founded German Society for Experimental Psychology (Deutsche Gesellschaft für experimentelle Psychologie), a position he held until 1927. From this power base Müller could also successfully fight for academic positions for his students (E. Haupt, personal communication).

The connections between psychological and social order become even more direct and transparent as we turn from association to perseveration. We have already heard that according to Müller and Pilzecker, we owe it to perseveration tendencies that we are able to become independent of others and think systematically. To this we now need to add a number of further observations.

First, there was some, however vague, link between perseveration and madness, inasmuch as Müller and Pilzecker had adopted the term 'perseveration' from psychiatry. Psychiatrists had spoken of perseveration in cases where patients were unable to free themselves from some memory or fantasy. Müller and Pilzecker distinguished between normal and pathological forms of perseveration: in pathological cases it was always the same presentation that returned, whereas in the normal individual, different presentations could persevere at different times (Müller and Pilzecker 1900: 60). Yet some similarity between the normal and the pathological cases obviously remained.

Second, this similarity between perseveration and madness was reinforced by another resemblance, that is that of perseveration and genius. As one of his students reported, in his lectures Müller would emphasise that '*Le génie, c'est la persévération*' (Kroh 1935: 158). The resemblance of genius and madness is of course another pervasive topos of Western culture; here it worked to link both genius and madness to perseveration.

Third, assuming that perseveration, like association, is both a psychological and a social phenomenon, what then are we to think of the 'perseverating individual'? What is this kind of individual like? The answer is not hard to find. Obviously, the perseverating individual will be strong enough to rise above the threshold of power even without allies and supporters, even without politicking and politeness. Naturally, to succeed, the perseverating individual will have a high degree of perfection, self-discipline, toughness, ambition, and restlessness. Müller sought to project all of these qualities to his students. For instance, it was an open secret that he himself was paying the water bills of the institute, and that 'he was never ready to make concession to those above. A proud sense of independence prohibited him from asking for funds' (Katz 1936: 238). His students were also told that he suffered from chronic insomnia, and the whole institute was expected to put up with the resulting bad temper (Katz 1972: 106). The members of the Göttingen Institute also knew that Müller was a man of perfectionism and self-discipline. Indeed, as a good drillmaster, Müller subjected himself to the tightest regime of all: he was 'possessed' by duty, 'a fanatic of objectivity' (Katz 1936: 238); he rewrote everything four to five times (Kroh 1935: 157); he opposed improvisation to the point of refusing to hold seminar discussions (Katz 1936: 237); he despised philosophical topics that could not be treated with precision – he warned his students about one of his colleagues by saying 'Be careful, he lectures on ethics'; and he fought in himself the tendency to work on metaphysical topics:

> And it may well have been only due to the most reckless discipline of the will, a discipline that he applied to himself in all areas of life, that he escaped the dangers that resulted from this side of his talent and work.
>
> (Kroh 1935: 187)

Clearly this was a man who possessed perseveration to an unusual degree.

Fourth, Müller went further by actually providing experimental proof of his unusual character. In Müller and Pilzecker (1900), the two authors went to some length to emphasise that of all thirteen experimental subjects, Müller was the one with the highest degree of perseveration. Deviating from an otherwise impersonal reporting of numbers and hypotheses, the authors described how Müller's unusual degree of perseveration expressed itself in everyday life. To make the case still more vivid, they contrasted Müller with his wife who, as will be recalled, had been the major experimental subject. They related that Frau Müller would be better than her husband at tasks that demanded primarily associations, such as learning a new language. Herr Müller, on the other hand, would be the one with the firm character; it would be he who would remember to carry out decisions made earlier. Moreover, he was always able to return quickly to his work after an interruption, and he had a hard time trying to stop thinking about scientific problems. Again, in these respects too, his wife was his opposite (1900: 69–75).

Fifth, Müller was not altogether unsuccessful in this self-portrayal, at least when judged by this obituary:

> He, who in a lecture dared to say: 'Le génie, c'est la persévération'; he to whom science owes the experimental discovery of perseveration in normal persons; – he, like few others, had the right to lay claim for himself and his œuvre to that peculiar form of genius that we call perseveration.
>
> (Kroh 1935: 158)

These then are the connections between the social order of Müller's institute – including the social order of his memory experiments – and the psychological order of his theory of the mind. The drillmaster's regime with its strict rules and humiliations could be justified as the key to the education of powerful and independent scientists. After all, such scientists needed a high degree of perseveration – self-discipline, restlessness, independence – and these virtues could only be learnt and maintained by subjecting oneself to a drillmaster's regime. Moreover, Müller's regime could also be seen as 'natural'. There was nothing unnatural about the purgatory, Müller's ubiquity, and the students' dependence: The social arrangement was similar to the mechanics of the mind, a mechanics that had proved its mettle in the struggle for survival. And finally, there could be no arguing with Müller's leadership position. The man with the highest degree of perseveration – a degree oscillating between madness and genius – had to be the natural master psychologist.

Given these various connections, Müller's hostile reaction to the Würzburg work starts to appear in a new light. Opposition to the sufficiency of association and perseveration was not just an attack on his psychological theory; it was by the same token an attack on his leadership and his claim to genius. The Würzburgers could be read as trying to repeal 'Müller's laws' in both the social and the psychological realm.

The interlocutor

It is much more difficult to do for the Würzburg school what I have done above for the Leipzig and Göttingen Institutes. Fortunately, though, these extra difficulties are significant and telling. For instance, whereas Wundt alone defined the Leipzig position, and Müller the Göttingen view, there was no one such dominating voice in Würzburg. It seems that Külpe never aspired to such authority. This fact makes it more complicated to identify *the* Würzburg view on issues such as psychological or social order. Or, whereas both Wundt and Müller spent many decades in one place – Wundt thirty-eight years in Leipzig, Müller thirty-five in Göttingen – Külpe moved from Würzburg to Bonn, and later to Munich, all within just twenty years. Of Külpe's more important colleagues, only Bühler followed him from Würzburg to the latter two places. Furthermore, and as mentioned before, many psychologists traditionally categorised as Würzburgers actually spent only short periods of time in the Würzburg Institute. And finally, debate and criticism flourished within the school.

In good part this diversity of voices finds its natural explanation in Külpe's style of leadership. If contemporary sources are to be believed, the social order of the Würzburg school was indeed a far cry from the Leipzig or Göttingen models. Külpe had trained with both Müller and Wundt, and thus he knew the pros and cons of the respective social arrangements. As Külpe's unpublished correspondence reveals, it even seems that he succeeded in getting the best of both worlds: Külpe did his PhD in Leipzig, but Müller both suggested the topic and provided much of the initial supervision in correspondence (Külpe *Nachlass*, Staatsbibliothek München, Suppl., Box 9). After he had risen to the position of Wundt's *Institutsassistent*, Külpe acted as something of a go-between for the two *Geheimräte*. For instance, in 1888 he supplied Müller with a list of all of the instruments used in the Leipzig Institute (ibid.).

One can only speculate on the young Külpe's reasons for moving from Göttingen to Leipzig. No doubt Wundt's broader philosophical interests and the better-equipped laboratory were strong incentives. Moreover, it seems more than likely, judging by reports on Külpe's character, that Külpe profoundly disliked Müller's draconian regime. As it happens, once in Leipzig, and once he had risen to *Institutsassistent*, Külpe was also deeply unhappy with the ways Wundt treated his students (Hellpach 1948: 175). Being Wundt's assistant, Külpe tried to be more supportive than the master; accordingly, the Leipzig students called Külpe 'the kind mother of Wundt's institute' (Kiesow 1930: 167). Külpe must have been eager to do things differently if he ever got his own institute. And he did when in 1894 he was called to Würzburg.

His long-term assistant, K. Bühler, described Külpe as someone who would 'capture one's soul'. Külpe would always have time, endlessly listen to students' and colleagues' ideas, and assist in developing them. Indeed, Bühler went so far as to say that the major fruits of Külpe's psychological research were not Külpe's own publications; these fruits were generously left to his students in discussions, supervisions, and introspective reports. Moreover, Külpe never imposed his ideas upon his students: 'each one of his independent students had *the widest possible room to move* both with respect to methods and with respect to theoretical interpretation. Külpe would go along with his students and provide the support they needed' (Bühler 1922: 249). Bühler thought that Külpe's running of his institute could not be better summarised than with Külpe's own concluding words to his *Introduction to Philosophy* (*Einleitung in die Philosophie*, 1910): 'The former monarchical constitution of science has given way to a democratic one' (Bühler 1922: 253).

Messer called Külpe his 'friend' and wrote that he chose to study experimental psychology with Külpe because he had been impressed with his personality (1922: 11–13). And Külpe's Munich colleague, C. Baeumker, also reported that Külpe helped others unselfishly in their research and that 'the best of [Külpe's] own work in the field [of thought psychology] appeared in the writings of his students'. Baeumker saw Külpe as a character who disliked thinking in terms of an 'either – or', and who tended more towards 'the conciliatory and balancing "not only – but also"'. Baeumker described the mode of co-operation around Külpe as similar to 'Plato's academy' (1916: 81, 88, 91).

Such words of praise of Külpe's style of leadership did not just come from friends and colleagues. They also came for instance from Marbe who fell out with Külpe shortly before the latter's death (Marbe 1961: 198). When Külpe took up the chair in Würzburg in 1894, Marbe joined him as *Privatdozent*. Marbe later explained that this decision was based on his friendship with Külpe, a friendship that started while they both were working in Leipzig:

> I owed most...to my association with the young *Privatdozent*, Oswald Külpe, who in my time was acting as first assistant in the Leipzig institute. Külpe had uncanny industry and was, even then, better read than I have ever become....Külpe, with whom I ate lunch regularly in Leipzig and whom I also saw much of otherwise...was always accessible for all the questions I asked and for the explanations of the problems I attacked ...in view of our friendly relations and of everything I owed him it was of course natural that I should seek my field of activity in the place in which he had gained a decisive influence upon instruction and investigation.
>
> (1961: 192–3)

Despite their falling out over Marbe's 1915 criticism of thought psychology, Marbe was unwilling to find fault in his former friend:

> The fact that he then turned away from me with his school was undoubtedly due to the circumstance that when Külpe was further developing the psychology of thought, I was no longer in Würzburg...and that he now yielded to other influences than mine.
>
> (1961: 199)

We can even use Wundt's testimony as further evidence for Külpe's unusually democratic and generous attitude towards school members. It seems that after Wundt had published his criticism of Marbe and Bühler, Külpe wrote to him, defending the work done in his institute. In his reply, Wundt regretted learning that Külpe was a supporter of thought psychology, and continued:

> I assumed that your attitude towards it [i.e. thought psychology] was based on your generous and conciliatory ethos. I thought that you merely wanted to grant freedom to the people working in this direction, and that you were ready to wait and see whether anything comes out of it.
>
> (Wundt, letter Wundt–Külpe, 26 October 1907, quoted in Hammer 1994, 244–6, p. 244)

Testimonials aside, we can also point to other data to bring out the fact that Külpe's leadership style differed considerably from Müller's and Wundt's. For example, unlike his older colleagues, Külpe welcomed, and co-operated with, men who had already done their PhD elsewhere. Marbe had done his PhD in Leipzig, and Messer in Giessen. By the time they worked on the psychology of thought, they already held temporary professorships in Würzburg and Giessen, respectively. Ach, too, entered as a finished PhD in medicine, although he chose to do another PhD with Külpe. Bühler held doctorates in both medicine and Catholic philosophy. Moreover, even though Bühler was no doubt Külpe's closest student, following him as his assistant from Würzburg to Bonn and later to Munich, he was by no means a disciple, and he referred to Külpe's work only rarely.

We need to note, furthermore, that none of the major Würzburger experimental studies in the psychology of thought came from Külpe's pen; this is not to deny, of course, that his work on abstraction (1904), his review of Ach (Külpe 1907b), and a number of theoretical papers published between 1912 and 1914 were influential (1912b, 1912c, 1914).

One might even question to what extent Külpe deserves credit for the Würzburg paradigm at all. Some historians of psychology seem to assume that because the Würzburgers were referred to as a 'school' and because Külpe was the head of the Würzburg Institute, the central ideas in the psychology of thought must have come from him (Lindenfeld 1978). One only needs to consult the original sources of the period, however, to gain a rather different perspective. Marbe claimed priority for situations of consciousness, Ach for awarenesses, and Bühler for thoughts (Marbe 1915; Ach 1910: III; Bühler 1907b: 330). Marbe and Ach both saw themselves as inventors of the method of systematic introspection (Marbe 1915; Ach 1910: 18). And Marbe even presented himself as the father of the whole psychology of thought movement (1961: 196–8). Nowhere did Külpe challenge any of these claims. Either he was not the man to fight over priority claims, or else he agreed that his work had not been the starting point of, or crucial in, the most famous project carried out in his laboratory. In either case, Külpe comes across as a man who lacked Müller's and Wundt's authoritarian style of leadership.

Groping for a label for Külpe's style of leadership, one obvious term suggesting itself is 'interlocutor'. After all, the interlocutor contrasts both with the recluse and with the drillmaster. The first does not talk, the second does not listen. The topos of the interlocutor, too, has a long history. There have always been opponents of the recluse model in particular, that is people who have insisted on the social character of knowledge. For such authors, thinking demands interaction with others, open debate, conversation on equal terms, and spiritual and emotional community. It is not difficult to see that in their attack on Müller and Wundt, the Würzburg psychologists made use of this alternative model.

The interlocutor model figured most prominently in the Würzburgers' characterisations of their new experimental design. Time and time again, they pointed to the equality and reciprocity between experimental subject and experimenter. For example, Ach spoke of the 'continuous close exchange of ideas' between experimenter and experimental subject and insisted on the 'intimate connection' between the two (1905: 9, 26–7). Bühler called the interaction between experimenter and subject a 'division of labour' and professed that – when acting as experimenter in this 'ideally beautiful co-operation' – he saw himself as no more than an 'editor' of his subjects' ideas. As experimenter he sought to select the tasks so as to accommodate, in a 'loving' (*liebevoll*) fashion, his subjects' tastes and predilections. He stressed that the choice of words for describing thinking processes had to be left to the experimental subjects and that the attempt to understand and relive their experiences needed empathy and 'intimacy' (1907b: 299, 300, 309, 313). And Külpe wrote that

> the peculiar human relationship between experimenter and experimental subject is at the centre of life in the psychological institute. A relationship of trust enclosed both of them. The social ties of reciprocal concern and willingness to sacrifice, of mutual understanding and of goodwill form a condition for the success of scientific work....Thus in the psychological institute an atmosphere of noble human relationships is fostered, a truly ethical atmosphere as it might have existed in the Academy of Plato, in the Lyceion of Aristotle, or the column halls of the Stoics.
>
> (1914: 1229)

Little surprise therefore that Külpe participated in almost all of the experiments done in his laboratory, and that in one of his writings he even presented such participation as a duty: '[In psychology] the directors of institutes must participate in the experiments if they then want to be able to make any kind of informed judgement of these experiments' (ibid.).

At least for one key member of the Würzburg school, Bühler, the egalitarian spirit of the Würzburg laboratory and its emphasis on the equality of experimenter and experimental subject went hand in hand with an egalitarian model of the human mind. This model was most clearly spelled out in Bühler's reply to Wundt's criticism (Bühler 1908a). Wundt had argued that the higher processes of thinking were much more complicated than the lower processes of sensation and perception, and therefore not open to experimentation.

Bühler denied the premise of this argument:

> Though it may be true to say that in research one should, wherever possible, progress from the more simple to the more complicated, it does not apply to this case. For our thought processes cannot be looked upon as the more complicated.

In Bühler's view the notion that thought processes are more complicated than processes of sensation or presentation was due to a mistaken interpretation of thoughts. One mistakenly had to treat thoughts as condensed sensations and presentations to look upon them as highly complicated entities. To give up this 'sensualistic bias' was tantamount to treating thoughts on equal terms with sensations, feelings, and presentations. In other words, in Bühler's opinion, Wundt conflated three claims: that a 'brain-teaser' (*Denkaufgabe*) might be hard, that the process leading to its solution might be complicated, and that the retrospective description of the experience of finding a solution might not be difficult. According to Bühler, it was at present anyone's guess what made one task harder than another. But there was no compelling reason to assume that a hard task demanded complicated processes of thinking, or that an easy task could be solved with simple mental processes. Bühler also introduced introspective evidence for his position:

> After an act of thinking we are often able to say with all certainty: What I have just experienced was not complicated at all; perhaps all one had to do was to make a somewhat unfamiliar connection; perhaps the contents present in consciousness consisted of little more than two thoughts and the connection between them. And yet, we are still fully convinced that we have achieved a respectable thought performance.
>
> (1908a: 109)

Furthermore, Bühler wrote that he found far from paradoxical the idea that 'the more we have to concentrate and exert our thinking the more slowly and the more simply the processes of thoughts take place'. Indeed, Bühler believed the idea was pretty close to the truth. He also disagreed with Wundt's claim that only simple processes could be observed retrospectively. In Bühler's view it was the other way around: while we are able to report our step-by-step solving of a difficult task, we are unable to say anything much about the processes that take place when we hear and understand single, isolated words (1908a: 110).

By rejecting the Wundtian assumption of thought processes as complicated and inaccessible to experimental psychology, Bühler weakened – at least to the satisfaction of himself and other Würzburgers – Wundt's argument that the study of higher processes called for a special kind of psychology, that is collective psychology. At least implicitly he also denied that one had to leave the study of thinking to those who had first undergone training in the experimental psychology of sensation, feeling, and presentations. That experimental psychology had no limits had already been Külpe's position in 1893 when he wrote that 'in principle there is no topic of psychological inquiry which cannot be approached by the experimental method. And experimental psychology is therefore fully within its rights when it claims to be the general psychology that we propose to treat' (Külpe 1893: 12, tr. Danziger 1979: 213). Later the Würzburg school also attacked Wundt's collective psychology more directly (see Chapter 5).[7]

It fits nicely with my argument that Bühler flatly rejected Wundt's claim according to which the presence of another person impeded thinking:

> Wundt's claim that the presence of the experimenter must necessarily have a negative effect on the thinking of the experimental subject does not accord with the facts....How does Wundt know that such interference *had to* occur in our case, even though the experimental subjects denied being aware of any such effect? There is no axiom from which Wundt could possibly deduce his claim; for it is no more than an unproved assumption on Wundt's part.
>
> (1908a: 90)[8]

As far as the advocate of the interlocutor model is concerned, the presence of others and involvement with them actually furthers thinking, not impedes it.

Külpe and his colleagues were of course quite aware that theirs was an attack not just on Wundt's theories but also on his social position and authority. In an article written on the occasion of Wundt's 80th birthday, Külpe insisted on Wundt's students' right to 'go their own ways' and went on to say that 'we do not allow any authority to prescribe for us what we are to tackle and what we are to teach' (1912d: 110). Marbe was more scathing, perhaps because he knew that after the publication of his study on judgements, Wundt had sought to exclude him from the editorial board of the *Archiv für Psychologie*.[9] Marbe alleged that Wundt's psychology was built on the pillars of social power alone:

> It is Wundt's method to make claims, to repeat these perhaps dozens of times, and thereby to eventually bring his students – which in this case should rather be called his disciples – to believe these claims. Under these circumstances it is easily understood that adherence to Wundt's doctrines of the higher mental processes extends no further than the influence of Wundt's authority. Wundt calls the experiments based on the method of systematic self-perception 'interrogation experiments'. With much more justification one could call his method...the 'authoritarian method'.
>
> (Marbe 1908: 358)

Marbe contrasted Wundt's 'authoritarian method' with the workings of the Würzburg laboratory, stressing that the new experimental design had made it possible for different psychologists to share a common and agreed upon observational foundation for all their different theories. All authors from within the Würzburg school, Marbe wrote,

> report investigations which I could at any time check and re-check with them. And this possibility will always, in the end, lead us to accept jointly or reject jointly their [theoretical] claims. All we have to do is experiment with a sufficiently large number of qualified observers. However different the protocols of the experiences might happen to be, it is impossible that two competent experimenters could get fundamentally different results.
>
> (1908: 353–4)

Compared with the Würzburgers' attack on Wundt's psychological and social order, their criticism of Müller was much more guarded.

In part this may have been due to the Würzburgers' genuine appreciation of great parts of Müller's work. But it seems also natural to surmise that some social–political factors played a role as well. Wundt largely confined himself to controlling and dominating his Leipzig institute and his house journal; he never went to conferences, and he kept a distance from the psychologists' professional organisations. As already mentioned, Müller invested much greater efforts in shaping German psychology at large; he was involved with the leading journal, and he was the chairman of the professional organisation. For the Würzburgers, it would have been much more difficult to attack someone publicly and personally with whom they regularly planned, organised, and met at conferences. It also would have been more risky to challenge a man of such power and influence within German psychology.

Nevertheless, the Würzburgers signalled a distance from Müller already by their preferred place of publication. None of the major Würzburg studies appeared in 'Müller's mouthpiece', the *Zeitschrift für Psychologie*. Instead they were published in the *Archiv für die gesammte Psychologie*. This journal was edited by E. Meumann, the founding father of experimental educational psychology.

Moreover, the Würzburgers rejected Müller's, as well as any other form of, associationism in no uncertain terms. For instance, Watt insisted that '*every theory*' that sought to make do with associations alone was incompatible with his experimental results (Watt 1905: 421). He also ridiculed associationism as the view according to which the mind was a house built from presentation blocks, held together by the 'mortar' of association (1905: 417). Ach maintained that the functions that Müller attributed to perseveration were in fact carried by determining tendencies. Rational and goal-directed thinking was not due to perseveration; it was due to determining tendencies. It was not perseveration that created independence from associations and external influences; it was determining tendencies: 'The influence of determining tendencies brings about an independence from arbitrary external stimuli and from the usual associative train of presentations.' (1905: 196). Messer made essentially the same point in terms of 'tasks' (1906: 31). On top of that, Messer even attempted an explanation for Müller's and other associationists' failings. It was the very preoccupation with nonsense syllables that had made associationists incapable of recognising the importance of meaning for mental life, and only if one recognised the importance of meaning could one also come to identify 'tasks' and 'thoughts' (1906: 113).

In some of Külpe's remarks one can detect allusions to Müller's regime and its justification. To begin with, Külpe insisted that associationism was a theory not of the normal, but of the pathological, mind. Külpe spoke of the 'complete myth' according to which 'organised thinking' was governed by laws of association alone. A thinking governed by association alone was 'disorganised thinking, the anarchy of *Ideenflucht* (flight of ideas)' (1912b: 209). Contemporaries will hardly have missed the irony of this turn in the argument; after all, Müller himself had alluded – *pro domo* – to the topos linking genius and madness. Külpe went further in his university lectures; here he approvingly quoted Wundt's idea that associationism explains the animal's soul at best (1920b: 6).

It is especially interesting to note that Külpe spoke of determining tendencies and tasks as the realm of freedom, and that he distanced this realm from the domain of association and perseveration. Associationism was bound to conceptualise the mind around the idea that '*it* thinks' – associationism was unable to capture the phenomenon of '*I* think' (1912b: 212). Put differently, associatonistic psychology had failed to notice the 'activity of the soul' (1912c: 1086), the 'great importance…[of] the spontaneous activity of the subject' (1920b: 195). As Külpe stated the Würzburg doctrine, between thoughts there existed 'peculiar, free relations', relations that could not be subsumed under laws of association (1912c: 1086). Perseveration and association were no more than 'means towards ends', and these ends were set by 'acts or functions, expressions of the personality, activity, and spontaneity' (1920b: 193).

Undoubtedly, anyone familiar with Müller's regime and his theory will have noticed that Külpe was challenging both. There was no freedom in Müller's institute, no freedom for his experimental subjects, and no freedom in his conception of the human mind. By insisting on freedom and spontaneity in the human mind, Külpe could be read as demanding from Müller freedom also for the student, and freedom for the experimental subject.

Külpe continued this line by emphasising the importance of thought psychology for education and character building. Thought psychology was crucial here because 'task' and 'duty' were closely related concepts: 'Psychologically speaking, every task that we take on becomes a duty' (1912c: 1106). Now, if task and duty were situated on the level of thinking, and if thinking was the realm of freedom and spontaneity, then surely Müller's educational regime could not produce human beings of strong character.

The interrogator

I have already suggested that the social order of Külpe's Würzburg – and later Bonn and Munich – Institutes allowed for a considerable amount of diversity of views and change of positions. Indeed, over time even the interlocutor model itself was partly displaced and complemented by what we might call the model of the 'interrogator'.

We can approach this aspect by noting the difference between Wundt's and Bühler's characterisation of the Würzburger experimental design. Whereas Bühler described it as a co-operation between equals, Wundt represented it as a process of interrogation or examination, a process in which the experimenter dominated the experimental subject. I now want to show that at least some quarters within thought psychology increasingly emphasised the very control and surveillance of experimental subjects that Wundt, rightly or wrongly, attributed to Bühler's experiments.

Bühler's characterisation of his experiments with Külpe and Dürr clearly invoked the interlocutor model. We have seen above how Bühler described the co-operation with his two subjects as equal and loving. Bühler did not interrupt his subjects; he let them speak for as long as they wanted and without imposing any terminology upon them; he selected the tasks so that they would be to his subjects' liking; and when publishing the results of his research he indicated which retrospective report came from which subject. Bühler felt little need to control his subjects' reports except by checking them for contradictory statements (Bühler 1907b: 306).

Already prior to Bühler's experiments, however, other experimenters of the Würzburg school had used a number of techniques that subtly weakened the cognitive and social authority of the experimental subject. For example, several of the Würzburg experimenters gave up the earlier custom of identifying the author–subject of introspective reports by name. Justified by the need for privacy (e.g. Westphal 1911: 434), this practice in fact downgraded the experimental subject's position: whereas the subject had earlier had the status of almost a co-author of the papers that incorporated his report, he was now reduced to playing an anonymous part within a set of data. And, most importantly, whereas earlier the subject had had the status of an eyewitness, whose name functioned as a signature and as an oath, his reliability and honesty soon needed to be safeguarded and indicated by other means. In his criticism of the Würzburg school,

Müller drew attention to this change and lamented its effects (Müller 1911: 5–6).

Moreover, two Würzburgers, Ach (1905) and Grünbaum (1908), used hypnosis, and one leading Würzburg experimenter, Messer, interrupted his subjects at various stages of the experiment to gain more detailed reports of the mental processes up to this point. Use of this method meant that the subject was unprepared for the intervention and left unaware of its intended use. The idea of interruption as a methodological tool was soon picked up and developed further by a member of Müller's Göttingen school (Baade 1913a, 1913b, 1918).

In some subtle ways, then, the Würzburg experiments contained the seeds for the interrogator model of experimentation from quite early on. A number of factors were needed, however, to speed up, or facilitate, the development of the interrogator model. I shall mention three of them here.

First, we need to note the Würzburgers' involvement in applied psychology – an issue to which I shall return in greater length in the next chapter. Applied psychology was an area of keen interest for several key members of the Würzburg Institute. *Inter alia,* they worked on forensic psychology, that is lie detection, on experimental pedagogy, and on psychopathology. The crucial point about applied psychology was that it was typically based upon the interrogator model: it worked with experimental or test subjects; such subjects were not trained psychologists, and certainly not social or cognitive equals of the psychologist. Many experiments in applied psychology were designed so as to lead to the development of standardised tests of the abilities of school children, or the honesty of suspects in criminal investigations. Because such tests already existed in applied psychology, it was but a small step to suggest their application to experimental subjects in the 'pure' psychology of thought as well.

A second factor that facilitated the shift towards the interrogator model in the psychology of thought was the persistent criticism, coming from many quarters, that self-observation was unreliable, arbitrary, and open to manipulation and suggestion. I have explained such criticism in the last chapter. On the one hand, critics felt uneasy about the fact that the Würzburg experiments had originally used only Würzburg psychologists as subjects. As Wundt had pointed out, this opened the door to *folie à deux,* or theory-induced error. On the other hand, however, if thought psychology was to follow the model of applied psychology and use lay persons as experimental subjects, it had to safeguard the trustworthiness of these lay subjects. Accordingly, one critic, Georg Anschütz, was willing to accept the results of systematic experimental self-observation only if a variety of different subjects had been used, if their reports had been subjected to statistical analysis, and if the experimenter had obtained information about the subjects' character (1911: 487).

Of the – present and former – Würzburg psychologists, only Bühler rejected such criticism outright (1909b). Külpe proved himself to be wide open to such criticism and suggestions, however. As seen in Chapter 1, Külpe eventually conceded that a battery of checks and tests were needed to control the experimental subject and the subject's reporting. The experimenter had to check subjects' ability to observe, their goodwill, their moods and dispositions, their knowledge and level-headedness. He also had to reckon with the possibility that some people are better in focusing on external rather than internal objects, that some have no sense for nuances, and that others are insecure and likely to utter contradictions.

Finally, I want to suggest, in part somewhat speculatively, that the Würzburg experimental design was influenced by psychiatry from the start, and that towards the end of the first decade of this century, the increased interest in 'psychopathology' further augmented this influence.

My evidence for an influence of psychiatry upon systematic experimental self-observation is somewhat circumstantial, but telling nevertheless. Ach, to whom we owe the first detailed characterisation of the method in print (1905), had worked with the famous psychiatrist E. Kraepelin before coming to Würzburg, and Külpe had studied not only with Müller and Wundt but also with the Leipzig psychiatrist and brain anatomist P. Flechsig (Geuter 1986: 139; Baeumker 1916: 76). Schultze, who did his PhD under Külpe and advocated systematic experimental self-observation, commented that the question-and-answer method used in the context of Würzburg thought experiments was all of a piece with the ways 'in which the psychiatrist records the case history [i.e. the anamnesis] of his patient'. He therefore suggested 'anamnestic method' as a label for the Würzburg type of questioning (Schultze 1906: 253). Grünbaum, who had followed Ach in using hypnosis within an experimental self-observation setting, also later turned to psychiatry (Grünbaum 1908; Geuter 1986: 171). And finally, F. Hacker's work on dreams, a work critical of Freud's earlier

Traumdeutung, was also carried out under Külpe's supervision (Hacker 1911).

Contemporary sources indicate that the psychologists' interest in psychopathology increased around the same time as the Würzburg school engaged in its famous experiments. As one reviewer of the psychological scene observed in 1908: 'psychopathology knocks impetuously at the doors of exact psychology; not only does it demand to get in, but some of its representatives even claim a dominating role for it: The pathological is supposed to illuminate the normal mental life' (Gutberlet 1908: 30). Interestingly enough, Külpe was one of the leading figures of this new movement within psychology. In his contribution to the debate over the proper institutional place for psychology (1912b), Külpe suggested that psychology find itself a new home in the medical faculty alongside psychiatry. As Külpe saw it, the psychology of the normal and the psychology of the pathological should and could illuminate one another (1912b: 190). To defend this thesis, Külpe discussed what he took to be the central recent methods and results of psychopathology. In each case, he tried to show how these methods and results needed to be improved by heeding the insights of thought psychology. And in a footnote at least, Külpe submitted that although Wundt's concept of 'interrogation method' did not apply to Bühler's experimental design, it well fitted the questioning method used by psychiatrists (1912b: 191).

If the overall gist of this chapter is roughly on the mark, then we should expect that a change in the social order of psychological experiments might also suggest changes on other levels. There is no evidence that the social order of Külpe's institutes in Bonn and Munich – places where he held chairs after 1909 – differed from the social order in Würzburg. I also have no reason to believe that Külpe gave up or modified his view that thinking and knowledge acquisition are essentially social activities. As concerns the internal structure of psychology as a field of knowledge, however, Külpe clearly raised psychopathology from the status of a marginal field to a much more prominent position. And at least according to one of his critics, Georg Anschütz, Külpe's various pronouncements on thought psychology implied that it was 'co-extensive with psychology proper, or at least its central area' (Anschütz 1912b: 198).

Whether or not this was in fact Külpe's view, it is worth pointing out that in 1912 Külpe published a sketch of the constitution of the human soul that easily justified the position that Anschütz attributed to him. The soul now had two strata: a lower stratum of thoughts, sensations, feelings, and presentations, and a higher stratum occupied by an actively governing ego. Külpe couched the relation between the two strata in unmistakably political terms by speaking of the 'monarchist structure of our consciousness':

> The ego sits on the throne and governs. It notices, perceives, and registers what enters into its kingdom; it deals with that intruder; it consults its experienced ministers, the principles and norms of its state, the contingent needs of the present, and it adopts a position vis-à-vis the intruder, to set him aside, to give him a useful form, or to govern against him. And the sensations and presentations have, in general, good cause to condemn as reactionary, harsh, and arbitrary the rule of this monarchic ego. They take their revenge during sleep and are up to mischief in our dreams. They then show what results from such anarchy.
>
> (1912c: 1090)

This is a powerful picture and a key passage in Külpe's œuvre. I shall discuss its importance for understanding Külpe's political philosophy in Chapter 5. In the present context, I shall confine myself to the following remarks. First, the hierarchical ego psychology presented in this passage cannot be found in any of Külpe's earlier writings. It clearly differed from Bühler's more egalitarian view of consciousness, with its rejection of the higher/lower divide. Second, there is some evidence that not all Würzburgers agreed with the shift. Marbe maintained that the link between the results of experimental thought psychology and this picture was tenuous, and that Külpe's way of presenting it 'was far from being stringent' (1915: 27). Third, Külpe's distinction between the acts of the ego on the one hand, and the correlates or products of this activity on the other, is clearly influenced by Brentano, Husserl, and Stumpf (Brentano [1874] 1924; Husserl [1901] 1984, [1911] 1987; Stumpf 1906). Fourth, Külpe's metaphor could also conceptualise psychopathological phenomena. This Külpe did explicitly in the aforementioned paper on the relationship between psychology and psychiatry. There he paralleled the dream state with states of mental disorder, and suggested that mental disorders are due to a lack of ego control (1912b). In using a philosophical model of the human mind for the purposes of psychopathology, Külpe did not stand alone: at least two other psychopathologists at the time, K.

Jaspers and T. Oesterreich, drew on Husserl for the very same purpose around the same time (Jaspers 1912; Oesterreich 1910). Fifth, Külpe now directed the main focus of the psychology of thought towards the activity of the ego itself (1912c: 1089). Given this new focus, and given the crucial importance of the monarchical ego for mental health, the psychology of thought could claim to be the main area of psychology and psychopathology. Sixth, and finally, I cannot abstain from the more speculative remark that the manner in which the ego was to control and suppress the potentially anarchistic level of feelings and sensations was reminiscent of the ways in which the psychiatrist was to control patients, the experimenter to check subjects, and the applied psychologist to test school children and criminals.

Summary

In this chapter I have made a first attempt to give a sociological account of Wundt's, Müller's, and the Würzburgers' different stances *vis-à-vis* thought psychology. The three parties had very different theories of the human mind. I have tried to show that these theories were intertwined with the social structures of the respective psychological institutes in general, and Wundt's, Müller's, and Külpe's leadership styles in particular. The division of cognitive and social authority between experimenter and experimental subject was another important aspect of these social orders.

In Wundt's case we found that key features of his psychology were shaped by the goal of justifying the social order in the Leipzig Institute: Wundt, the solitary master psychologist, studied the higher, solitude-demanding processes of thinking; his students and assistants studied the lower processes of sensation and perception. In Göttingen too a match between social and psychological order had been achieved: Müller's theory of the mind stipulated a rule-bound struggle between presentations over passage into consciousness. In the light of this theory, Müller could justify both his own privileged leadership position and his strict 'drillmasterly' educational regime. In Würzburg we found an 'egalitarian' theory of the mind co-existing with a fairly egalitarian social order in the institute and in the experimental situation. We also noted that, for various reasons, some Würzburgers later adopted less egalitarian models of experimentation and of the mind.

More was at stake in the controversy over thought psychology than the reliability of self-observation, the existence of non-depictive thoughts, and the limitations of associationism. And this 'more' included positions of institutional leadership, power, and authority. The Würzburgers' attacks on Wundt's and Müller's psychological doctrines were – by the same token – assaults on hierarchical structures that they found unacceptable; Wundt's and Müller's defences of their respective psychologies were – by the same token – attempts to shield and justify these very same structures.

4 Purist versus promiscuist

Introduction

The Würzburgers' thought-psychological studies were published at a time when the institutional and intellectual borders between psychology and other fields of study were still very much in flux. Indeed, psychology was not even an independent discipline yet. Psychologists such as Wundt, Müller, and Külpe held professorial chairs in philosophy, and Wundt had earlier trained and taught as a physiologist. As is to be expected under such circumstances, different psychologists had diverging views on the relationship between psychology and other academic subjects, and thus also contrary positions on the proper institutional location for psychology. In this chapter, I shall explain the links between these disagreements and the controversy over thought psychology.

The simple dichotomy between *the purist* and *the promiscuist* [*sic*!] suffices for capturing the most prominent turn-of-the-century positions regarding the relationship between psychology and other disciplines. The *purist* insisted that psychology be both conceptually and institutionally independent of other disciplines. The *promiscuist* was happy to allow permeable borders with respect to at least some other fields. I shall argue that Wundt was a psychological purist with respect to logic, physiology, and the applied sciences. The Würzburgers, however, were fully committed to promiscuity in all three directions. Müller was open to physiology, but showed little active interest in developing links between his psychology, the applied sciences, and contemporaneous logic or epistemology.

I shall concentrate my analysis on Wundt and the Würzburgers and make no more than passing references to Müller. Only Wundt and the Würzburgers explained their positions on interdisciplinarity and related matters repeatedly in print and at some length.

Psychology and physiology

Wundt's purism

It is not difficult to show that Wundt pursued a purist strategy *vis-à-vis* physiology. Almost all the central pillars of Wundt's theoretical edifice were in fact border stones marking the realm of psychology as distinct from that of physiology.

Wundt did not count psychology among the natural sciences. Psychology was *the most fundamental* field of the humanities (*Geisteswissenschaften*) (1889: 44). To understand why psychology was not a natural science, we need to turn to what Wundt regarded as *the* basic dichotomy of human experience. Wundt suggested that an unbiased study of human experience revealed that 'every experience contains *two* factors that in reality are inseparable: the objects of experience and the experiencing subject'. Natural science would study only the former, whereas psychology would investigate both the former and the latter:

> Natural science seeks to determine the properties and the reciprocal relations among *objects*. Thus natural science abstracts...from the subject....Psychology cancels that abstraction and thus it investigates experience in its immediate reality. It thereby reports on the relations between the subjective and the objective factors of immediate experience and informs also on the genesis and the interrelations of the different contents of immediate experience.
>
> (Wundt 1896: 12)

In other words, Wundt proposed that both psychology and the natural sciences were empirical sciences, but that psychology studied 'the given' in its immediacy, whereas the natural sciences looked at it merely 'as a system of *signs* on the basis of which one has to form hypotheses of the real nature of the objects' (1896: 23).

Wundt also contended that causality in psychology was qualitatively different from causality in the natural sciences (Wundt 1894: 43). Four characteristics distinguished mental from physical causality. First, whereas for the natural sciences cause and effect were 'separate experiences, *disjecta membra*', in that the causal connection between two events 'comes only from the conceptual connection and treatment of experience', in psychology the connection between psychological elements was not a matter of theory but 'a fact of immediate consciousness' (1894: 43, 108; tr. Mischel 1970: 6). Second, mental causality could not be reduced to physical causality because the explanation 'of psychological processes is everywhere shot through with value determinations'. Such value determinations never occurred in causal explanations provided by natural science (1894: 98; tr. Mischel 1970: 8). Third, psychological causality was distinct from physical causality because 'the formation of mental products which indicate a conscious purposive activity, in which there is a choice between various possible motives, requires a real consideration of purpose' (1894: 117; tr. Mischel 1970: 8). In other words, Wundt held that the psychologist, in explaining human action and behaviour, inevitably had to bring in the goals and purposes of the agent. And fourth, physical causality presupposed a 'substance-substrate': that is, to give a causal explanation in the natural sciences, one needed to refer to the properties of some underlying and continuing substrate. Mental causality was different. No underlying substrate was assumed, and all processes and events were 'depictive' (*anschaulich*) and thus observable (1908b: 260–2).

Wundt called the assumption that psychology dealt only with depictive elements, their compounds and their unique mental–causal relations 'the principle of mental actuality' (1908b: 260). The principle also included the idea that 'every mental content is an *event* (*actus*); there are no constant objects of the kind that natural science must presuppose in its realm – within our immediate inner experience' (1894: 101).

The principle of actuality was the link between Wundt's insistence on a separation between psychology and the natural sciences, on the one hand, and his rejection of the unconscious on the other hand. The assumption of unconscious presentations and feelings ran counter to the idea that the realm of psychology was the realm of the immediately given; after all, unconscious presentations and feelings were inferred, not observed (1896: 34; 1903a: 327). To allow for unconscious processes meant to accept psychological hypotheses concerning unobservables, and this in turn amounted to an assimilation of psychology to physiology and other natural sciences. This move needed to be resisted.

Wundt was distinctly unimpressed by what he regarded as the standard argument for unconscious processes and called the unconscious 'a concept steeped in mysticism' (1896: 34). It simply was not true that the unconscious was needed as something of a storage space for presentations that had passed through consciousness and that were now waiting to be mobilised or reproduced. Presentations were never stored and lifted into consciousness as such; the fact that they could never be recalled exactly indicated that what seemed like a 'recalled' presentation was in fact a new–newly created–presentation (1903a: 327–9).

Another consequence of 'the actuality theory of the soul' was 'the heuristic principle of psychophysical parallelism' (1903a: 772). The actuality assumption implied that the character of *mental* elements, compounds, and relations was principally different from the nature of *physical* elements, compounds, and relations. And this incommensurability of the mental and the physical realms made 'psychophysical causality' inconceivable. All one could conceive of were two causal chains – the chain of mental causality and the chain of physical causality – running in parallel; all one could envision was that certain mental events 'regularly co-occurred' with certain physical events. But 'regular co-occurrence' was not enough to establish a causal link. For a causal link one also needed general laws under which the events could first be subsumed, only then to be incorporated 'in the general context of experience' (1908b: 252).

To put it differently, the principle of psychophysical parallelism stated that 'mental events can only be caused by mental events, and physical events can only be caused by physical events' (1908b: 256). This was because 'cause and effect always presuppose a homogeneous whole of which they are parts. But no whole is homogeneous if its parts belong to completely different viewpoints of experience', that is to the viewpoints of psychology and the natural sciences (1897: 380).

Wundt never forgot to mention that the principle of psychophysical parallelism was merely a 'heuristic' idea, not a 'metaphysical' dogma. By this he meant that the scope of the principle was to be decided by scientific enquiry. Thus the principle did not claim that every physical event was accompanied by a mental event, or even that every mental event was accompanied by a physical event. The principle solely denied that between the two levels there could ever be more than mere co-occurrence (1903a: 773).

Wundt trusted that the joint work of psychology and physiology already allowed one to formulate a close-to-certain conjecture on the scope of the psychophysical principle:

> There is no elementary mental process, no sensation and no subjective emotional excitement that is not paralleled by a physiological process, or complex of physiological processes. Now, because all mental processes consist of such elements, it is obvious that the principle of psychophysical parallelism is a universally valid heuristic principle for mental contents.
>
> (1903a: 776)

The parallelism on the basis of elements did not imply an isomorphism of combinations and connections, however. Assume that we have three elementary mental processes, *m1*, *m2*, and *m3*, and that these processes are connected in way *mc*. On the basis of the above, we can then say that almost certainly there are three physiological processes (or complexes thereof), *p1*, *p2*, and *p3*, such that *p1* regularly accompanies *m1*, *p2* accompanies *m2*, and *p3* accompanies *m3*. We can also say that *p1*, *p2*, and *p3* must be connected in some way or other for it to be possible that *m1*, *m2*, and *m3* are connected in way *mc*. But there is no licence for assuming that the way in which *p1*, *p2*, and *p3* are connected, say *pc*, is in any way similar, or analogous, to *mc*. For instance, as Wundt insisted, the fact that ideas were linked by association did not justify the assumption that there were 'association fibres' in the brain (1903a: 776). Or, the other way around, even if physiology were ever to understand the brain 'as clearly as the mechanism of a pocket watch', this physiological description of neural combinations and connections would not tell the psychologist much about mental processes (1903a: 777).

To make these sorts of claims was of course tantamount to denying that psychology could in any form be reduced to physiology. The form of reductionism that Wundt never tired of attacking was 'psychophysical materialism', a position he attributed to three of his former students, Külpe (1893), Meumann (1904), and H. Münsterberg (1889, 1891). Psychophysical materialism differed from both 'pure materialism' and psychophysical parallelism. Psychophysical materialism was unlike pure materialism in so far as it rejected the idea that *all* mental events were caused by physical events in the brain; according to psychophysical materialism, *elementary* mental events were not so caused. In this respect, psychophysical materialism sided with psychophysical parallelism. It parted company with the latter, however, by assuming that every causal explanation of the sequence (combinations and connections) of mental events had to refer to *physical* causal relations that existed between the *physical* accompaniments of the mental events. In Wundt's words:

> According to this view, psychology has to solve two tasks. The psychological task is purely descriptive: Describe the sensations that can be observed during the working of certain senses and neural apparatuses. The physiological task is causal-explanatory: Show how, on the basis of the connections between physiological functions, complex mental processes...can arise out of sensations.
>
> (1908b: 149)

Wundt insisted that this view implied the demise of psychology as an independent discipline:

> Because the first task...obviously is merely auxiliary with respect to the second, it is clear that this position aims to turn all of psychology into a part of the physiology of the sense organs and into a part of the physiology of the nervous system.
>
> (1908b: 149)

Wundt's way of combating this position drew on all of the above: psychophysical materialism contradicted psychophysical parallelism, the dualism of causality, the dichotomy of *Natur-* and *Geisteswissenschaften,* the principle of actuality, and all of the arguments given in support of these more general claims (e.g. Wundt 1896). Moreover, Wundt judged psychophysical materialism to be a position based on wishful thinking: the kind of brain physiology needed for carrying through the psychophysical materialists' programme was, even by their own admission, a thing of the far-off future (Wundt 1904: 358).

Wundt's purism with respect to physiology, however, did not rule out that there could be some limited co-operation between the two disciplines. This much was already suggested by Wundt's early label for experimental psychology: 'physiological psychology'. The 'physiological' in 'physiological psychology' referred first to the fact that experimental psychology often used experimental settings invented by physiologists. Indeed, for the psychologists, the *means* often were physiological, but not the ends: 'The experiment remains a psychological one as long as...the purpose is self-observation under varying conditions' (1908b: 222). Second, modern psychology differed from older philosophical approaches in that it aimed to identify regular co-occurrences between mental and physiological processes. For reasons mentioned above, such a programme could only concern very elementary mental processes, but it was an important part of psychology nevertheless (1908b: 223). And third, 'physiological psychology' also signalled the possibility of provisionally inserting elements from one causal sequence, say the physiological, into the other causal sequence, say the mental. This seemed natural when one science – in this example psychology – was as yet unable to close a causal gap with its own devices and theories. But here again it had of course to be remembered that 'the heterogeneous elements of explanation must always only be looked upon as proxies for the homogeneous elements that, for the time being, or forever, are, or remain, unknown' (1908b: 257).

Despite this defence of 'the physiological' in psychology, Wundt thought of physiological psychology as no more than a 'transitory' field of study. This impure field of enquiry would no longer be needed once psychology had properly developed its own experimental tradition, and once physiology would be able to provide all the information about bodily processes that the psychologist might need (1896: 21; 1908b: 224). From here on, psychology and physiology could run in parallel, without ever having to cross over into each others' terrain.

Külpe's promiscuity

Külpe did not share Wundt's goal of trying to keep psychology separate from physiology. Nor did he adopt any of Wundt's more specific measures for achieving such separation: Külpe argued that psychology was not a *Geisteswissenschaft*; he rejected mental causality; he refused to side with Wundt's version of psychophysical parallelism; and he contradicted Wundt's principle of actuality.

The split between Wundt and Külpe over the relationship between psychology and physiology preceded Würzburg thought psychology by a decade. As shown in an important paper by K. Danziger (1979), Külpe's repudiation of Wundt had already happened in Külpe's *Grundriss* of 1893.

Külpe's central resources in his repudiation of Wundtian purist psychology were Richard Avenarius's and Ernst Mach's views on science and psychology (Avenarius 1888, 1890; Mach 1896). Avenarius and Mach rejected the metaphysical dualism between the mental and the physical, claiming that 'experience' (*Erfahrung*) showed no such division. The two positivists nevertheless allowed for the possibility that experiences could be investigated from two different points of view: that is, either as *dependent* on, or as *independent* of, the particular physiological system to which they belonged. The first viewpoint was that of the physical sciences, the second that of empirical, or scientific, psychology. In their conception of a scientific psychology, the notion of a physiological system was central for Avenarius and Mach. They demanded that psychological explanations be based on physiological principles, and that mentalistic concepts – that is, concepts presupposing the notion of an active ego-subject – be excluded from psychology. As Mach and Avenarius saw it, such concepts had no basis in experience. Both men saw it as the task of science to provide the most economical description of the interrelations among experiences, and Mach conceived of scientific laws as stating functional relationships between observables. Mach also held that the sciences were to be thought of as one hierarchical structure in which less general sciences were situated below more general ones. As progress towards greater thought economy meant the formulation of ever more general theories and laws, less general sciences were to be reduced to more general ones. With regard to psychology this meant that it should ultimately be reduced to physiology and biology (Danziger 1979: 210–12).

Because Külpe adopted the position of these positivists, he demanded that psychology relate the facts of experience to the 'corporeal', physiological or biological individual or organism. Only in this way, Külpe wrote in 1893, could psychology become a *natural science*: 'The objects of psychological enquiry would never present the advantages of measurability and unequivocalness, possessed in so high a degree by the objects investigated by natural science, if they could be brought into relation only with the mental individual' (1893: 4; tr. Danziger 1979: 209). For Külpe, mental processes had to be explained by physiology.

As Külpe wanted psychology to become a natural science, he had no sympathies for any position that restricted the applicability of the experimental method. And this in turn meant that Külpe did not share Wundt's scepticism concerning the applicability of the experimental method to the study of higher mental processes. Encouraged by Ebbinghaus's study of memory, Külpe wrote that 'in principle there is no topic of psychological inquiry that cannot be approached by the experimental method. And experimental psychology is therefore fully within its rights when it claims to be the general psychology that we propose to treat' (Külpe 1893: 12; tr. Danziger 1979: 213). This was of course one of the seeds out of which the Würzburg thought psychology could later grow.

From about the turn of the century, Külpe increasingly distanced himself from positivism, and he subjected Mach's philosophy to detailed criticism in his little book *Contemporary Philosophy in Germany* (*Die Philosophie der Gegenwart in Deutschland*, 1902; here 17–24). But this did not amount to a return to the Wundtian position, for in the same book, Külpe also attacked Wundt's principle of actuality – the cornerstone of all of Wundt's purist moves. Against the Wundtian claim 'that consciousness and mental facts coincide', Külpe set the Nietzschean aphorism: 'I insist on the phenomenality of the inner world as well as the outer world: Everything of which we are conscious is thoroughly made up, simplified, schematised, interpreted' (quoted from Külpe 1902: 100). Külpe marshalled a number of arguments against Wundt.

First, Wundt's principle was plausible only for one's own mental experience, but not for the mental experience of someone else. In Külpe's view such privilege for the first person was unscientific. Second, the principle was plausible only for one's own experience in the present. It was not credible, however, for one's remembered experiences. And third, Külpe pointed to his own abstraction experiments in which subjects were often unable to report facets of their experiences that clearly they had to have seen. For instance, of a given configuration of letters, subjects might recall only the letters themselves, but not their colour (1902: 101–2).

For these and similar reasons, in Külpe's view, it was Nietzsche and not Wundt who was closer to scientific psychology. After all, perception was always interpretation ('apperception') in the light of earlier experience; inner perception could not be an exception to this rule:

> 'To experience a mental process', 'to perceive it', 'to be conscious of it', 'to apperceive it' – all these are therefore synonymous expressions....But since this is so, mental appearances too are of course 'simplified' or 'made up', as Nietzsche says. And they have this nature since we experience them and know of them only in the form in which apperception has modified them; and this form cannot be calculated *a priori*.
>
> (1902: 103)

Külpe's rejection of the actuality principle undermined anew the Wundtian distinction between psychology and the natural sciences;

the data of psychology no longer had the special immediacy that according to Wundt, distinguished them from the data of the natural sciences. If psychology, like the natural sciences, only dealt with *appearances*, then it too needed to go beyond the given. And this 'beyond' was either physiology, or else some version of the unconscious. Külpe now was no longer as insistent as he had been in 1893, that this 'beyond' be that of physiology. Whichever form one chose, from Wundt's perspective it amounted in any case to an opening of the floodgates to physiology.[1]

The debate over the relationship between psychology and physiology and the controversy over thought psychology

Wundt's 1907 criticism of the Würzburg studies did not contain any direct references to his and Külpe's disagreement over the status of psychology as a natural science or a *Geisteswissenschaft*. Nor did Wundt refer directly to the issue of whether psychology would – wholly or in part – reduce to physiology. Nevertheless, it seems natural to assume that the debate over the relationship between psychology and physiology was one of the subtexts of the controversy over thought psychology.

First, the Würzburgers' psychological investigation, and postulation, of non-depictive conscious entities (situation of consciousness, awareness, thought) violated Wundt's principle of actuality. According to this principle, all mental contents were conscious, and all conscious contents were depictive. We have seen that for Wundt the principle of actuality was the most important bulwark against both psychophysical materialism and the idea of psychology as a natural science. In ignoring the principle of actuality, the Würzburgers implicitly rejected Wundt's purism *vis-à-vis* physiology. No doubt, part of Wundt's hostility was due to the fact that he recognised what the thought psychologists were doing.

Second, pretty much the same observation can be made concerning the unconscious. Again, the assumption of unconscious mental processes ran counter to Wundt's principle of actuality. Ach, Watt, Messer, and Bühler all worked, more or less explicitly, with the idea of a psychological unconscious. They used introspective reports as data for hypothetically determining the workings of the unconscious mind. Ach followed the tradition from Herbart to Müller and spoke freely of the threshold of consciousness, of the unconscious, and of presentations that moved back and forth between the conscious and the unconscious realm (Ach 1905, *passim*). Watt invoked Külpe's argument against Wundt in going beyond the introspectively given (Watt 1905: 425–7). Messer relied less on Külpe than on a conception developed by T. Lipps (1901). According to Lipps, both Wundt and the physiologically inclined psychologists, such as Münsterberg (1889, 1891) and the early Külpe (1893), had worked with an oversimplified picture of the human 'mind-brain'. In trying to explain the sequence of conscious contents, both camps had tried to make do with just two levels: the level of conscious contents, and the level of physiological processes. Wundt then explained the sequence of mental contents by stipulating a mental causality between them, Münsterberg first by relating conscious contents back to their underlying physiological processes, and second by referring to the causal relations between these physiological processes. What was missing, Lipps thought, was an appreciation of a third level, the level of the unconscious. Conscious contents were the endpoints of mental processes taking place unconsciously. Mental causality held between these mental processes. Lipps submitted that this conception was common to him and to Müller.

Again, we can see why this talk of the unconscious was unacceptable to Wundt. Any use of the unconscious was in breach of the principle of actuality, and thus a way of opening the floodgates that protected psychology from the natural sciences. Here it did not matter whether or not the unconscious was itself conceptualised as either physiological or psychological. Even if it was conceived of as psychological, the damage was done. For once psychology was no longer restricted to studying the immediately given; its main distinctive feature, the feature that separated it from the natural sciences, had been lost.

And third, the very fact that the Würzburgers used experiments in the study of higher processes was of course itself an indication that they were on a different tack from Wundt. It is true that Wundt never directly argued that psychology was a *Geisteswissenschaft* because the higher – and thus the most important or valuable – mental processes were studied by collective psychology, a field that relied heavily on the methods of *Geisteswissenschaften* such as philology or history. Wundt did not need such an argument because he could argue for the inclusion of psychology in the humanities on other grounds. Nevertheless, drawing this conclusion was undoubtedly much too natural and plausible for Wundt's colleagues and contemporaries to miss. But then the employment of experiments with respect to the higher, thinking processes appears in a new light. This employment weakened the case for Wundt's construal of psychology as a *Geisteswissenschaft*.

Pure and applied psychology

Wundt's purism

Wundt was a purist also with respect to the distinction between pure and applied psychology. While he did not seek to rule out applied research altogether, Wundt insisted that – at least for the time being – the communication between pure psychology and areas of application remain one way only. Wundt felt particularly worried about the application of psychology in pedagogy. This application, for which Wundt's former student, E. Meumann, introduced the label 'experimental pedagogy' (Meumann 1907c), attracted considerable attention at the beginning of the century. Wundt's opposition to the increase of applied research in this area went so far that in 1903 he withdrew his support for Meumann's *Archiv für Psychologie* – the journal in which the Würzburgers published most of their studies – because 'in the issues published so far, a total of 873 pages have been devoted to education, but only 715 pages to all other areas of psychology' (letter by Wundt to Meumann, quoted from Bringmann and Ungerer 1980: 70).

Wundt discussed applied psychology both in his *Logik* (1908b) and in a longish criticism of Meumann's 1908 book *Intelligence and the Will* (*Intelligenz und Wille*) (Wundt 1910c). In the *Logik,* Wundt distinguished between 'general', 'applied', and 'practical psychology'. Practical psychology was the ability to empathise with others, to understand them, and to be able to manipulate them for one's own ends. This was the psychology of the politician, the historian, the medical doctor, or the teacher. Wundt did not believe that either general or applied psychology had contributed much to the development of this kind of knowledge. Practical psychology was a mixture of tacit skill and common sense, and as such it grew out of practical experience (1908b: 297–8). Wundt also held that scientific as well as applied psychology deserved little credit for recent, undeniable advances in history and the social sciences as well as pedagogy. Almost all of what these fields owed to psychology they owed to practical psychology (1908b: 298).

Wundt did not rule out that one day practical psychology might learn some lessons from scientific psychology. For instance, in the realm of pedagogy one could modify some psychological–

experimental tools and settings and use them for evaluating school children's achievements. More important than this piecemeal use of scientific psychology, however, would be a more holistic approach; that is, the practitioners would undoubtedly profit most from acquainting themselves with the results of general, theoretical psychology. Proper knowledge of modern scientific psychology would free them from the shackles of 'vulgar' folk psychology and thus help them understand mental life in its diversity and complexity (1908b: 301). Wundt submitted that most of the *Geisteswissenschaften* too would gain enormously from a familiarity with (his) psychology. For the *Geisteswissenschaften*, the results of collective psychology would be particularly important. But again, the reception of scientific psychology should be holistic rather than technological. The aim should not be to find some isolated psychological techniques, but rather to raise one's general understanding of psychological processes (1908b: 301).

In Wundt's view, there was good reason to assume that psychology could never be applied in practice as successfully as the natural sciences. 'The singular character' of mental events made it impossible 'that special rules can be derived from general psychological principles, rules by means of which individual mental processes can be changed according to predetermined goals' (1908b: 298–9).

In his 1910 paper 'On Pure and Applied Psychology' (*Über reine und angewandte Psychologie*, 1910c) Wundt again emphasised the importance of collective psychology for the *Geisteswissenschaften*, but now there was an additional twist to his argument that is worth noting. Wundt submitted that collective psychology had a twofold character: On the one hand it was a central part of general, theoretical psychology. On the other hand, it also was *the most important* area of application for pure psychology: in at least some respects, collective psychology was application of pure psychology to collective phenomena (1910c: 7–9).

Wundt was much less enthusiastic about experimental pedagogy as an area of application for psychology. Applied psychology had three possible tasks *vis-à-vis* pedagogy: first, the 'practical–technical' task of finding, for instance, the most efficient methods of memorising; second, 'the practical–theoretical' task of determining differences in talents, or studying differences in achievement between students of different ages or sexes; and third, 'the purely theoretical' task of clarifying the general development of the child's mental life (1910c: 10–11). Wundt regarded the third task as the most important and insisted that the first two could best be tackled only on the basis of solutions to the third problem. The third problem in turn presupposed pure psychology proper: the development of the mind could only be adequately described and explained if one already possessed a theory of the endpoint of this development, that is a theory of the adult mind (1910c: 11).

Wundt believed that contemporary applied psychology – and especially experimental pedagogy – had chosen a different path, and that this path would harm all of psychology. To begin with, writers in this area had adopted the mistaken view that psychology was a natural science. To this inadequate conception they had then added the further error of assuming that natural science had always been involved in applied research. As a consequence they had rushed into applications, starting with 'practical–technical' and 'practical–theoretical' problems. And to make matters even worse, they had tacitly come to believe that the data gained in and through the work on such practical problems could even be the foundation for general, and pure, theories of the mind.

We already know that Wundt did not think of psychology as a natural science. But he also opposed the conception of natural science that he suspected lay behind some psychologists' rush into applications. Wundt claimed that the development of successful technologies presupposed decades if not centuries of undisturbed purely theoretical natural-scientific work. Natural science had no direct obligation to serve practice, and the aims of practice were best served by ignoring them: 'The following general principle is still undisputed today: Science exists primarily for its own sake, and it best serves the purposes of practice if it lets itself be guided primarily by the problems of purely theoretical knowledge' (1910c: 13).

Wundt's greatest fear, however, was that the emphasis on application would invite the construction of poor 'pure' psychology. Even psychologists who spent most of their time working on applied questions would sooner or later wish to develop general theories of the mind and its development. But in so doing, they would be prone to draw – primarily or exclusively – on the data they already knew well; that is, on the often incoherent data that had been collected with special, practical–technical interests in mind. Not surprisingly, the resulting theories could therefore always be shown to be overgeneralisations based on one-sided material; to be permeated by 'vulgar' folk-psychological concepts and ideas; and to be filled with contradictions (1910c: 17–18).

Wundt referred only in passing to applied psychology in his criticism of the Würzburgers' experimental research on thinking. Nevertheless, his suspicion that the latter was infected by impure practical motives was easy enough to detect. For instance, Wundt coined the expression 'interrogation method', and he suggested that it was derived from, or modelled on, practices in school examination and the courtroom (1907c: 303, 305, 340 349). In light of the above, it is clear why Wundt had to reject borrowing such methods. It meant an intrusion of pedagogy and legal practices into psychology. There also were, in Wundt's criticism, several allusions to a proximity between hypnotic experiments and Würzburg practices (1907c: 302, 310–11). For Wundt, hypnosis was a method of the medical doctor, or, more specifically, of the psychiatrist. As Wundt had already argued in 1892 in a book-length study, hypnosis belonged 'in the hospital room, not the psychological laboratory, and the inducement of hypnotic sleep…is justifiable only on medical grounds' (1892a: 12). Using hypnosis within psychology was to cross the boundary between psychopathology and pure psychology in the wrong direction. And finally, Wundt suspected that the Würzburgers' systematic experimental self-observation would become popular and successful with educationists and applied psychologists. On the one hand, the method was so easy to use that it just *had to* appeal to applied psychologists. On the other hand, the method could easily be used in the classroom – asking the children for their thoughts – and it could readily be combined with the use of questionnaires, a method that met with increasing interest amongst applied psychologists at just that time (1907c: 359–60; Stern and Lipmann 1908; Baerwald 1908; Schlesinger 1908). For Wundt, the use of questionnaires was the epitome of a method foreign to psychology, and he therefore vigorously opposed their use (Dorsch 1963: 38).

Summa summarum, Wundt's attack on the Würzburgers' thought psychology was also a defence of the purity of theoretical psychology with respect to practice.

The Würzburgers' promiscuity

The Würzburgers engaged in a large variety of applied projects. Külpe's interest in psychopathology went as far back as his student days, when he studied with the Leipzig psychiatrist and brain anatomist P. Flechsig (Baeumker 1916: 76). Külpe emphasised the importance of psychopathology for normal psychology from his early *Grundriß* (1893: 16–17) until his late paper on the relationship between psychology and medicine (1912b). In the latter text he stressed that a real dialogue between psychiatry and psychology was necessary, and that both needed one another: 'The modern development of psychiatry tends towards a thorough psychological comprehension and foundation; and normal psychology feels the strong need to take into account pathological observations and explanations' (1912b: 190).

Külpe's intellectual promiscuity was not restricted to psychiatry, however. In the same context, as well as elsewhere (1912c), he praised the existence of the many channels of communication that had opened between modern thought psychology on the one hand, and the philosophical, natural, and moral sciences (*Geisteswissenschaften*) on the other hand (1912b: 188–90). And Külpe's work as a supervisor and head of school reflected his stated convictions. Amongst the dissertations and other studies that Külpe supervised and evaluated in Würzburg, Bonn, and Munich were a considerable number of applied investigations. These dealt with such topics as 'the influence of the length of work and breaks from work upon the intellectual abilities of school children' (Friedrich 1897), 'the psychological analysis of self-education' (Cordes 1898), 'the influence of various medicines on the ability to comprehend' (Ach 1899), 'the partial and complete achievement of the school child' (Mayer 1903), lie detection (Wertheimer 1906), 'qualitative types of work' (Pfeiffer 1908), 'the ability of school children to abstract' (Koch 1913), 'on the development of the ability to abstract in female pupils' (Habrich 1914), and 'the ethical judgement of young people' (Roth 1915). Of Külpe's assistants, E. Dürr published first a thought-psychological study of the will in the major forensic journal, *The Courtroom* (*Der Gerichtssaal*) (Dürr 1906), and then followed this with two books addressed to the teaching profession. One treated attention and was based on lectures to school teachers; the other was an introduction to pedagogy (1907, 1908a). When Dürr became professor of philosophy in Bern in 1906, he was replaced by Karl Bühler who from around 1910 onwards developed an increasing interest in child psychology (Bühler 1911, 1918).

Two men who co-operated with Külpe but were not his students also must be mentioned as further proof of the strong inclination of Würzburg psychologists towards applied research. Messer wrote three books on education, one on the role of apperception in teacher–student, and teacher–headmaster interactions (1899), one on language learning (1900), and one on how to educate good citizens (1912a). Moreover, Messer's *Sensation and Thought* (*Empfindung und Denken*, 1908) was primarily aimed at teachers. Here as well as elsewhere Messer followed the example of Külpe and emphasised the importance of thought psychology for applications in pedagogy (1908: 184). And, in a paper on 'the importance of psychology for pedagogy, medicine, jurisprudence, and economics', Messer not only expressed his enthusiasm about the many applications of psychology, but also predicted the emergence of a new professional, 'the practical psychologist' (1914a: 190).

Finally, the standard history of applied psychology in the German-speaking world, F. Dorsch's *History and Problems of Applied Psychology* (*Geschichte und Probleme der angewandten Psychologie*, 1963), treats Marbe as one of the six major 'pioneers' of the field. And rightly so. Marbe was the first psychologist in Germany to act, in 1911, as an expert witness in court (1945: 38). In the following year, he published a much-cited review of applications of psychological knowledge in natural science, medicine, linguistics, philology, literary studies, aesthetics, history, pedagogy, jurisprudence, economics, and philosophy (1912a). Subsequently, applied research became ever more central both to Marbe's own work and to the Würzburg Institute, where he succeeded Külpe in 1909. This work ushered in books on forensic psychology (1913b, 1926a), the psychology of accidents (1926b), and the psychology of advertisement (1927).

Psychology and logic

Wundt and the Würzburgers also differed in their respective views concerning the relationship between psychology and philosophy. Their disagreement in this area was of course part of a wider debate in German philosophy at the time, that is the debate over psychologism. I have written about the latter controversy at length elsewhere (Kusch 1995a). I shall here introduce only a few central points.

Psychologism

From the 1880s onwards, a number of German logicians, epistemologists, and mathematicians started to attack what they eventually came to call 'psychologism', meaning, roughly, an excess of psychological thinking in key areas of philosophy and mathematics in general, and logic in particular. To the antipsychologicists [*sic*!], the following kinds of arguments seemed indicative of psychologistic tendencies in the writings of their contemporaries:

1 Logic studied certain 'laws of thought'.
Psychology is the study of all kinds of laws of thought.
Therefore, logic is a part of psychology.
2 Normative disciplines must be based upon descriptive–explanatory disciplines.
Logic (epistemology) is a normative discipline with respect to human thinking.
Psychology is *the* descriptive–explanatory discipline with respect to human thinking.
Therefore, logic (epistemology) must be based upon psychology.
3 Everything is either a mental or physical entity.
Numbers, meanings, and logical laws are not physical entities.
Therefore, numbers, meanings, and logical laws are mental entities.
4 Logic is a theory about judgements, concepts, and inferences.
Judgements, concepts, and inferences are human mental entities.
Therefore, logic is about human mental entities.
5 The touchstone of logical truth is self-evidence.
Self-evidence is a human mental experience.
Therefore, logic is about human mental experience (and thus a part of psychology).
6 We cannot conceive of an alternative logic.
The limits of conceivability are mental psychological limits.
Therefore, logic is relative to the human species, and thus a psychological phenomenon.

In some cases there was considerable dispute over the question who actually held these views; we need not enter into these debates here. It seems not altogether unfair, however, to attribute argument 1 to Lipps (1880, 1893) and Heymans (1894), 2 to Wundt (1880), 3 to Husserl (1891), 4 to Jerusalem (1905), 5 to Elsenhans (1902), and 6 to Erdmann (1892, 1907).

Arguments against these and other allegedly psychologistic authors were provided by many writers, amongst them Frege (1884, 1893, 1894) and Husserl (1900), as well as the neo-Kantians H. Cohen (1902), P. Natorp (1887), H. Rickert (1904, 1913a), and W. Windelband (1884). Of these and the many other critics, Husserl was undoubtedly the most successful. Although he was himself accused of being a closet advocate of psychologism more often than any other philosopher at the time, no other antipsychologistic text was cited and discussed as often as Husserl's *Prolegomena* to his *Logical Investigations* (*Logische Untersuchungen*, [1900] 1975), the text which contained his attack on psychologism. Here is my reconstruction of some of Husserl's central arguments:

1 If logical rules were based upon psychological laws, then all logical rules would have to be as vague as the underlying psychological laws. *But,* not all logical rules are vague. *Therefore,* not all logical rules are based upon psychological laws ([1900] 1975: §21).
2 If laws of logic were psychological laws, then they could not be known a priori. They would be more or less probable rather than valid, and justified only by reference to experience. *But,* laws of logic are known a priori; they are justified by apodictic self-evidence, and valid rather than probable. And *therefore,* laws of logic are not psychological ([1900] 1975: §21).
3 If logical laws were psychological laws, they would refer to psychological entities. *But,* logical laws do not refer to psychological entities. And *therefore,* logical laws are not psychological laws ([1900] 1975: §23).
4 All forms of psychologism imply species relativism: logic is construed as being relative to the human species, that is to human psychology. Species relativism is an absurd doctrine: if truth varies with different species then one and the same judgement could be true for one species and false for another. But this contradicts the meaning of truth. And, to make truth relative to the constitution of a species makes truth temporally and spatially determined. But this conflicts with the notion of truth. Truths are eternal ([1900] 1975: §§36, 38–40).
5 It is a psychologistic prejudice to think that prescriptions meant to regulate psychological events must be psychologically grounded. This overlooks the distinction between (a) laws that *can be used for setting* norms on how to reason and (b) laws that *are* norms on how to reason. Logical laws are laws of type (a), and they have no relation to mental processes ([1900] 1975: §41).
6 It is a psychologistic prejudice to think that logic is concerned with ideas, judgements, inferences, and proofs, and that all these are psychological phenomena. If this were true then the same reasoning would turn mathematics into a branch of

psychology. But this is utterly implausible and has been refuted by Frege and Natorp, amongst others ([1900] 1975: §45).

7 It is a psychologistic prejudice to think that self-evidence is a psychological phenomenon. Logical principles, such as the principle of non-contradiction, say nothing about self-evidence and its conditions ([1900] 1975: §50).

A good deal of the psychologism controversy concerned the validity and originality of Husserl's arguments, the question of whether his accusations against specific authors had been fair, and the issue of whether Husserl had himself relapsed into psychologism either in his very 'refutation' of it, or else in later parts of his *Logical Investigations*.

The debate over the systematic ties between logic and psychology was at the same time a debate over the question of whether the new, experimental psychology still belonged within philosophy departments. This was hinted at in Husserl's book, where he lamented the fact that too much contemporary philosophical effort was going in the wrong direction:

> the best scientific energy is directed at psychology – psychology as an explanatory natural science that should not interest philosophy any more than should the sciences of the physical processes...I am pleased about the otherwise promising development of scientific psychology, and I take a strong interest in it – but not as someone who expects any kind of *philosophical* clarification from scientific psychology.
>
> ([1900] 1975: 214)

Other authors had been, and were, more outspoken. Indeed, the allegation that experimental psychology had nothing to do with philosophy 'proper' had accompanied experimental psychology from early on. Thus a reviewer of the first volume of Wundt's journal *Philosophische Studien* thought the title inappropriate, and suggested changing it to 'physiological' or 'psychophysical studies' (Horwicz 1882: 498). Another author, K. Güttler, demanded that experimentalists leave philosophy departments. One of his main arguments was that no one could cope with the overburdensome task of being both a philosopher and an experimental psychologist (1896: 22–7).

The philosophical purists also took their 'plight' to widely circulating dailies and monthly magazines. For instance, R. Lehmann wrote in *Die Zukunft* in 1906 that the general growth in philosophical interest squared badly with the tendency to fill ever more philosophical chairs with experimental psychologists. Lehmann asked rhetorically whether the resulting decline in academic philosophy would 'continue until the present generation has reached rock bottom and until a new generation will come forth' (1906: 487). And another writer, M. Frischeisen-Köhler, calculated in *Die Geisteswissenschaften* that between 1873 and 1913, the experimentalists' share of philosophical chairs had risen from 7.7 per cent to 22.7 per cent (Frischeisen-Köhler 1913: 371).

The pure philosophers went public whenever 'yet another' philosophical chair had 'fallen' to an experimentalist. Already the appointment of Ach to 'Kant's chair' in Königsberg in 1906 aroused a lot of anger amongst pure philosophers (Marbe 1913a: 28–9; Lück and Löwisch 1994: 68), but the proverbial 'final straw' was the appointment, in 1912, of Müller's student E. Jaensch, to H. Cohen's chair in Marburg. When a student protest had no effect, Cohen's former colleague, Natorp, went public and wrote an article in the daily *Frankfurter Zeitung*. He complained bitterly of the fact that again a professorial chair in philosophy had been 'surrendered to a special science which has no more to do with philosophy than any other special science' (Natorp 1912: 2). At the same time, Natorp, Husserl, Rickert, Windelband, the neo-Kantian A. Riehl, and the neo-Fichtean R. Eucken drew up a petition, collected signatures for it, and submitted it to all German-language universities and ministries of education. The petition was signed by 107 philosophers in Germany, Austria, and Switzerland, demanding that no more philosophical chairs go to experimental psychologists. The last paragraph contained the key suggestion (tr. Ash 1980b: 407–8):

> In the common interest of both disciplines it must be taken carefully into consideration that philosophy retains its position in the life of the universities. Experimental psychology should therefore be supported only by the establishment of its own professorships, and everywhere where previously philosophical professorships are occupied [by experimental psychologists] new chairs of philosophy should be created.

Finally, before returning to Wundt and the Würzburgers, we need to register that Husserl's as well as many other philosophers' purism was unidirectional. That is to say, while the philosophical purists denied that experimental psychology had anything to contribute to philosophy, they in turn tried to contribute to psychology in some way or form. This was particularly true of Husserl. He envisaged a new philosophical discipline called 'phenomenology', which would, amongst other things,

> prepare the ground for *psychology as an empirical science*. It analyses and describes (especially as a phenomenology of thought and knowledge acquisition) the experiences of presentation, judging, and coming to know, experiences that in psychology must find their genetic explanation, that is, their exploration according to interrelations subject to empirical laws.
>
> ([1901] 1984: 7)

Husserl believed that experimental psychology needed his help to be able to rise from its current '*de facto* unscientific' level ([1911] 1987: 20). Experimental psychology found itself in this sad state because it had neglected the tasks both of a 'direct and pure' descriptive analysis of consciousness and of a clarification of key concepts. Lacking such analysis and such clarification, the descriptive and explanatory concepts of experimental psychology were no more than 'coarse class concepts'. For the same reason, experimental work lacked theoretical guidance, and its results remained without explanation ([1911] 1987: 18). As is well known, the framework that Husserl recommended for all of psychology was based centrally on the concepts of 'acts' as 'intentional experiences'. Thus, for instance, the analysis of meaning presupposed three types of acts: the acts of perception that were directed at the physical sign; the sense-giving acts that attached a meaning to the sign, that is the '*meaning-conferring acts* or the *meaning-intentions*'; and the acts of perception or imagination that presented or 'gave' the object referred to or meant, that is '*meaning-fulfilling acts*' (LU 1975 [1900] I: 44). The meaning-intending act was directed towards an *intended* object, but this object was not yet given within the meaning-intending act itself. Only an act of meaning-fulfilment, an act that fulfilled the meaning 'delineated' by the meaning-intention, could ultimately 'give' the object referred to (cf. Kusch 1989: 67).

Wundt's purism

Turning back to Wundt, it seems that his psychological purism *vis-à-vis* logic and philosophy was something of a mirror image of Husserl's logical purism with respect to psychology. Wundt never tired of deploring the intrusion of logic into psychology; until a change of heart in 1913, he desired institutional independence for psychology; and he believed that logic could do with a little help from psychology.

It was of course the last-mentioned idea that led many 'pure' philosophers to brand Wundt a proponent of psychologism (Dubs 1911: 119; Geyser 1909a: 259; Heidegger [1913] 1978: 79–90; Husserl [1900] 1975: 69, 80, 184; Moog 1918: 360; 1919: 101; 1922: 110–129; Natorp 1901: 277; Spranger 1905: 14; Ssalagoff 1911: 161; Windelband 1884: 248). Wundt placed logic between psychology and all other scientific disciplines. Logic was a normative discipline that sought to identify those combinations of ideas that were generally valid. Whereas psychology studied how humans *in fact* thought, logic investigated how one *had to* think to obtain scientific knowledge. Wundt always was committed to the idea that the psychological study of logical thought was the necessary first step in the development of any scientific logic. Although he did not regard this psychological study as a part of logic proper, he emphasised nevertheless that only a psychological study of thought could identify the special and unique features of logical thinking (1906: 1, 11).

Wundt suggested that psychological analysis could pinpoint three features that distinguished logical thought from all other types of thinking: spontaneity, self-evidence, and universality. First, as concerned spontaneity, Wundt held that logical thinking was experienced by humans as a free inner activity, that is as an act of willing. Thus logical laws of thought had to be understood as laws of the will. Second, logical thought had the special character of an inner necessity, a character that led one to ascribe immediate certainty to the combinations of ideas produced by logical thinking. Third, and finally, the laws of (logical) thought had universality in two ways: they were evident for all reasoners, and their applicability was not restricted to any particular realm of objects (1906: 75–6, 84–5).

The two last-mentioned features of logical thought, self-evidence and universality, allowed one to distinguish more clearly between psychological and logical laws. Logical laws of thought were all those rules 'which contain regularities regarding that which is self-evident and universal in our thought'. Psychological laws lacked both self-evidence and universality. Moreover, while both psychological and logical laws of thought were arrived at through generalisation from, and observation of, factual thinking, only 'logical laws of thought are, at the same time, *norms* by means of which we approach factual thought and test its correctness'. However, this opposition between logical and psychological laws did not mean that no psychological concepts entered the vocabulary of logic. Wundt believed that because logical laws of thought were identified within human thinking, and because logical thought was usually intertwined with other kinds of mental activities, the formulations and explanations of logical laws would always and inevitably contain psychological concepts (1906: 88–9).

Wundt did not believe, or accept, however, that his position amounted to some sort of 'psychologism'. In his own treatment of 'psychologism' in his paper 'Psychologism and Logicism' (*Psychologismus und Logizismus*, 1910b), Wundt accused Brentano and Husserl of being amongst the worst 'psychologicists', Husserl because of his tacit reliance on the concept of self-evidence in his criticism of psychologism, and Brentano because of his theory of judgements (according to which judgements consisted of presentations) (1910b: 526–36, 612).

As Husserl's mirror image, Wundt of course had to agree with the former that modern psychology had no place within philosophy departments. Indeed, in 1896, Wundt wrote that no one could doubt any more that 'psychology…is on its way to turning itself from being a part of philosophy into an independent, positive science' (1896: 2). And in 1908, Wundt regretted that 'many still take psychology to be a "philosophical" discipline'. This impeded progress within psychology and thus it seemed to be 'at present the primary task of psychology to resolve a union that so damages its independent development'. Looked at 'from a logical point of view', psychology had no more to do with philosophy than with physics or history (1908b: 19).

Finally, just as Husserl and other 'pure' philosophers tried to cleanse logic of psychological ideas, so also Wundt aimed to exorcise logical concepts and conceptions from the realm of psychology. So while Husserl was fighting 'psychologism', Wundt saw himself as combating 'intellectualism' and 'logicism'.

Intellectualism and logicism had their roots in 'vulgar psychology' (1908b: 151). 'Vulgar psychology' was Wundt's term for what today is called 'folk' or 'common-sense psychology'. Folk psychology tended to overestimate the importance of logical and intellectual mental processes. In so doing, it committed two fallacies. On the one hand, by regarding logical reasoning as *the* central paradigm for *all* mental processes, folk psychology conflated whole and part. Committing this mistake was facilitated by the fact that in most people's mental life, logical processes stood out from, and seemed independent of, all other processes. The naïve observer would conclude, mistakenly, that logical processes were the most important parts of mental life, and that all other parts could be reduced to them. On the other hand, any interpretation and classification of, and reflection on, mental life used some logical tools and criteria. This was the point where the folk psychologists would make a further mistake. They would project the logical properties of their tools onto the mental processes they were trying to classify. And thus they ended up with giving special privilege to those mental contents, that is presentations and judgements, that seemed 'most logical'. To bring out that intellectualism was rooted in reflection, Wundt sometimes called it 'the psychology of reflection' (1908b: 151).

The intellectualism or logicism of 'vulgar' psychology had also shaped much of scientific psychology and physiology. Wundt tried to show its influence upon faculty psychology, associationism, Herbartian psychology, the physiology of perception, and Brentano's psychology.

Faculty psychology was a child of intellectualism for two reasons. First, it aimed for a simple, 'logical' subsumption of the diversity of human mental life under a few concepts. And these concepts were usually taken from naïve ordinary language. Second, amongst the faculties, faculty psychologists usually favoured the more logical ones, and sought to assimilate others to them. Thus feelings were analysed as forms of diffuse knowledge, or volitions were construed as caused by intellectual motives. Both moves distorted mental life (1908b: 151–2; 1910b: 554–5).

Associationism betrayed its logicistic roots by striving for as few laws of association as possible, by classifying associations according to logical relations and principles, and by overemphasising presentations. Herbartian psychology shared the last-mentioned mistake with associationism (1908b: 152–5; 1910b: 557–63).

Even the sense physiology of Helmholtz and the young Wundt was not free of intellectualism. This was not surprising as such, because in their attempt to distance themselves from philosophical disputes, the physiologists fell back on common sense. And in psychological questions common sense was nothing but intellectualistic folk psychology. The most important intellectualistic heritage within physiology was the assumption of 'unconscious inferences'.

Rather than study mental experiences as such, and in their immediacy, the physiologists replaced these real experiences with 'logical constructions'. And to protect their arbitrary constructions from refutations by experience, they placed them under the threshold of consciousness. In Wundt's view, the physiologists had taken a fiction for reality:

> processes of logical thinking...presuppose a highly developed consciousness. To situate such thought processes...already on the level of the unconscious life of the soul is thus a fiction at best, a fiction that might illustrate how processes [of perception] *might* occur *if* they were logical acts of thinking. But of course, they are not acts of this kind.
>
> (1910b: 566)

Wundt was most hostile towards the alleged intellectualism and logicism of Brentano, Stumpf, and Husserl. All three men were guilty of trying to replace psychological analysis by a logical–semantical study of words, and all three tried to assimilate presentations to judgements. For instance, although Brentano thought of judgements as being composed of presentations – this made him a psychologicist after all – he also, in Wundt's view, mistook each individual presentation for an implicit judgement. And he did so by construing presentations as intentional states of mind (1910b: 568).

In Stumpf's case, Wundt disliked the theory of feelings. For instance, Stumpf had suggested – contrary to Wundt – that not all affects had several phases. Stumpf's example had been the affect of fright 'that can be over in one second without different phases being distinguishable' (quoted from Wundt). Wundt's reply was scathing:

> I have never observed an instance of the affect of fright that really was over in one second...I therefore am inclined to think that Stumpf has never carefully studied this affect. All he has done is reflect on the conventional meaning ('momentary affect') of the word 'fright'. One can hardly blame him for that. For what he does here is but a necessary consequence of his general method.
>
> (1903a: 240)

Wundt was even more critical when it came to evaluating Husserl's suggestions and claims. In Husserl's case Wundt did not just attribute an inability to keep logic out of psychology; in Husserl's case Wundt attributed, as it were, a logical imperialism of sorts: 'His programme is not just to ban psychologism from logic; his programme is to ban psychologism from psychology'. Wundt suspected that Husserl wanted 'to turn psychology into logic', or 'into an applied logic'. Wundt also spoke of 'Husserl's…recklessly developed logicism that seeks to turn psychology into a reflective dissection of words and concepts' (1910b: 516–19, 569).

Wundt showed no inclination to accept Husserl's claim that phenomenological analysis clarified important concepts for both psychology and logic. The concepts that Husserl selected for his study were 'the scientifically unchecked concepts of vulgar psychology, and these concepts do not become in the least bit more psychologically scientific by this detour through purely logical conceptual analysis' (1910b: 579).

On Wundt's reading of Husserl, 'the phenomenology of thought' did not study phenomena of thought as these were given to consciousness. Instead, phenomenology resolved all contents of consciousness 'into logical acts of thinking' or into 'linguistic forms'. With respect to the alleged mental acts, Wundt remarked that Husserl acted no differently than the physiologists had done when stipulating inferences in the unconsciousness. Like the inferences of the physiologists, so also Husserlian acts were not observable. And with respect to the 'linguistic forms', Wundt lamented that these were not studied from the perspectives provided by the psychology, or history, of language. Instead Husserl investigated them from the viewpoint of a logical metareflection on the results of grammar. Unfortunately, this procedure ignored the basic insight that meanings were not stable, and thus disqualified phenomenology as a foundation for empirical psychology. Husserl's phenomenology was 'a psychology without psychology' (1910b: 573–6, 580, 603–4, 607).

Finally, it deserves to be noted that Wundt's opposition to the philosophy of Brentano and his followers included an attempt to exclude them from the major psychological journals. When in 1903 Wundt advised Meumann on how to select the editorial board for the new *Archiv für Psychologie,* he recommended that A. von Meinong – another well-known representative of the 'Austrian school' – be left out:

> As far as your inquiry about Meinong is concerned, I wish to urge caution towards him and others like him. The time does

> not appear to have come to make the *Archiv* into a forum for [the expression] of all imaginable viewpoints in psychology. As the situation now stands, it appears best to limit the scope of views that are to be expressed in the *Archiv*, to those suggested by the names of the contributing editors...including Lipps, Höffding, and Jodl. This scope is broad enough but excludes all the more or less scholastic reflection psychology of Brentano...as well as the truly speculative psychologists. If you accept Meinong's writings, the journal will lose its character completely and nobody will know what to think of it.
> (9 October 1903; quoted from Bringmann and Ungerer 1980: 64)

The Würzburgers' promiscuity

As already implied by the above, the Würzburgers' thought-psychological studies started to appear at a time when the relationship between psychology and logic, or psychology and philosophy, was *the* most important issue amongst German-speaking philosophers and psychologists. Faced with the choice between Wundt's psychological purism concerning logic, on the one hand, and Husserl's logical purism *vis-à-vis* psychology on the other hand, a number of key Würzburgers tended to side with the latter. That is to say, they allowed – *pace* Wundt – for the possibility that logic and 'pure' philosophy might influence experimental psychology, while insisting that logic owed nothing to psychology. No wonder then that, as seen in Chapter 2, the neo-Kantians were so positive in their reaction to the Würzburgers.

The issue of the relationship between logic and psychology was of course explicitly the topic of Marbe (1901). Marbe set himself the task of testing the psychological claim that there was a specific quale that characterised all judgements. Brentano and Wundt were amongst those who had made proposals on what this quale might be like. As seen earlier, Marbe's subjects were unable to find these qualia in cases where, by the standards of logic, they had formed judgements; that is, in cases where they had said or done something to which the concepts of true and false could be applied. Marbe therefore concluded by calling for a greater independence of logic from psychology. Logic was meant to answer the question of how our mental contents corresponded to real and ideal objects; psychology queried the quality of mental contents and their dependence upon the brain. Answering the latter set of questions could never amount to solving the first-mentioned problem: 'Currently logic is to a considerable extent nothing but a non-methodological psychology of judgement. In the future, logic will have to construe itself as non-psychologically as possible' (1901: 98). Marbe insisted that experimental psychology could and should continue to study judgements; for instance, it could tackle problems such as what kinds of judgements gave rise to what kinds of feelings and presentations. Psychology could also test and refute the naïve psychological assumptions that Marbe found endemic amongst contemporaneous 'logicians' such as Sigwart or Wundt (1901: 95–7). But the clarification of the true essence of judgements fell to logic.

Marbe continued this line of argument in later writings, and even against fellow thought psychologists. For instance, in a review of E. Dürr's book on epistemology (Dürr 1910), Marbe essentially repeated his plea for an independence of logic from psychology (Marbe 1912b), and later he accused Watt, Messer, and Bühler of 'psychologism'. These authors had insisted that psychologists were, after all, able to give a psychological criterion of judgements, namely the task. To Marbe, in so doing, his former colleagues had narrowed the scope of what one means by a judgement; according to him there were many judgements (in the logical sense) where no prior tasks could be identified (1915: 11). Nevertheless, Marbe insisted that logic and experimental thought psychology stay in close contact. Paradoxically, logic needed experimental thought psychology as a bulwark against psychologism. It was only experimental study that could unmask and refute mistaken psychological claims in the realm of logic (1912a: 69–73).

There can be no doubt that Wundt regarded Marbe's 1901 intervention as something of a betrayal. Here was, after all, a former Leipzig student and experimental psychologist, who sided with *the* enemy of experimental psychology, that is Husserl, and accused Wundt of psychologism – alongside Brentano. The extent of Wundt's anger can be measured by his reactions. In the fifth edition of the *Grundzüge* (1903a) he called Marbe's book a 'maltreatment of psychology'. The judgements formed by Marbe's subjects were not 'original judgements' but mere 'artefacts of the experiment' (1903a: 580). And Marbe was of course one of the two main targets of Wundt's 1907 criticism, together with Bühler. Wundt was active also behind the scenes. For instance, Wundt repeatedly let Marbe's superior, Külpe, know what he thought of the author of *Experimental-Psychological Investigation about Judgement*. In 1905, Wundt wrote to Külpe that Marbe was a 'fine lecturer' but 'a scholastic psychologist....I can't make any sense of his "judgement experiments"' (letter Wundt–Külpe, 12 December 1905, quoted from Hammer 1994: 237). And in 1907, after the publication of his criticism of Marbe and Bühler, Wundt wrote to Külpe:

> First of all, I am very sad to learn from your letter of the extent to which you identify yourself with the interrogation method. Up to now, I have always thought that Marbe was the inventor of this utterly reprehensible method....It made sense to me that Marbe was the inventor of this method. For I take him to be a man who occasionally might perhaps construct a sensible instrument but who lacks all talent for being a psychologist.
> (26 October 1907, quoted from Hammer 1994: 244)

It is significant that Wundt wrote his extensive criticism of the Würzburg experiments not after the publication of Marbe (1901), nor after Ach (1905), Watt (1905), or Messer (1906). His 'On Interrogation Experiments' appeared only after Bühler had published the first of his three articles on thoughts. Indeed, Wundt concentrated most of his attack on Bühler. In light of the issues raised in this subsection, this finds a natural explanation. For with Bühler's papers, the very 'scholastic reflection psychology of Brentano' and Husserl, the type of psychology that Wundt had tried to exclude by not accepting Meinong on the editorial board, entered the *Archiv*. And it did so with a vengeance.

Bühler began his first paper with praise for Husserl's methodology in *Logical Investigations*:

> Husserl has recently developed an original and very fruitful method, a kind of transcendental method. In general terms, he assumes that the logical norms can be fulfilled and then asks himself what this allows us to infer about the processes that must be regarded as the bearers of these law-fulfilling events.
> (Bühler 1907b: 298)

Bühler went on to claim that his study of 'the *hic et nunc* of what is experienced while thinking' would prove Husserl's assumptions true. In the main body of his text, Bühler made repeated use of Husserl, Brentano, and Stumpf. He used Husserl's terminology for analysing parts and moments of thoughts, adopted Husserl's classification of acts, and used Husserl's distinction between meaning-intention and meaning-fulfilment (1907b: 299, 329, 347).

Indeed, his reliance upon Husserl was so substantial that one contemporary critic commented that 'Bühler's...experiments are...an attempt to check and confirm Husserl's phenomenology in an experimental way' (Aster 1908: 62). Bühler also deplored 'excessive' criticisms of 'the Austrian psychologists' and rejected the suggestion that theirs could be called a 'reflection psychology' (1907b: 358). Later, when replying to Wundt's criticism, Bühler used Stumpf's dichotomy of appearances versus functions (1908a: 113). Stumpf's paper with this title was reviewed very favourably by Bühler in 1908; Bühler treated Stumpf as an ally against all those who denied the existence of non-depictive thought (Bühler 1908b); and Bühler knew Stumpf well: before coming to Würzburg in 1906, Bühler had spent a term studying with Stumpf in Berlin (Wellek 1968: 199).

Despite the fact that Dürr already had written positively about Husserl in 1903 (Dürr 1903), Bühler later claimed that it had been he who 'in the winter of 1905' introduced the work of 'Brentano and his school' into Würzburg thought psychology – in part 'against Külpe's hesitation and resistance' (Bühler 1922: 252). If so, then Bühler must have been very persuasive indeed, and especially with respect to Külpe himself. Recall that Külpe in 1907 introduced the distinction between functions and contents, a distinction that differed from Stumpf's dichotomy (between functions and appearances) in all but its name (Külpe 1907b). In 1910 Külpe praised the 'descriptive psychology' of the 'astute Brentano' (1910a: 69), and his epistemological *opus magnum, The Realisation* (*Die Realisierung*, 1912a), contained numerous positive references to, and uses of, Stumpf (10, 117–18), Meinong (10), and Husserl (15, 16, 71, 73, 87, 127). Külpe praised Husserl's œuvre 'as one of the outstanding testimonies to the independence of thought' (1912a: 127). In his lectures, Külpe accepted Husserl's antipsychologism (Külpe 1923b: 90–1), although he suggested that a close link between thought psychology and logic was still advantageous. But in this relationship it was logic that set the tasks for psychology, not vice versa (1912c: 1103). In light of all this, it no longer comes as a surprise that Stumpf was eager to win Külpe for a professorship in Berlin. In a letter to Külpe, Stumpf described the prospect of 'a co-operation with you' as 'beautiful and fruitful' (letter to Külpe, 30 October 1908; Külpeana V, 18).

Even if Bühler was crucial in introducing phenomenology into Würzburg, he was no doubt helped by Messer. There were only a few references to Husserl's work in Messer (1906), but Messer did adopt, for instance, Husserl's categories of meaning-intention and meaning-fulfilment (1906: 85). Within two years of completing his thought-psychological study, Messer had converted to phenomenology almost completely: Messer's *Sensation and Thought* (1908) was as much an introduction to Husserl's *Logical Investigations* as it was an introduction to modern psychology. Husserl was the most-quoted author of the book, and there was hardly any chapter that did not start from phenomenological ideas and terminology. Not surprisingly, Messer also warned of the dangers of psychologism and adopted Husserl's brand of antipsychologism (1908: 179). In later writings Messer defended Husserl against Wundt's attacks: 'Husserl's phenomenological investigations…are not "mere verbalistic", "mere grammatical" and "scholastic" analyses. No one should discredit such important and thorough investigations with labels like "writing-table psychology"' (1912b: 126). Curiously enough, Messer even felt the need to defend Husserl against Husserl himself. Messer argued in two papers (1912b, 1914b) against Husserl's claim according to which phenomenology was not part of psychology. Like the phenomenologist, the psychologist too aimed for '*general* knowledge'; and, like the psychologist, the phenomenologist had better be interested in coming to know 'reality'. Messer concluded that phenomenology could not be separated from psychology, and that it was, as 'pure psychology', psychology's most basic part (1912b: 123–4). And in an update on his evaluation of Husserl's work published in 1914, Messer still missed in Husserl's work the proof 'that modern psychology doesn't know of the immanent analysis of essences'. Messer now suggested that especially the 'psychology of thought' could claim competence in this area:

> With this method [i.e. the method of systematic experimental self-observation] the phenomenological method – as Husserl himself, his students, and his followers use it with respect to psychological objects – harmonises only too well.…One notices precious little of the abyss that according to Husserl separates phenomenology from *all* psychologies.
>
> (Messer 1914b: 64)

While Messer granted Husserl that his 'reductions' indeed set phenomenology apart from psychology, he felt that 'this separation is only of purely theoretical significance, and is in fact completely irrelevant for the practice of research'. And thus Messer upheld his earlier verdict that phenomenology *'is psychology – indeed it is psychology's most basic part'* (1914b: 66).

Conclusion: a place for psychology

The different stances – purist versus promiscuous – that Wundt and Külpe took with respect to physiology, the applied sciences, and philosophy came out most dramatically around 1912 and 1913, when the 'pure philosophers' were organising their petition. I can summarise the differences between the purist and the promiscuist by considering these two texts.

In 1912 Külpe published his longish paper, 'Psychology and Medicine' (*Psychologie und Medizin*, 1912b). In this paper Külpe suggested a distinction between philosophical psychology and another branch of psychology that – because of its empirical and experimental character – should become a separate natural science. While Külpe did not deny the existence of many close ties between experimental psychology and philosophy, he argued nevertheless that 'the combination of psychology (as a special science) with philosophy is beyond the capacity for work, the talent and the inclination of a *single* human being'. In Külpe's view, the continuation of the status quo could only lead to 'dilettantism' in philosophical questions. Külpe went so far as to express understanding for the pure philosophers' opposition to this 'invasion of specialists' (1912b: 266–7). Having discussed a number of psychopathological studies allegedly lacking in thought-psychological sophistication, and having emphasised the need for a thought-psychological component in the training of medical doctors, Külpe suggested that experimental psychology should lobby for chairs in the medical faculty (1912b: 190–263).

In light of the above, these statements make sense. First of all, the idea that experimental psychology should move to the medical faculty was natural to make for a man who saw psychology as closely tied to physiology. Second, Külpe's main argument for why psychology should move to the medical faculty was that psychiatrists as well as ordinary medical doctors needed psychological knowledge. Put differently, for Külpe, psychology ought to be situated where its application could make the greatest possible difference. And third, Külpe had opened himself to the concerns and ideas of pure philosophers to such an extent that he was willing to go along with their wishes and separate pure and experimental psychology.

Read in this light, it was actually surprising that Külpe did not support and sign the philosophers' petition – as Messer did. Indeed, the pure philosophers were themselves startled. Rickert sent Külpe the text of the petition with the following note attached to it: 'Dear colleague! It would be great if you too could sign. I hope it is clear that we are not trying to say anything *against* psychology' (Külpeana V, 13). Rickert also referred to the proximity between Külpe's and the pure philosophers' positions in the daily *Frankfurter Zeitung*, and expressed his astonishment about the fact that Külpe had not signed (1913b: 2).

Perhaps the explanation lay with the fact that Külpe and some other experimentalists were working on their own petition. In Külpe's *Nachlaß* of around the same time the following draft petition can be found. It is worth quoting in full:

> The German Society for Experimental Psychology deems it its duty to present the state government with the following proposal: Experimental psychology and philosophy should not be represented in personal union by one and the same person. Experimental psychology should be represented at every university as an independent scientific discipline with its own professorial chair and its corresponding institute. The main reasons are the following:

1 Experimental psychology has the character of a special science, with respect to its methods and results, its subject matter, and its perspectives. Experimental psychology has the character of an empirical science and thus its relation to philosophy is no closer than the relation of any other empirical science to philosophy.

2 Since Fechner founded experimental psychology in 1860, its area has grown to such an extent that its proper representation needs its whole man. In general it exceeds the capacity for work, the talent, and the interest of one researcher to be productive not only in psychology but also in philosophy.

3 A continuation of the current situation has the danger that Germany will be passed in an area into which it itself has introduced experimental practice. In other countries (North America, England, Belgium, Italy) experimental psychology has already become independent, and wonderful institutes have been created there.

To put our proposal into practice, three ways are prima facie possible:

1 Make psychology one of the preparatory subjects within the medical curriculum. O. Külpe has explained this possibility in his treatise on psychology and medicine (Leipzig: Engelmann 1913).
2 Make psychology one of the subjects in which high-school teachers are examined. In this case, because of its close ties to pedagogy and the *Geisteswissenschaften*, it should be both a special main subject and a general cultural subject (*allgemeines Bildungsfach*). In both cases, psychology must be separated from philosophy, the discipline within which psychology currently is practised.
3 Institute a combination of both ways.

Under any circumstances, it is desirable that psychology as a special science remains within the philosophical faculty, and that wherever there exists a division into sections, it is put in the mathematical and natural-scientific section.

The Society is, at all times, willing to embark upon the project of a more detailed development of a plan for the position of psychology within the system of education. The Society will act in this matter as soon as the State Government has expressed its willingness to consider the present proposal.

(Külpeana V, 13)

This petition was never sent and it is difficult to determine its origins and fate. What is striking is that Külpe was directly mentioned, and that the goals of the petition were in effect indistinguishable from those of the philosophers. They also clearly reflected the promiscuous tendencies that characterised the Würzburgers' thinking about the relationships between disciplines.

Wundt's intervention in the public debate over the institutional future of psychology both reflected and summed up his overall purist agenda. To begin with, Wundt's *Psychology Struggling for Survival* (*Die Psychology im Kampf ums Dasein*, 1913a) defended and attacked on two fronts: against the philosophers' 'get psychology out of philosophy' stance, and against some psychologists' desire to 'get philosophy out of psychology' (1913a: 2–3). Of special interest to us is the latter topic; here Wundt took issue with Külpe. How Wundt saw the battlefront comes out most clearly in a letter to his son Max (2 February 1913):

> A great agitation has started amongst the philosophers; and it is meant to lead to a separation of psychology from philosophy. The psychologists of the party 'Müller & co. alias Külpe' are blowing the same horn, and their melody is only slightly different.
>
> (Wundt Archiv, Leipzig)

Wundt was unimpressed with Külpe's plea for a psychological education of medical doctors. In his view, the only kind of psychology the general practitioner needed was 'practical psychology…which, like every ingenious talent, is partly innate, partly acquired through practice'. Wundt was ready to admit that the psychiatrist needed more than just this practical talent, but he saw no reason for moving empirical psychology into the medical faculties. Instead he suggested that not all of psychology need be an academic, that is university-based, subject, and that psychiatrists might receive their psychological training in asylums (1913a: 11, 15).

Wundt's main worry was that an institutional split between philosophy and psychology would soon turn experimental psychologists into mere artisans. By his count, almost half of all of the psychological literature extended into metaphysics and epistemology. And 'the most important questions of psychological education are so closely linked to epistemological and metaphysical questions that one cannot imagine how they could ever vanish from psychology'. Wundt also rejected the American model of having separate psychological institutes as being ill-suited to Germany: an independent psychology would give exclusive attention to practical application, and this would not fit the emphasis on theoretical work typical of the German university system (1913a: 18, 24, 26).

Wundt found that both the organisers of the philosophers' petition and Külpe shared the same restricted picture of psychology; for both psychology equalled experimental psychology. This was the point where for Wundt the struggle over the institutional location of psychology linked up with the debate over the proper psychological study of thinking. That psychology should be shifted to the medical faculty, or that it was a natural science, was plausible only as long as one believed two tenets: on the one hand, that thought could be studied experimentally, and without *collective psychology*, and on the other, that an experimental study of thought was a natural-scientific study. Both assumptions were wrong.

Psychology was 'both a part of the science of philosophy and an empirical *Geisteswissenschaft*; and its value for both philosophy and the empirical special sciences resides in its being the main negotiator between them' (1913a: 30, 32).

Wundt thus reaffirmed and argued anew a number of purist manoeuvres that he had held for some time. First, psychology was not a natural science and thus there was a principled gap between psychology and physiology. Second, this gap would be the widest for those parts of psychology that studied the higher and highest mental processes: *collective psychology*. To challenge the notion that collective psychology was essential for a study of thinking – as the Würzburgers had done – was in fact to weaken the independence of psychology. Third, the institutional position and nature of psychology should never be defined by its possible applications. Wundt now went further and even suggested that independent psychology departments would not be able to withstand the temptations of applying their knowledge. And fourth, and finally, to prevent psychology from losing its independence from physiology, applications, and bad philosophy, it needed to remain within philosophy. With this insistence on a place for psychology within philosophy, Wundt deviated from some of his earlier remarks on the need for separate departments for psychology. But the deviation was on the surface only. Wundt sought to keep psychology pure – and in an impure academic world this purity needed a slight dose of impurity in its defence.

5 Collectivist versus individualist

Introduction

Different stances regarding the new thought psychology were informed by different social, ethical, and political philosophies. I shall try to make my case for the main antagonists of the controversy. Broadly speaking, Wundt was a 'collectivist', whereas Bühler, Külpe, Marbe, and Messer were 'individualists' of sorts. Wundt's opposition to introspection was part and parcel of his general hostility towards individualism; the Würzburgers' advocacy of self-observation was part of a conscious rejection of collectivism.

The collectivism of Wundt's *Völkerpsychologie*

Up to this point, I have used 'collective psychology' as a translation for *Völkerpsychologie*. This translation is not literal of course. Literally, *Völkerpsychologie* means 'peoples' psychology'; the genitive of 'peoples'' being a *genitivus objectivus*. Using 'collective psychology' as the English translation brings out the collectivism of Wundt's *Völkerpsychologie*. But it loses the ethnic or nationalistic emphasis on the *Volk*. Because, as we shall see shortly, this emphasis was important for Wundt, but contested by Marbe, I had better stay close to the original term. I shall use the original German word in this chapter.

The first extensive programmatic outline for a *Völkerpsychologie* was sketched by the Herbartian philosopher M. Lazarus and the linguist H. Steinthal in 1860, in the leading article of their newly founded journal *Zeitschrift für Völkerpsychologie und Sprachwissenschaft* (Lazarus and Steinthal 1860). Wundt's own *Völkerpsychologie* was indebted to these authors, although his was a critically modified version of theirs. In his view, *Völkerpsychologie* should confine its research to three subject areas (1888b): language, myth, and custom. *Völkerpsychologie* studied universal–general features of the human mind, and such features could only be found in these three areas (1888b: 27). Moreover, Wundt was unwilling to reduce *Völkerpsychologie* to a mere application of individual psychology (1908b: 227–8). He emphasised that

> the conditions of mental reciprocity produce new and specific expressions of general mental forces, expressions that cannot be predicted on the basis of knowledge of the properties of the individual consciousness....And thus it takes both individual psychology and *Völkerpsychologie* to constitute psychology as a whole.
>
> (Wundt 1908b: 227)

First and foremost, *Völkerpsychologie* was needed for collecting objective data about psychological processes that could be reliably and objectively studied through neither introspection nor experiment: 'it is precisely at that point where experimental method reaches its limit that the methods of *Völkerpsychologie* provide objective results' (1908b: 227). These methods were the 'comparative-psychological' and the 'historical-psychological': the first compared phenomena of different cultures; the second compared different, successive stages of one and the same cultural phenomenon (1908b: 242).

According to Wundt's programme, *Völkerpsychologie* dealt both with small collectivities such as the family, the group, the tribe, and the local community, and with large communities such as nations, peoples, states, or humankind as a whole. Amongst collectivities of various size and character, however, the *Volk* had a special position. It was this special position that justified calling the second main branch of psychology *Völkerpsychologie*:

> In the overall development of mental life...*Volk* is the main unifying concept. All other [collectivities] are related to it. Families, classes, clans, and groups exist only within the *Volk*. The concept '*Volk*' does not exclude these other communities; on the contrary, it includes them. This is because the concept '*Volk*' picks out *Volkstum* [the traditions of a people, national traditions] as the main and decisive factor for the most fundamental creations of any community.
>
> (1913b: 5)

Wundt's emphasis on *Volk* was significant – and it is important for us here because Marbe criticised the special position of *Volk* within a psychology concerned with collective phenomena (see below). No doubt, the stress upon the *Volk* in Wundt's psychological thinking was an expression of the political aspirations that were common in the third quarter of the nineteenth century (Oelze 1991: 80–1). This was the period when many German intellectuals supported German national unification. Wundt himself was actively involved in the politics of his home state of Baden, and from 1866 until 1868 he even served as representative for a Heidelberg district in the Baden diet. Also in 1866, Wundt was invited to serve on the peace commission following the war between Prussia and Austria. His enthusiasm for German unification was probably instrumental in earning him this invitation (Diamond 1980: 42). In later years, Wundt's German nationalism became militant and hostile; during the First World War Wundt argued for the superiority of the German mind over French 'sophisticated egoism' and British 'reckless materialism' (1915: 35, 45; cf. Kusch 1995a: 220–1).

Wundt conceptualised the relationship between the individual and the collective in a number of different ways: On different occasions, the two *relata* were 'the individual' and 'the community' (1908b: 294), 'the individual will' and 'the collective will' (1897: 390; 1912 III: 33; 1918: 306–15), 'the individual spirit' and 'the collective spirit' (1897: 614), 'the individual organism' and 'the collective organism' (1897: 616–19; 1908b: 296), 'the individual personality' and 'the collective personality' (1908b: 197; 1897: 624), the 'individual consciousness' and the 'collective consciousness' (1907a: 246), and finally, 'the individual soul' and 'the soul of a people' (*Volksseele*) (e.g. 1888b; 1911: 8).

To justify talk of collective souls and the like, Wundt drew once more on his 'principle of actuality'. This principle was thus crucial not only to his psychophysical parallelism and his argument against physiological materialism; it was also the central pillar of his *Völkerpsychologie*. As Wundt reformulated the principle in this context, it stated that any system of lawful interconnections between conscious experiences qualified as a soul. Both the conscious experiences of singular individuals and the conscious experiences of groups of individuals showed such systematic interconnections. And thus both could be called 'souls' (1911: 9–10).

Although the 'soul of a people' (*Volksseele*) did not exist 'outside of the individual souls', it was at least as 'real' as the latter (1908b: 294; 1911: 10). Moreover, it was as correct to say that the *Volksseele* was the 'product' of individual souls as it was adequate to maintain that individual souls were the 'products' of the *Volksseele* (1911: 10). Furthermore, the *Volksseele* could continue to exist over time, even though 'its' individual souls – that is, the souls of the individuals participating in it – came and went (1912 III: 36). When speaking of the difference between collective and individual wills, Wundt even called the collective level more real than the individual one:

> All effects of the collective will are incomparably more powerful than those of the individual will....If then one wants to measure the degree of reality by the effects, then one would have to accept that the collective will is undoubtedly the more real.
>
> (1897: 390–1)

Wundt also extended the concepts of 'organism' and 'personality' to cover collective phenomena. By organism he meant

> a composite [and independent] unit. This unit consists of parts. Each of these parts is two things at once. First, it is a unit with properties similar to the more complex composite unit. Second, it is a unit that serves the whole as an *organ*.
>
> (1897: 616–17)

The individual organism was no more than a 'preliminary stage' (*Vorstufe*) of the collective organism. In the individual, the organs had no self-consciousness of their own, and their co-operation was hardwired. This meant that the individual organism possessed only a '*restricted*' form of independence. In the collective, on the other hand, the hardwiring was missing, and the functioning of the whole was based upon the parts' self-conscious actions. This made the collective organism '*free*' (1897: 619). Not every collectivity deserved the title 'organism', however. To qualify, the members of a collectivity had to share 'all important purposes in life...to a high degree'. This sharing was so strong that no individual could be the member ('organ') of more than one collective organism (1897: 623). Finally, an organism became a personality once it possessed autonomy, consciousness of its own actions, and the ability to choose. In the case of collective personalities, decisions resulted from the interactions between a great number of individual persons. Such interactions were governed either by natural, innate disposition, or else by norms. In the former case, Wundt spoke of a 'natural community', in the latter of a 'cultural community' (1897: 628).

Wundt's ethics

Wundt was also a prolific writer on ethics. His *Ethik* went through several editions during his lifetime, and it came in three volumes. For present purposes, it is worth noting that ethics and *Völkerpsychologie* were closely connected enterprises: *Völkerpsychologie* was the 'portico of ethics' (1912 I: iii). The link between the two was twofold. On the one hand, *Völkerpsychologie* provided the social ontology upon which all *Geisteswissenschaften* and philosophy had to build. On the other hand, *Völkerpsychologie* was also the science of the phylogenetic development of ethical norms and values. In this role, *Völkerpsychologie* provided the empirical starting point for ethics. It was the task of ethics to clarify and purify those norms and values towards which the development of culture seemed to be leading. The moral philosopher was thus refining morality from within, and did not stand to the side or beyond the historical process. Wundt saw this position as being of one piece with 'the ethics of speculative Idealism' (1912 I: v; 1887: 8). This ethics included the ideas of a moral progress of humankind (1897: 304), and the moral value of the state.

The core of Wundt's ethics comprised six main norms: two concerned the individual in isolation, two the individual's relations to others and the community, and two the individual's relationship to humankind:

> The individual norms…
>
> Think and act in such a manner that you never lose your self-respect.…
>
> Fulfil the duties – towards yourself and others – that you have taken upon yourself.
>
> The social norms…
>
> Respect your neighbour as you respect yourself.
>
> Serve the community to which you belong.
>
> The human norms…
>
> Look upon yourself as a tool in the service of the moral ideal.…
>
> Dedicate yourself to, and sacrifice yourself for, the purpose that you have recognised as your ideal task.
>
> (1912 III: 152–8)

In cases of conflict between these norms, social and human norms took precedence over individual ones, and human norms over social ones. There could be a conflict between the two social norms, too. The statesman who sacrificed individuals to further his country's interests acted primarily on the second social norm, whereas the 'soft and feminine characters' tended to favour compliance to the first. Wundt sided with the statesman's perspective (1912 III: 157). Elsewhere in his ethics, Wundt reminded his readers that the state 'serves ideal purposes of absolute value, purposes with respect to which the life of the individual has no value at all' (1912 III: 102).

These 'ideal purposes of absolute value' had nothing to do with human well-being or happiness. Instead, the ultimate goal of morality was the unceasing 'production of intellectual creations':

> On what basis do we judge individuals and peoples of the distant past...? We do not judge them on the basis of the amount of happiness that they enjoyed themselves, or the amount of happiness that they created for their contemporaries. Instead, we judge them on the basis of what they have done for the future development of humankind....Such ultimate purpose can only be the production of intellectual creations [*die Hervorbringung geistiger Schöpfungen*].
>
> (1912 III: 85–6)

Happiness itself was without moral value. And ways of organising social, economic, and political life were subject to moral evaluation and assessment only indirectly: They became moral issues only in so far as their proper handling was an inevitable auxiliary means for the production of outstanding intellectual deeds. One of the key means for this production was nationalism. The creation of ever more, and ever better, cultural works presupposed the competition between peoples and their states. To strive for a single 'world-state' was thus ultimately immoral (1897: 634, 657–8, 661).

Wundt's political philosophy

Wundt had some specific proposals on how best to organise society and the state. To promote the ultimate moral goal, society had to be divided into two classes: an 'active' and 'higher class of society', containing the 'active intellectual bearers' of culture; and a 'passive' and 'lower class of society' that adopted and preserved whatever the 'active intellectuals' handed down. Wundt insisted that membership in one of the two classes should be based upon merit. Obviously, because this order of society, that is this division into two classes, was meant to promote culture, 'it need[ed] no justification on the basis of what good it might do to the individual' (1912 III: 259–60).

Wundt's views on the overriding importance of culture, or intellectual products, fitted well with both his early political activity in the *Bildungsverein* (Educational League) and the goals and aspirations of the *Bildungsbürgertum*. Wundt was active in the *Bildungsverein* during the early 1860s, lecturing to workers. The Educational League was a movement of the liberal and bourgeois intelligentsia. It had no concrete programme on how to improve the workers' working and living conditions. Its main remedy for all grievances was to try to improve the workers' education and degree of culture (*Bildung*) (Ungerer 1980: 103–4). Wundt withdrew from the Educational League when around 1865 it became more radicalised (Diamond 1980: 42). It is also striking to note that Wundt's highest moral value – an endless increase in culture and *Bildung* – was but an extrapolation of the ideology of the *Bildungsbürgertum*, an ideology adopted by the German bourgeoisie after the failed revolution of 1848. Robbed of the opportunity to influence the political scene, the bourgeoisie withdrew into the ideal of inward cultivation, an ideal according to which the increase of *Bildung* was *the* single overriding value (Nipperdey 1990: 382–9).

We already know that according to Wundt the state stood above the individual (cf. 1920a: 336). To this we must now add that the state was also of higher moral value compared with society. By 'state' Wundt basically meant a collective organism and personality in the senses introduced above (1897: 628; 1920b: 347). Society differed from the state in being 'everywhere governed by centrifugal impulses' (1912 III: 305). On the one hand, society tended to separate physical neighbours from one another by emphasising differences of birth, wealth, profession, interest, or intellectual ability. On the other hand, society was also the realm of numerous smaller or larger interest groups, associations, clubs or unions; these often set their own interests before the common good. In short, society provided 'an obstacle to the complete unification of all people in the state' (1912 III: 305).

But this obstacle could be overcome. The state could discipline and tame society, and indeed, this was the state's calling. The state had to 'reform society'. The state had to turn 'the most important associations of society' into mere organs of the state (1897: 633). The state

> intervenes and thereby removes the contingency that characterises the emergence of associations in society. The state removes this contingency by adjusting the associations to its plans – and these plans are guided by the moral purposes of the whole. Society as such lives in the present, but the state is directed towards the tasks of the future, and it puts the passing energies of society's life into the service of lasting goals.
>
> (1912 III: 307)

The state educated the citizens and thereby achieved both a softening of divisions in society and an orientation towards the ultimate moral goal of creating more intellectual goods (1912 III: 307–8).

The best form of government, and the form of government towards which history had advanced, was some kind of combination of democracy, oligarchy, and monarchy. The modern state was democratic in so far as the citizens elected representatives to one or more chambers of parliament; it was oligarchic in so far as it was *de facto* governed by the two classes of the civil servants and the professional politicians; and it was monarchic in so far as no state could do without a monarch, or monarch-like figure at the top. Here Wundt counted elected presidents as 'monarch-like' figures. No state could do without such a figure, because it represented the coming together of the collective and the individual (1917: 311). While Wundt supported a parliamentary system, he did not think that a single chamber best served the promotion of the national interest. He advocated a three-chamber system in which one chamber represented the population's support for different political parties, one chamber different regions, and one chamber the main professions (1920a: 376).

Wundt's criticism of individualism

Up to this point, I have concentrated on summarising Wundt's main collectivistic contentions with respect to social ontology, ethics, and political philosophy. To tie all this back to his scepticism concerning introspection in general, and the Würzburg-type thought experiments in particular, we need to turn to his criticism of individualism.

We already know from earlier chapters that Wundt opposed materialism and intellectualism in the realm of psychology. To this we must now add that both of these 'sins' were closely tied to socio-ontological and politico-philosophical individualism. As Wundt saw it, materialistic psychologists tended to favour individualism. First, they reduced the mental life of the individual to being 'a combination of atomistic sensations'. Second, they denied the existence of mental causality. Third, they therefore could not conceive of the possibility of mental phenomena that were more than, and irreducible to, the sensations. And fourth, because they held this reductionism for the individual mental life, they also had to maintain it for the case of collective mental phenomena (1908b: 292).

Intellectualism led to individualism via a different route. In line with its exaggerated emphasis on reasoning, intellectualism treated every social phenomenon as something that had to be planned, debated, and agreed upon. It therefore was prone to construe the state or other collectivities as emerging from contracts between pre-social individuals (1908b: 292). In Wundt's view, 'no conception of the state misunderstands its essence more radically than the individualistic theory according to which the state emerged from a real or fictive social contract' (1912 III: 325). This theory was wrong not least because it distorted the true relation between society and the state. For the contract theorist, the state was not the power above, but the artefact of, society. Rather than correctly conceive of the state as an organism, the contract theorist looked upon the state 'as a machine that can be construed according to different plans'. Moreover, because the state existed by an agreement between self-interested individuals, it was unlikely that the tasks of the state would go much beyond the protection of life and property of the individual (1912 III: 326). Indeed, the state itself would probably be 'fleeting and contingent' (1897: 612). Ultimately, the contract theory was most dangerous, however, because it naturally led to two further ideas: the 'universal state' and 'anarchism'. If the state was nothing but an agreement amongst self-interested parties then there was no reason to deny the possibility or desirability of a world-state. But such a world-state would again be so thin, so restricted in its rights *vis-à-vis* its citizens, that it was tantamount to having no state at all. And that was anarchism (1917: 336).

In Wundt's view, individualism was *the* central premise of a host of politically dangerous positions: Marxism, social democracy, utilitarianism, and capitalism. In the case of Marx and Lassalle, Wundt deplored their psychology as well as their ethical stance. Allegedly, their materialistic psychology led them to embrace individualism and determinism, and they construed the human being as 'a logical machine that reacts...to external influences' (1912 II: 245). Their psychological individualism was most perspicuous in their internationalism; after all, internationalism denied the importance of collectives such as the family, the tribe, or the nation (1912 III: 286). And finally, psychological individualism led to ethical and political individualism. The 'key motive' of Marxism and social democracy was 'egoism'; both aimed exclusively for the 'satisfying of the natural needs of life' of the individual (1912 II: 246).

Utilitarianism was just another form of egoism: 'one's own, that is, the individual's, feelings of pleasure are without moral value; it follows that the same holds also for the feelings of pleasure of the many or of all. Utilitarianism is an extended egoism' (1912 III: 81).

Wundt's opposition to socialism did not make him an ardent advocate of capitalism, however. Yet another product of individualism, capitalism was a positive historical phenomenon only in so far as it had emerged 'as a powerful, indeed indispensable means of support for culture' (1912 III: 251). But one could already see that it would not play this role for long. Rather than create wealth to support *Bildung*, capitalists tended to reverse this order and support culture only where it improved their social standing or their wealth. Capitalism thus needed reform – but such reform could only come from 'a collectivistic mode of thinking' (1912 III: 257).

Finally, forms of individualism or collectivism – socio-ontological, metaphysical, psychological, ethical, political – were all interconnected. Wundt made this point most forcefully in comments on Descartes and Herbart. Descartes' metaphysical ideas 'reflected' the ethical views of his time: 'Descartes' atomistic concept of the soul is the true reflection of the psychological and ethical individualism of the Enlightenment' (1912 III: 33). And Herbart's assumption of a conflict-free, collective soul was influenced by 'the higher and educated German middle classes of his time, the conservatively thinking civil servants and the academics' (1912 II : 215).

Collectivism and thought

In light of the above, it seems that Wundt had even more reason to attack thought psychology than what I have suggested in Chapters

2 to 5. Given the ways he viewed individualism, Wundt had no choice but both to conceive of the Würzburgers as advocates of individualism, and regard their *psychological* theories as *politically* dangerous.

We already know that Wundt suspected the Würzburgers of intellectualism and of psychophysical materialism. In this chapter we have seen that for Wundt intellectualism and materialism led straight to socio-ontological individualism. And this form of individualism in turn gave rise to egoism, utilitarianism, Marxism, social democracy, and capitalism. All of these positions were of course unacceptable to Wundt.

As far as Wundt's likely perception of the Würzburgers is concerned, their denial of the need for a collective psychology must have appeared as a very clear symptom of individualism. The Würzburgers tried to study thinking with the help of the very method – introspection – that Wundt had marked as the primary tool of *experimental* psychology. And experimental psychology was of course the heartland of *individual* psychology. In advocating introspection as *the* decisive method for thought psychology, the Würzburgers in fact insisted that all of psychology could throughout be based upon individualistic premises. There can be no doubt that Wundt himself made the link between this thought-psychological individualism and political individualism in its full sense. For instance, in his autobiography he reported that trying to understand thinking on the basis of individual consciousness had always seemed to him as hopeless as the attempt to understand the state 'as a purely individual discovery' (1920a: 218).

Seen from Wundt's perspective, the Würzburgers had also failed to appreciate properly the links between association, apperception, society, and the state. A certain homology between psychological associationism and political individualism was a legacy of British associationism. Classical associationism had established

> a metaphorical homology between...the structure of society...and the structure of the human mind....just as societies were considered to be formed by the combination of separate and independent persons, so individual minds could be thought of as formed by the association of separate mental elements.
>
> (Danziger 1990b: 347)

Wundt sought to destroy psychological and political associationism. And he did so by trying to establish a new metaphorical homology between the relationship of state and society on the one hand, and the interaction between apperception and association on the other. The state selected from, and mediated between, the conflicting interests in society just as apperception selected from, and mediated between, the conflicting associations. As the state was the higher instance with respect to society, so also apperception was the higher realm *vis-à-vis* association. The state, like apperception, was the vehicle of reason, culture, and *Bildung*.

Lest my interpretation of this point appears a bit too speculative, I hasten to add that Wundt was himself quite explicit about the analogy between the individual and the state. Thus he spoke of the importance of the Platonic idea according to which 'the state is the "human being writ large"' (1912 III: 29), and elsewhere he wrote that 'there are but…two *real* persons: the individual and the state' (1912 III: 326–7). Indeed, one might say that for Wundt the state was the individual writ large, and the individual the state written small. Individual psychology provided nearly all of the metaphors for the collective in general, and the state in particular: the concepts of the collective will, the collective soul, the collective organism, the collective consciousness, and the collective personality were modifications and extensions of the already existing concepts of the individual will, the individual soul, the individual organism, the individual consciousness, and the individual personality. Of course, once these concepts had been extended and altered in this way, they could then be used again for the individual; the collective now became the metaphor for the individual. Thus Wundt wrote, for example, that what appeared to be a single and individual will was in some sense already a collective will, that is a combination of conflicting motives (1897: 414; 1918: 327).

Needless to say, in the Würzburgers' thought-psychological studies up to 1907, Wundt could find neither parallel developments nor adoptions and adaptations of his metaphors and ideas. Most importantly, the Würzburgers had little sympathy for apperception. This could appear to weaken the metaphorical homology between a strong state and a strong will. The cases of Ach (1905) and Watt (1905) were particularly clear in this respect. Rather than work with the assumption of a choosing and selecting will (i.e. apperceptive processes), Ach and Watt introduced new kinds of deterministic regularities that governed the relationships between presentations. By Wundtian standards they thus failed to capture the true sense of the individual personality. There was no government, no state, in Ach's and Watt's societies of the mind.

We might also note a striking parallel between the two classes of the Wundtian society and the two levels of psychologists in the Wundtian research school. The division between the 'active' and the 'passive' class – or, as Wundt also said, between 'the leading spirits' and the rest of society – resembled the division of labour in the Leipzig Institute in which the master psychologist acted as if students needed someone to think on their behalf. I have argued above that the Würzburgers attacked the social order of Wundt's institute; now we have some reason to suspect that for Wundt the disagreement ran deeper than was suggested there.

Finally, Wundt's view on the relationship between the leading spirit, or spirits, and the passive rest can perhaps also clarify one point that may have aroused some reader's curiosity. The point I have in mind is the seeming contradiction between Wundt's collectivism on the one hand, and advocacy of the solitude topos on the other. In light of the doctrine of 'leading spirits', this might be a possible solution: although a leading spirit was always part of the collective, there was no need – after the initial upbringing and training – for him or her to seek a reciprocal dialogue with others. Indeed, one might even argue that such dialogue might endanger the activity of the leading spirit and pull it down to the level of the passive masses. Of course, psychologists such as the Würzburgers – if then they could be regarded as leading spirits at all – might fall into this trap more easily than the Leipzigers. After all, only the former interacted regularly with the passive masses of school children or teachers.

Uniformity and individualism: Brönner and Marbe

Was Wundt right about the Würzburgers' commitment to individualism? And what did the Würzburgers make of Wundt's collectivism in psychology, ethics, and political philosophy? It is difficult to provide a straightforward answer to these questions. Very few of the Würzburgers wrote about social and political issues, and even those few who did usually expressed their views only long after the main thought-psychological studies had been published. Nevertheless, and with these provisos in mind, it can be shown that at last some key Würzburgers rejected both Wundt's social ontology and aspects of his ethical and political thinking. One might therefore suspect that their individualistic way of approaching the study of thinking was linked to a more principled individualistic philosophical commitment. And this suspicion is not new. At least one observer of the German psychological scene, the Swiss C. Sganzini, reported in 1913 that the '*Funktionspsychologie*' of Külpe and the Würzburg school now represented the main alternative to Wundt's interpretation of 'collective-mental processes'. For the 'younger schools of psychologists and sociologists', there existed only 'individual-psychological processes'. They accepted that some of these individual-psychological processes were 'social'. But to call an individual-psychological process 'social' did not mean for them that it was part of some larger collective process or soul; a psychological process was 'social' only if it was 'caused by stimuli coming from other humans' (1913: 235).

One of Sganzini's main pieces of evidence was a dissertation written under Marbe's supervision in Würzburg and published in the *Zeitschrift für Philosophie und Philosophische Kritik* in 1911. The author was W. Brönner, a man who had not published thought-psychological studies himself. His study was entitled 'On the Theory of Collective-Mental Phenomena' (*Zur Theorie der kollektivpsychischen Erscheinungen*, 1911).

Brönner claimed that Wundt was wrong to apply the concepts 'soul', 'spirit', or 'consciousness' to collective-psychological phenomena. Wundt's main argument had been that both in the case of the individual and in the case of the collectivity, one could speak of an 'interconnection' (*Zusammenhang*) between mental states. Brönner objected that Wundt had not explained the quality and the laws of such interconnection for the case of the collectivity. The connections between mental states of the individual could be captured, for instance, by the laws of association. But nowhere had Wundt explained 'how we are to understand the possibility that some conscious process in person *A* causes, by means of association, another conscious process in another person *B*'. Not only had Wundt failed to provide such analysis for the case of association, he had failed to do so for *any* of the psychological laws about the interconnection between mental states of the individual (1911: 9).

Brönner did not deny that different individuals could have similar mental processes, or similar mental states. He also granted that individuals could influence one another, and thus change one another's minds. But he denied that this created a mental interconnection *between* individuals that was anything like the mental interconnection *within* a single individual. Only in the former was there an objective physical event that separated the two minds. It was produced by one mind and perceived as an objective stimulus by the other (1911: 10).

According to Brönner, Wundt was also wrong to hold that language, myth, and custom were different from, and more than, the sum of what individuals produced as their language, myth, and custom. Wundt was right only under the counterfactual assumption of a totally isolated individual without any society around him or her. The sum of what a thousand completely isolated individuals would produce as their language 'would indeed be different from, and less than' the language spoken by a people. If, however, one started from individuals who had been raised and were living in a community, then one could no longer maintain Wundt's thesis. In their case, the sum was – at least as much as – the whole. The language, myths, and customs of a people were 'nothing but the sum of the languages, myths, and customs of the individuals, minus the peculiarities of some individuals and groups' (1911: 11).

Moreover, Brönner was unconvinced by Wundt's claim that the *Volksseele*, or a collective spirit, was 'real'. To refute Wundt on this point, Brönner distinguished between 'the empirically real' and the 'hypothetically real' on the one hand, and 'the physically real' and 'the psychologically real' on the other. An entity was empirically real if it was observable; it was hypothetically real if it was to date unobserved, in principle observable, and currently part of the best explanation of an empirically real phenomenon. Physical reality was made up of events in space and time, and psychological reality consisted of conscious processes. Brönner argued that collective spirits and *Volksseelen* fitted into none of these categories. As psychological entities they obviously were neither empirically nor hypothetically *physically* real. But neither could they be empirically or hypothetically *psychologically* real: conscious processes could be observed only in self-observation, and self-observation could only take the conscious processes of the individual self-observer as its objects (1911: 12–13).

Finally, Brönner also rejected Wundt's proposal that the effects (*Wirkungen*) of the collective souls attested to their reality (*Wirklichkeit*). Here the effects in question were language, myth, and custom. As Brönner saw it, this argument overlooked that '"reality" and "having effects" are completely different concepts'. All told, Wundt's theorising reminded his critic of mythological forms of reasoning (1911: 14).

Brönner sought to offer a systematic alternative to Wundt's collectivism. Central to this alternative was the thesis of the 'uniformity (*Gleichförmigkeit*) of mental events'; that is, that 'under the same or similar conditions different persons have the same or similar mental experiences'. Thumb and Marbe (1901) had found that different experimental subjects tended to react to the same stimulus words with the same or similar reaction words. Other studies had confirmed this result under a variety of conditions; they had also shown that suggestion could increase the uniformity of reactions (Brönner 1911: 33).

'Collective-mental phenomena', that is the very phenomena that had led Wundt to introduce entities such as collective souls, could be explained on the basis of the uniformity assumption and interaction between individuals. Interaction was a key factor; it 'increases the uniformity of the conditions under which the mental life of a plurality of individuals occurs'. Over and above the interaction, the uniformity of conditions, and the natural uniformity of mental lives, however, there was no need to stipulate any further entities or processes. To follow Wundt's example and do so meant to turn concepts into real essences, and thus to fall back into the practices of scholastic realism or German idealism (1911: 36–40).

Marbe was clearly satisfied with his student's way of dealing with Wundt's collectivism. In his *Outline of Forensic Psychology* (*Grundzüge der forensischen Psychologie*, 1913b) and his *Uniformity of the World* (*Die Gleichförmigkeit in der Welt*, 1916) Marbe endorsed Brönner's position. In the latter book Marbe dealt with the topic in great detail (1916: 20–147).

Like Brönner, Marbe made 'the uniformity of human mental life' the key concept of his analysis. This uniformity was of great importance in everyday life; it made other people's behaviour and their mental states predictable and understandable. Marbe distinguished mental uniformities that seemed tied to specific cultures and time periods from mental uniformities that held for humans of different cultures. An example of the latter was humans' tendency to favour the numbers 0, 5, 8, and 2 (in that order) when in circumstances where any number between 0 and 9 had to be named or written down. Marbe compared here information about the (estimated) age of the deceased on Roman gravestones; modern-day experimental subjects' estimations of the lengths, in centimetres, of drawn lines; and the rough ages given by 'uneducated Negroes in Alabama'. In all three cases, the most frequent last digits of the numbers given were 0, 5, 8, and 2. Marbe reasoned that this 'obvious uniformity' was 'not a mere product of culture but a fact that was grounded in the mind of the human being' (1916: 54). The ways people spoke, on the other hand, were culturally dependent uniformities (1916: 68). The distinction between culturally dependent and independent uniformities was close to that between 'primary' and 'secondary uniformities'. Primary uniformities existed independently of interaction and suggestion, whereas secondary uniformities were created and strengthened by interaction and suggestion (1916: 123).

An investigation of secondary uniformities was tantamount to a study of 'social organisations in the widest sense of the word'. Social organisations increased the degree of conformity of conditions for their members, and thus also augmented the degree of uniformity in their thinking, feeling, and acting. As examples, Marbe cited 'socialism, the brilliant organisation of the [Catholic] *Zentrum* party,...religious orders...Catholic schools and seminars...and clubs for military training'. Marbe indicated a personal political preference by deploring the fact that 'liberal parties of various strands have too few of such organisations; this is the main reason for their negligible influence' (1916: 124).

It was necessary for humans to ascribe their individual mental experiences to an 'I' or a 'self' (*Ich*). Accordingly, this way of talking was ubiquitous in everyday life, and psychological science too had to make use of it. Often the 'I' would be equated with the 'soul'. Marbe had no objection in principle to this occasional substitution of the 'soul' for the 'I', but he clearly regarded soul-talk as much more problematic. He was particularly concerned that talk of a soul might mislead speakers into believing that the soul was something like a 'real carrier' of the experiences (1916: 115). Things got worse, however, once people started attributing their joint mental experience first to a 'we' and then to a collective soul. Whereas I-talk was irreducible, we-talk was not. And whereas an individual soul, as a bearer of individual mental life, was merely 'an illusion', the assumption of a *Volksseele* was 'actually a farce' (1916: 115).

One central line of Marbe's criticism – a line not present in Brönner's paper – concerned the adequacy of the emphasis on *peoples* in Wundt's *Völkerpsychologie*.[1] The force with which Marbe belaboured this point deserves attention, especially if we remember that Marbe's book was published halfway through the First World War; that is, at a time when Wundt and others were busy arguing for the superiority of the German *Volksseele* over all others (Kusch 1995a: 220–1). Indeed, the relevant passage is worth quoting at some length:

> Do we not also find within political parties a peculiar connection of presentations and feelings, just like we find within the whole *Volk*? And do not such connections sometimes transcend national boundaries? Given Wundt's definition of the collective consciousness, might we not also speak of the collective consciousness of conservatism, of liberalism, or of international social democracy? Speaking of the collective consciousness of social democracy can hardly be dismissed on the grounds that under the influence of the World War, collective consciousness seems shattered. For just like international associations, so also peoples and states sometimes break apart. Is it not true that great numbers of persons have placed in the past, and are placing now, the international consciousness of social democracy above the national consciousness? And is it not true that in their case the international consciousness is more developed than the national one?
>
> (Marbe 1916: 117)

Of course, in this passage, Marbe did not endorse social democracy, and he did not write critically or contemptuously about the national collective consciousness. But the mere fact that Marbe wrote neutrally and in a matter-of-fact style about social democracy already signalled a considerable degree of intellectual distance from the Wundtian and other contemporaneous forms of nationalism.

Marbe did not only insist that international collective phenomena fulfilled Wundt's criteria for a collective soul. He also believed that much smaller social groups would qualify as well: for instance, professional organisations, parishes, and interests groups. This immediately raised a tricky question, however: In how many different collective souls could a given individual soul participate (1916: 118)?

Marbe suspected that Wundt had been misled by the fact that in the case of some wholes, namely 'collective objects' (*Kollektivgegenstände*), the whole possessed properties not possessed by any of the parts. But from this it did not follow that the whole had a reality over and above the parts: 'Think of a house. Although a house has a plethora of properties not possessed by any of its parts, no one claims that the house is an additional reality, additional to its components' (1916: 119).

Moreover, the assumption of *Volksseelen* and of other collectives was likely to push scientific research in wrong directions. If no *Volksseelen* were assumed, then the psychologist or sociologist would have to study the emergence of collective phenomena out of complex interactions between individuals. If *Volksseelen* were acceptable, then the psychologist could confine him- or herself to declaring the collective phenomena an outflow of the collective soul and leave it at that. That was indeed what Wundt had done. Wundt's *Völkerpsychologie* tended to ignore the very interaction that allegedly had created its phenomena. To make this point more forcefully, Marbe drew on the criticism that the linguist H. Paul had marshalled against Wundt's treatment of language. According to Paul,

> he [Wundt] treats language only from the perspective of the speaker.... And he pays no attention to the learning of language and its consequences. Thereby he has made it impossible for himself to understand properly the conditions of the development of language. As Wundt has it, changes of language do not originate in the individuals; instead they flow, with a peculiar natural necessity, from the common *Volksgeist*.
>
> (H. Paul, quoted in Marbe 1916: 121)

Another argument used by Marbe against Wundt involved animal psychology. According to Marbe, it could no longer be denied that at least higher animals, such as partridges, possessed intellectual capacities. Thus – presumably by Wundt's own standard – one also ought to accept that partridges had souls. But given that partridges interacted with one another, and that their mental states were interconnected, one could then also argue that partridges could be looked upon as a people, and that they possessed a *Volksseele*. In Marbe's view, this consequence was the endpoint of a *reductio ad absurdum* (1916: 122).

To sum up, it seems obvious enough that Brönner and Marbe had, by 1911 and 1916 respectively, formulated an individualistic alternative to Wundt's socio-ontological collectivism. Unfortunately, the available evidence does not tell us to what degree Marbe had such individualistic commitment in 1900. And thus it is impossible to show that a self-conscious commitment to some sort of socio-ontological individualism was one of the causes of the Würzburger thought psychology. Nevertheless, the following three factors at least lend some plausibility to this claim. First, Marbe used a method, that is introspection, that his former teacher Wundt had clearly labelled as individualistic. Second, at the very same time that Marbe worked on thought psychology he also began his investigations into uniformity, and here too with a book-length study; Marbe's famous study of judgement (Marbe 1901) and his book with Thumb on the uniformity of associations (Thumb and Marbe 1901) appeared in the same year. And third, in his autobiographical writings, Marbe never spoke of a conversion to individualism, or a sudden realisation that his thought-psychological work implied a commitment towards individualism. Instead, Marbe presented his work on thought psychology and his investigations into uniformity as closely related projects (1945, 1961).

Bühler on Wundt as 'the Diogenes in the barrel'

Bühler too published extensive critical discussions of Wundt's *Völkerpsychologie*, focusing on Wundt's treatment of language in particular. Although these texts come from a much later period – the mid-1920s and the early 1930s – they still throw an interesting light on Bühler's and Wundt's views. Bühler's main accusation was that Wundt's collectivism was actually an extreme form of individualism; Wundt had no account of interactions amongst individuals, and thus his only way of conceptualising collectivities was as individuals writ large. At least a vague anticipation of this criticism can be found in the seventh of Bühler's *Habilitationsthesen* of 1907. There Bühler stated that the Wundtian 'parallel between *Volksseele* and individual soul is without any empirical foundation' (Bühler 1907c: 2).

In his *Crisis of Psychology* (*Die Krise der Psychologie*, [1926] 1929), Bühler demanded that the psychology of language adopt the following criterion of progress: 'The dubious notion of the *Volksseele* must be made redundant; but one must not lose sight of these facts that led [Wundt] to introduce the notion in the first place' ([1926] 1929: 32).

Bühler alleged that Wundt had never solved the problem of how to relate individual and collective psychology to one another. Wundt's concepts of 'collective soul', 'collective consciousness', and so forth, appeared only in Wundt's programmatic pronouncements on *Völkerpsychologie*; they did not, however, play much of a role in Wundt's detailed, two-volume analysis of language. When it came to providing a detailed psychological account of language, Wundt emerged from his collectivistic cocoon as an extreme individualist: He analysed language predominantly as the outer expression of the inner experiences of an isolated speaker, and he failed to take account of the referential (object-related) and pragmatic (hearer-related) aspects of language. Wundt's conception of language was thus utterly non-social; indeed, it was 'solipsistic' ([1926] 1929: 32).

In his *Theory of Expression* (*Ausdruckstheorie*, 1933), Bühler extended this criticism by arguing that Wundt's model of the human being was 'Diogenes in the barrel', an individual physically isolated from others (1933: 135). Indicative of Wundt's failure to capture the social aspects of language was first and foremost the following: he construed the relationship between the collective soul and collectively shared language to be directly parallel to the relationship between the individual soul and the individual's language. Collectively shared language expressed the collective soul as the individual's language expressed the individual soul. Clearly, the isolated, non-social individual provided the model for the collective. Because the model contained no social interaction, the modelled phenomenon could not be based upon social interaction either ([1926] 1929: 32).

As an alternative, Bühler proposed 'locating the origins of semantics in the community rather than in the individual'. This meant recognising that language was not primarily a *product*, or *expression*, of the community; instead language was one of every community's most important 'constitutive factors'. And to study language from this perspective was to investigate how language functioned as social individuals' main tool of 'reciprocal control and regulation' (*gegenseitige Steuerung*) ([1926] 1929: 39).

Külpe and Messer on Wundt's ethical and political ideas

Although Marbe and Bühler challenged Wundt's social ontology, they did not directly comment on his ethical and political ideas. There is of course some indirect evidence that they did not share Wundt's vision of a *Volk*-bound, corporate state. In his 1916 book, Marbe hinted that his sympathies lay with political liberalism. And Bühler's later teaching activity in the 'Red Vienna' of the interwar period attested to at least some – however limited – tolerance of, and sympathy for, a social-democratic agenda (Gardner and Stevens, 1992).

More direct comments on Wundt's ethical and political writings can be found in Külpe's and Messer's writings. Messer dealt with ethical and political issues in his book *The Problem of Civic Educating* (*Das Problem der staatsbürgerlichen Erziehung*, 1912a), and later in his *Ethik* ([1918] 1925). The first book outlined a programme for the education of good citizens. Messer shared with Wundt the concern that social democracy might increasingly endanger the state, and he also feared that 'ethical individualists' might undermine belief in 'the moral significance of the state' (1912a: 156, 158). Unlike Wundt, however, Messer put much emphasis on the notion that the interests of humankind should always be placed above those of the state. It might seem that Wundt had suggested something similar when he had situated 'the human norms' above 'the social norms' (1912 III: 152–8). But in Wundt's case this order between the state and humankind was somehow without deeper significance because he never dealt with possible conflicts between the interests of the state and the interests of humankind. It was as if such conflict could not really arise. Messer insisted that it could. And he sided 'with Kant and our classic literature' in maintaining that in cases of conflict the morality of humankind should always be placed above the *raison d'état* (1912a: 146).

In his postwar *Ethik,* Messer challenged Wundt on two further grounds. On the one hand, he rejected Wundt's 'evolutionary theory', according to which the defining criterion of moral actions was the 'furthering of the development of culture'. In Messer's opinion, this was either false or circular: it was false if by 'culture' one meant science, art, or the economy; any of these could be advanced for moral as well as for immoral reasons. If one thought of 'culture' as 'moral culture', however, then one had failed to give a non-circular account of morality and its criteria ([1918] 1925: 78). On the other hand, Wundt went wrong also by completely tying the individual to his or her state. This contradicted the Christian moral doctrine that each human being was 'a child of God'. Wundt fell back into the pre-Christian, ancient-Greek view according to which the individual was nothing more than a citizen ([1918] 1925: 104). Messer insisted that the individual had both the right and the duty to break the state's laws if these laws turned out to be immoral by the individual's moral criteria ([1918] 1925: 107).

In both his criticism of Wundt's ethical evolutionism and his insistence on the individual's right to resist the state, Messer had been preceded by Külpe. For instance, already in 1910 Külpe wrote that 'a revolt against the state can have moral significance in circumstances where the state follows tendencies that harm culture, that are egoistic, and that run counter to the comprehensive interests of humankind' (1910a: 321). And in his *Ethics and War* (*Die Ethik und der Krieg,* 1915a), Külpe wrote that 'cultural values are not *per se* moral values. And it would be wrong to say that it is primarily, or exclusively, the morality of a people that enables it to create cultural values' (1915a: 31). Furthermore, Külpe challenged Wundt's Hegelian assumption that the correct morality could be read off historical development: 'He can show us how morality has developed – but he can do so only after we have already determined what is moral. But he cannot determine what is moral on the basis of historical-genetic considerations' (1910a: 295).

Külpe's discussion of ethical 'individualism' and 'universalism' is also noteworthy here. Whereas ethical individualism held that the object of the moral will was the individual, ethical universalism gave this position to the collective. Individualism was further divided into egoism and altruism, depending on whether oneself or another subject was the purpose of the action. Universalism had as many subforms as there were communities: 'thus one can speak of a *social*, a *political*, a *national*, and a *human* universalism' (1910a: 317). Counting Wundt amongst the '*human* universalists', Külpe distanced himself from his former teacher and emphasised the need for 'a unification of the individualistic and the universalistic tendencies'. Universalism was wrong in so far as it was based on a mistaken ontology. Communities could 'never be the *immediate* object of the will' because communities only ever had '*ideal* existence'. Obviously, this contradicted Wundt's claim that collectivities had not just ideal but *real* existence. Külpe's compromise formula between universalism and individualism stated that 'the community is the ultimate, but only the ideal object of the moral will; the individual is the proximate and real object of the moral will' (1910a: 320).

'The ego sits on the throne and governs'

In light of the material presented in this chapter, it is well worth returning once more to Külpe's earlier-quoted comparison between the soul and the state.[2] We can now see that such comparison probably had a deeper significance. After all, Külpe was no doubt aware of the way in which Wundt paralleled the state and the soul. Any use of this system of metaphors was therefore also an indirect comment on its alternative uses.

Here is the passage once more, only now I shall also quote the two following sentences:

> The ego sits on the throne and governs. It notices, perceives, and registers what enters into its kingdom; it deals with that

intruder; it consults its experienced ministers, the principles and norms of its state, the contingent needs of the present, and it decides to take a stand toward the intruder, to leave him aside, to give him a useful form, or to rule against him. And the sensations and presentations have in general good cause to condemn as reactionary, harsh, and arbitrary the rule of this monarchic ego. They take their revenge during sleep and are wont to make mischief in our dreams. They then show what results from such anarchy. But in every ego resides an ineradicable high-handedness, a self-satisfaction that the ego acquires and uses within its realm. And thus we can understand why the ego is quite unwilling to submit to another will.

(1912c: 1089–90)

A first striking difference between Külpe's and Wundt's respective mind/state talk was this: whereas there was anarchy, conflict, revolt, suppression, and ruler's arrogance in Külpe's vision of soul and state, Wundt's theories of both emphasised harmony and fusion. Wundt's state acted as mediator and educator of society's conflicting groups; it transformed society into an organism that supported its (i.e. the state's) ethical goals. And within the individual soul, apperception did likewise to associations. In both cases, the hierarchies were stable, legitimate, and unchallenged.

Second, Külpe's description of the mind/state focused on the *interaction* between individuals and small groups. There was talk here of the king's 'dealing with' intruders, of 'consulting' others, and of 'handing out' tasks. The state here was the result of an outcome of the struggle between individuals that try to influence and guide one another. This contrasted with Wundt's account of the state where the individuals disappeared, or were – in the Hegelian sense – *aufgehoben* in the collective soul. Külpe's use of the metaphor was close to the theme of interaction insisted upon by Brönner, Bühler, and Marbe.

Third, it should be noted also that the last two sentences gave the whole passage something of a dialectical twist. In the bulk of the passage, state politics was used as a metaphor for the soul. But in the last two sentences, the politics of the soul became itself an explanation for one key feature of state politics: that is, of why individuals were always prone to challenge any form of state rule. Individuals tended to reject suppression, but, ironically, they did so because they wished to be as high-handed outside their mind as they acted within, or they wished to be as high-handed outside as the ruler was already. Introspecting one's own mind thus taught one a political lesson of individual freedom and self-determination!

Fourth, it seems a natural conjecture that for Wundt and Külpe the metaphorical exchange between mind and polity had different starting points. I have suggested before that for Wundt collective phenomena were modelled on the individual; the collective soul was the individual soul writ small. This does not seem to hold for Külpe and the Würzburgers; as the summaries of Brönner's, Bühler's, Marbe's, and Külpe's texts make clear enough, these men were consciously trying to avoid the Wundtian move of modelling the collective on the individual. But while disallowing the Wundtian move, Külpe did not have qualms about making the collective the model for the individual. The result was that whereas Wundt tended to 'psychologise', or naturalise, the state, Külpe, at least in this passage, politicised the soul. For Wundt, the state was a bit like the mind of a tranquil grown-up academic: serene, free from inner turmoil, and dedicated to the production of cultural works. For Külpe, the mind was a bit like the Wilhelminian state around him: The monarch did 'consult with his experienced ministers', but otherwise there were many who 'condemned' his regime as 'reactionary, harsh, and arbitrary' (cf. Nipperdey, 1992: 621–757). Unfortunately, Külpe did not indicate whether either mind or state could change.

Finally, it is perhaps worth pointing out that neither Wundt's nor Külpe's visions map easily onto party-political programmes and ideals of the time. Neither of them was a supporter of social democracy or Marxism, of course, and it is unlikely that they would have supported either the Catholic *Zentrum* or the primarily agrarian Conservatives. Perhaps one can risk the conjecture that Wundt sympathised with conservative *Nationalliberalismus* – which by the turn of the century emphasised the importance of the state and the nation, and the need for education. This much is suggested also by the fact that the Heidelberg division of the *Nationalliberale Partei* warmly congratulated Wundt on the occasion of his 80th birthday ('We regard you as one of us and thank you for your achievements on behalf of the national and the liberal cause', Wundt Archiv, 15 August 1912). Some of the Würzburgers' pronouncements, on the other hand, suggest at least a critical distance from *Nationalliberalismus* (cf. Nipperdey, 1992: 521–36). But other than that, little can be said. Wundt and the Würzburgers were like most German academics at the time: they did not actively engage in the political debates of the day.

Summary

In this chapter I have argued that Wundt and some of the key Würzburgers had different social, ethical, and political philosophies. Wundt's opposition to thought psychology was partly an opposition to the individualism of the Würzburgers' introspective methodology, partly a defence of collectivism. This collectivism insisted on the ontological priority of the *Volksseele* with respect to the individual, on the ethical priority of cultural works with respect to human happiness, and on the political priority of the state with respect to the citizen. By a complex network of analogies and implications, Wundt had linked introspection to individualism, and the latter to materialism, intellectualism, social democracy, Marxism, utilitarianism, capitalism, and anarchism. Small wonder, therefore, that Wundt tried to stop thought psychology in its tracks.

Needless to say, the Würzburgers did not accept the link between methodological individualism and the various political movements. But it is true to say that some of them did favour an individualistic approach. Brönner and Marbe developed this socio-ontological individualism at considerable length. I suggested that this form of individualism might already have informed Marbe's 1901 study of judgements, that is the study with which the thought-psychological movement began. At the same time as he was working on thought psychology, Marbe also wrote his first book on the 'uniformity' of mental processes. And this concept became Marbe's systematic, individualistic alternative to Wundt's collectivism.

Bühler's, Messer's, and Külpe's brands of individualism were less systematically developed. Bühler accused Wundt of lacking an account of interaction, and Messer and Külpe opposed key elements of Wundt's ethical and political collectivism. Finally, I contrasted Külpe's and Wundt's ways of using mind and state as metaphors. It turned out that Wundt naturalised the state, whereas Külpe politicised the mind.

6 Protestant versus Catholic

Introduction

In this chapter, it remains for me to argue for the importance of a fourth social variable in the debate over the Würzburgers' thought psychology: religion, or more precisely, the 'conflict of confessions' (*Konfessionenstreit*) in turn-of-the-century Germany. I shall try to make the following ideas plausible. First, Wundt's opposition to the Würzburgers' thought psychology was linked to his hostility towards Catholicism in general, and to his dislike for neoscholasticism in particular. Second, Wundt's metaphysics as well as key assumptions of his epistemology and psychology were – by the standards and criteria of the time – Protestant in spirit if not in letter. Third, Külpe, although like Wundt a Protestant, took up philosophical positions that – to his contemporaries – tasted of Catholicism. These philosophical views can be shown to be involved in the psychological work of the Würzburgers. Fourth, of other key Würzburgers, Bühler, Marbe, and Messer all had strong Catholic backgrounds. At early stages of their respective academic careers, all three men had intended to apply later for professorial positions in 'Catholic philosophy'. Although Bühler, Marbe, and Messer were apostates by the time they carried out their famous thought-psychological experiments, a certain sympathy towards Catholicism and neo-Thomism was still visibly present in their writings.

Academic philosophy and the *Konfessionenstreit*

The sour relations between the Catholic and Protestant confessions at the turn of the century were due both to political events in Germany and to developments within the churches.[1] During the last third of the nineteenth century, most German states had sought to limit Rome's influence over German Catholicism. Unsurprisingly, such attempts led to conflicts, that is the so-called *Kulturkämpfe* (cultural struggles), with the rank and file of the Catholic church. The struggle was particularly bitter in the predominantly Protestant Prussia. Bismarck was suspicious of Catholicism not least because Catholics regarded a German unification without the Catholic Austria as unsatisfactory, and because they expressed their dissatisfaction with being ruled by Protestants from Berlin. Catholics also often voiced internationalistic views, placing the international church above the national state. Prussia therefore tried to destroy political Catholicism, that is the *Zentrum* party; to restrict ecclesiastical jurisdiction; to increase the state's control over schools; and to secure its say in the training and appointing of Catholic priests. By 1878 eight bishop and 1,200 priest positions were vacant, and the Jesuits had been expelled from Prussian territory. It was only when Bismarck started to need political Catholicism as an ally against the growing social democracy that the pressure on the Catholics was eased, and many of the earlier decrees revoked. But the damage was not easily repaired. The opposition between Catholicism and Protestantism had been deepened considerably, and the Catholic population remained ambivalent towards the nation-state for a long time to come (Grane 1987: 201–3; Nipperdey 1992: 364–81).

Meanwhile, the Catholic church was struggling to define its position towards modern culture. In 1870, during the reign of Pius IX, popes had been declared infallible. This move had antagonised many progressive forces within the church. For the following pope, Leo XIII, the solution to the problem of adjusting to modernity lay in a return to medieval scholasticism. In his encyclical *Aeterni Patris* of 1879, Leo XIII declared Thomas Aquinas to be '*inter scholasticos Doctores omnium princeps et magister*'. St Thomas was seen as the crucial resource not least because he had emphasised the importance of reason, and because he had been the major Aristotelian of the medieval churchmen. Thus it was hoped that his work would provide the modern theologian and Catholic philosopher with the tools and the spirit to engage successfully with modern natural science. Finally, G. Sarto, who became Pope Pius X in 1903, attempted to recommit the church to its traditional values and hierarchies by launching his attack on 'modernism' within the church (1907). From 1910 onwards, all Catholic theologians had to swear the 'antimodernist oath' committing themselves to strict obedience to the pope in all doctrinal matters (Grane 1987: 207–8, 212–15).

German Protestantism did not engage in similar painful and divisive exercises. It did not fight running battles with the state, and its doctrinal disputes were less heated. Its leading thinkers, such as A. Ritschl, A. von Harnack, and E. Troeltsch, were united by their anti-Catholicism, their belief in progress, their conservative national liberalism, their belief that the reformation marked the normative origin of modernity, and their advocacy of 'personality' as *the* pivotal cultural, ethical, and religious concept (Graf 1989).

Needless to say, the *Kulturkämpfe* and the *Konfessionenstreit* had their repercussions within academia in general, and within philosophy in particular. As M. Weber wrote in 1908, 'The "freedom of science" exists in Germany within the limits of ecclesiastical and political acceptability. Outside these limits there is none' (quoted from Ash 1995: 19–20). The 'ecclesiastical limits' not only severely disadvantaged Jewish scientists throughout Germany; they also applied to apostates, and to Catholics and Protestants. Catholics would have better chances in Bavaria, Protestants in Prussia or Saxony. The principle *cujus regio, illius et religio* still had some force in turn-of-the-century German academia. For instance, when the Catholic Carl Stumpf was appointed to a professorial chair in Berlin in 1894, he had to be made palatable to the Protestant royal household as 'in confessional matters…a man of mild views', as shown by the fact 'that he married a Protestant and let his children be raised in the evangelical [Lutheran] faith' (quoted from Ash 1995: 33). When Ach was appointed to Kant's chair in Königsberg in 1906, there were suspicions that he owed the appointment in part to his Protestant faith (Lück and Löwisch 1994: 68). And as late as 1920, Husserl sought to improve Heidegger's chances for an appointment in Marburg by assuring Natorp that Heidegger was no 'Catholic philosopher', and had 'cut himself loose completely from dogmatic Catholicism' (Schuhmann 1994b: 139).

At the time, 'Catholic philosopher' did not simply mean a philosopher who happened to be a member of the Catholic church. Instead it meant a full commitment to neo-Thomism, and either holding, or potentially qualifying for, one of a half-dozen 'Catholic' professorial chairs in philosophy. The existence of such chairs was usually an outcome of compromise formulae at the end of the *Kulturkämpfe*. Many Protestant 'pure' philosophers were unhappy about these chairs. For instance, in 1909 the *Frankfurter Zeitung* gave its front page to an article by P. Hensel. The main theme of Hensel's article was that the career prospects of young – Protestant! – philosophy graduates were endangered from two directions at once: by Catholic philosophers and by experimental psychologists. Hensel reported that about a quarter of German professorial chairs (i.e. ten) were reserved for Catholic philosophers. He also claimed that in an increasing number of German philosophy departments, there were two professors, one 'psychophysicist', and one 'Thomist'. Allegedly, these two would always get along well; the Thomist would leave questions of fact to the psychologist, the latter would leave questions of value and religion to the former. To Hensel's taste, this made for an unholy alliance against true philosophy, that is German idealism (Hensel 1909). Hensel was followed four years later by M. Frischeisen-Köhler who engaged in a little statistical exercise to prove that pure philosophy had increasingly lost ground to both the neo-Thomists and the experimental psychologists (Frischeisen-Köhler 1913: 371):

	1892–3	*1900–1*	*1913*
Pure philosophers	32	29	27
Experimental psychologists	3	6	10
Catholic philosophers	4	7	7
Total number of full chairs	39	42	44

In claiming that German idealism and neo-Thomism were incompatible philosophical viewpoints, Hensel did not stand alone. Indeed, the conflict between Catholic and Protestant philosophy was usually couched as the opposition between Kant and St Thomas, or Kant and Aristotle. Interestingly enough, philosophers (and theologians) from both sides of the religious divide concurred in this. Here is an especially clear statement from C. Gutberlet, one of Germany's leading neo-Thomists of the time:

> When I gave my first speech, from this same spot, on the occasion of the academic celebration of Saint Thomas Aquinas, I spoke about the relation between Saint Thomas and Kant. I characterised Saint Thomas as the philosopher of objectivity, as a Catholic philosopher, and the thinker from Königsberg as the philosopher of Protestantism. I sought to present the difference between them as a conflict between two opposed *Weltanschauungen*. Although this deep opposition may not always be obvious in the case of these two standard-bearers,

> the development of Kantianism and Protestantism up to the most recent past highlights this divergence. The subjectivism that Kant inaugurated has led, in the short period since my first lecture, with logical consequence to the denial of all objectivity; 'the dead are making headway'. In many quarters, pure phenomenalism, immanentism and psychologism are now presented as the highest and ultimate wisdom.
>
> (1911: 147)

The Catholic theologian A. Ehrhard – until 1898 professor in Würzburg, then in Vienna – regarded neo-Thomism as a necessary 'reaction' to the 'antimetaphysical character of modern philosophy insofar as it has been shaped by Kant'. Ehrhard called Kant the 'philosopher of Protestantism' and the 'gravedigger of Christianity' at that (Ehrhard 1902: 184, 248). After all, Protestantism, and thus also Kant, was shaped by a desire for revolutions, by 'extreme subjectivism', and by German nationalism – Ehrhard reported that German-ness (*Deutschtum*) and Protestantism were frequently equated. But no religion could be built and survive on such principles (1902: 116–20).

The most forceful statements from the opposite camp came from R. Eucken and F. Paulsen. In his paper 'Thomas Aquinas and Kant: A Fight of Two Worlds' (*Thomas v. Aquino und Kant: Ein Kampf zweier Welten*, 1901), Eucken insisted that St Thomas and Kant were more than just the representatives of two opposing confessions; the difference between the two now marked 'the clash of two whole *Weltanschauungen*, the fight of medieval and bounded thinking with modern and free thinking' (1901: 1). Eucken sought to defend Kant against the neo-Thomist charge of subjectivism. As Eucken saw it, Kant's epistemology was not subjectivistic, relativistic, or psychologistic. Kant had sharply separated logic, ethics, and aesthetics from psychology in general, and individual sentiments in particular. While granting that St Thomas had some historical importance, Eucken denied that the schoolman's work had much contemporary significance (1901: 11). Eucken also took on the neo-Thomist accusation according to which the philosophy of Kant and German idealism led inevitably to denying the existence of God:

> Let us work...as friends of Kant for the true substance of life....If we do this then we are quite prepared to live with the Thomists' accusations of a lack of faith and of a subjectivism. For properly observed...the lack of faith and the subjectivism is

> not on our side, it is on their side. They lack faith in the spirit because they will allow for the spirit's reality only if it is embodied; they lack faith in the power of the spirit in history because they are unable to see the positive sides of the great changes which the modern era has brought;...and it is a subjectivistic endeavour – never mind how many millions are standing behind it, and how strong an organisation – to try to stop the great train of spiritual advance, and thus to turn back the wheel of world history.
>
> (Eucken 1901: 18)

Paulsen belaboured these same themes at great length both in his *Introduction to Philosophy* (*Einleitung in die Philosophie,* 1901a) and in his *Militant Philosophy: Against Clericalism and Naturalism* (*Philosophia militans: Gegen Klerikalismus und Naturalismus,* 1901b). Because Külpe and Messer regarded Paulsen and Wundt as intellectual bedfellows (Külpe 1910a: 276–80; Messer 1916: 80), and because Wundt referred positively to Paulsen's work (1908b: 146), it is worth following his train of thought more closely.

In the *Einleitung,* readers were informed that the following five traits characterised contemporary philosophy in Germany. First, it followed Kant in being 'phenomenalistic and positivistic'. Second, it was 'idealistic and monistic' because it believed that the 'essence of reality' could only be discovered in 'inner experience'. Based on this inner experience, philosophers had come to the conclusion that

> [physical] reality...is the appearance of a *spiritual, universal life,* and [that] this life is the realisation of a *unified meaning,* of the activity of *a will that realises ideas*. And a trace of this will is given to us in our own will.

Third, philosophy had shifted from an 'intellectualistic' to a 'voluntaristic' viewpoint. This viewpoint had first emerged with Kant's emphasis on the 'primacy of practical reason' but it had since started to influence all parts of philosophy, including ethics and metaphysics. Paulsen added that philosophy was not alone in making the shift from intellectualism to voluntarism. Both psychology and Protestant theology were doing the same. Fourth, philosophy increasingly had adopted 'evolutionary and teleological' perspectives. And fifth and finally, philosophy strongly emphasised the importance of history (1901a: VI–VII).

Paulsen explained that the voluntarism in Protestant theology amounted to a reinterpretation of the role of dogma for the religious believer. Dogmas were no longer taken to be theoretical truths and claims about the universe or God; they were not addressed to the understanding or to reason. Instead, dogmas were now thought of as directed towards the will, and as giving to the will 'aim and direction'. This meant a return to the Lutheran position, a position that rejected scholastic philosophy, and its 'false unity of belief and knowledge'. It was tantamount to 'putting ecclesiastical life on the firm basis of the gospel' (1901a: 13).

In Paulsen's view, Kant held the Lutheran position. Kant's criticism of metaphysics implied a separation of religious belief from theoretical knowledge. Moreover, Kant's voluntarism was obvious from the emphasis on the will in his practical philosophy. Indeed, voluntarism was the 'true central point of Kantian philosophy: a *Weltanschauung* receives its deepest and most decisive impetus not from the understanding but from the will, that is from practical reason' (1901a: 347).

Recognising the importance of the will was inseparable from rejecting the idea of a soul-substance. Put differently, to recognise the importance of the will was to accept Wundt's 'actualistic' conception of the soul. The restless activity of the will was incompatible with any conception of the soul as a permanent, unmoving substance. Ironically, given later uses of the scholastics' term '*actus purus*', Paulsen employed this term to characterise what he regarded as Wundt's most important insight:

> [for Wundt] the soul is not a persisting thing, but living activity, not *substantia* but *actus*. Via its name – which reminds one of the *actus purus* of Scholastic philosophy – Wundt's 'actualistic' conception of the soul refers back to the Aristotelian entelechy: The soul is activity, not passive being like matter; God is absolute activity, pure energy, *actus purus*, that is, eternal thinking of that absolute thought that is reality itself.
>
> (1901a: 391)

Like Wundt, Paulsen also accepted psychophysical parallelism; he followed Wundt in regarding this principle as a consequence of the actuality principle (1901a: 97).

As the title – *Philosophia militans* – already promised, in his second major book-size statement of his religious and metaphysical views, Paulsen was more polemical, and more openly anti-Catholic. For instance, in the first chapter he insisted that the 'Protestant

Germanic peoples of England, Germany, the Netherlands, and the North...were enjoying a stable development, whereas the Catholic and Romanic peoples had fallen from a state of limitless absolutism into a state of revolutionism' (1901b: 24).

For present purposes, the second chapter of Paulsen's book, a chapter entitled 'Kant, the Philosopher of Protestantism' (1901b: 29–84), is particularly telling. To begin with, Paulsen reminded his audience of an 1899 letter by Pope Leo XIII to the French clergy. In this letter the pope had warned Catholicism of Kantianism as a modern philosophy

> that, under the pretext of freeing human reason from all prejudices and illusions, denies reason the ability to reach any knowledge beyond the nature of its own functions; [a philosophy] that therefore sacrifices, on the altar of scepticism, all those proofs that traditional metaphysics had given as the basis for the demonstration of the existence of God, the spirituality and immortality of the soul, and the objective reality of the external world.
>
> (Leo XIII, quoted in Paulsen 1901b: 32)

Paulsen read this statement by the pope as proof that Catholicism was unable to grasp the true importance of Kant's work. He also suggested that Kant could easily have been persuaded that the main critical thrust of his work was indeed directed against Catholicism and the 'mediaeval-Catholic school philosophy' (1901b: 33).

Paulsen submitted that there were three ways of thinking about the relationship between reason and religious belief. '*Rationalism*' believed that reason on its own could produce a 'system of absolute truth', a system that could also function as a religious belief. This was the position of Plato, Aristotle, and Hegel.

'*Semirationalism*' assumed that although some religious truths came to humans via divine revelation, human reason could vindicate the religious beliefs at least in their broad outline. One could thus distinguish between 'natural' and 'revealed' religion. Semirationalism was the position of Thomas and the Catholic church. Thomism was important to modern Catholicism because of its 'conciliatory' nature. It reconciled reason with theology by giving reason some limited role to play. At the same time, the 'highly complex dialectical system' of Thomism served the useful function of both 'training and exhausting reason' (1901b: 37).

Finally, '*irrationalism*' was the view that reason was incapable of dealing with any question of faith, and that revelation was the only source of religion. This was the position first formulated by Luther, and since then advocated by both Kant and modern Protestant theology. Its adoption also meant a 'return to the true "Christianity" of Jesus' (1901b: 40). Paulsen found three main parallels between Luther and Kant. First, both men insisted on the 'autonomy of reason'. In Luther's case this expressed itself in his stance towards all earthly authorities; Luther did not accept any of them in questions of faith. Second, both Luther and Kant were 'anti-intellectualistic' and 'voluntaristic'; they did not believe that speculative, theoretical reason could support religious faith through proofs and demonstrations. And third, both emphasised autonomy in moral questions: 'Neither reason nor conscience are bound by an external, human authority' (1901b: 43–7).

Such autonomy was of course anathema to Catholicism. The key principle of Catholicism was 'absolute authority', and such a principle could only 'kill individuality'. Or, put still more strongly, Catholicism tried to turn humans into mere 'automata'. It tried to make superfluous, and ultimately to destroy, the individual's conscience and reason. The endpoint of all this was likely to be 'perfect absolutism', or upon closer scrutiny, 'idiotism' (1901b: 78).

Wundt's views on Catholicism, Protestantism, and German mysticism

Turning from Paulsen back to Wundt, the first things to register in this context are that he was the son of a Protestant parish priest, and that in his memoirs he reported early experiences with the religious divide. For instance, the 13-year-old Wundt suffered 'from the unhappy confessional conflict' when he had to spend a year in a Catholic *Gymnasium*. The school was attended by very few Protestants, several of the teachers were Catholic priests, and a large number of the Catholic pupils intended to train for the priesthood. It seems that Wundt was badly bullied by some his Catholic teachers (1920a: 39–41).

In his later life, Wundt was always highly critical of Catholicism, scholasticism, and neo-Thomism. Catholicism was 'the religion of restriction' (*Religion der Gebundenheit*), and the Catholic believer was 'the prisoner of his church. He does not think for himself; he blindly follows authority and the power of tradition'. The very existence of Catholicism in his own time and age proved to Wundt that people were still steeped in '*superstition*, that is, the belief in miracles and magical effects' (1912 III: 219–20).

The main reason why Catholicism had survived to the present, Wundt insisted, was its lack of coherence. A mishmash of doctrines and ceremonies, Catholicism had always been willing to assimilate whatever primitive forms of worship and ritual it had happened to encounter. Strictly speaking, Catholicism thus was unified only as an organisation, not as a faith. Instead, it was 'an encyclopaedia of all religions'. Within it, one could find ancient 'soul worship', 'tutelary gods', 'rudiments of an age-old fetishism', 'nature worship', 'the godly figure of Christ, surrounded by a wealth of magical ceremonies deriving from ancient mysteries', and 'special cults surrounding the mother of God, the apostles, and the saints'. All this amounted to a 'polytheism that even the Greek religions could not have matched'. In more recent centuries, Catholicism had learnt to make compromises *vis-à-vis* science, too. Wundt predicted that it would not be long until official Catholicism would even accommodate its Modernist critics (1923: 540–1).

Wundt did not think much of scholasticism either. Its essence was compilation and eclecticism. The work of Thomas Aquinas was a case in point. Wundt chastised Thomas also for having been one of the earliest advocates of the contract theory of the state. Wundt suspected that this idea derived from an ethical individualism, which in turn was the direct result of Thomas' s assumption of an individual soul-substance (1897: 612; 1909: 165–6). In Wundt's view, not only was Thomas's system dubious in its metaphysics or its psychology, it was also dangerous in its politics.

It attested to the intellectual backwardness of nineteenth- and twentieth-century Catholicism that it had to go back 600 years to find a philosopher congenial to its thinking and answering to its needs (1912 II: 238). In Wundt's view, such 'Catholic philosophy' did not deserve anyone's support and respect. In any case, since the Renaissance all truly Christian philosophy had been Protestant (1920b: 365–6). Modern philosophy since Descartes and Bacon had been 'a Protestant science' (1920b: 366); and the Reformation had paved the way for science by 'liberating the minds' (1909: 172).

With respect to contemporary neoscholasticism, Wundt distinguished between 'official' and 'free' versions. The official version was the neo-Thomism advocated by the pope and faithfully repeated by the so-called 'Catholic philosophers' of the universities. By 'free Neoscholasticism' Wundt meant the philosophy of Brentano and his followers. We have seen in Chapter 4 that Wundt disliked their 'logicism' and 'intellectualism'; now we can add to this that Wundt also wished to emphasise a *religious* divide between himself and the Brentano school. Free neoscholasticism arose because of developments in the *Zeitgeist* of the second half of the nineteenth century. Many intellectuals were fed up with Kant and Herbart, and none of the new thinkers, such as Spencer, could fully convince. In short, 'one needed something new, but was incapable of producing anything new oneself'. The 'natural' solution was a return to the past. While few Catholic intellectuals were ready to submit to the dogmas of official Thomism, many nevertheless sought peace with the church. The compromise formula was Brentano's philosophy. Although Brentano 'was quite unable to develop an independent philosophy', his resurrection of medieval Catholic thinking fitted the needs of the time. It remained close enough to the official dogmas for individuals to feel secure and calm; it advocated what people of the time valued highly, namely philosophical, argumentative acumen; and it avoided conflicts with church and state. And, at the same time, it allowed its followers to believe in their independence and freedom (1910b: 599–600).

Although Wundt was occasionally critical of Protestantism, too, his overall assessment of its past and future was fairly positive. Wundt claimed for instance that the Reformation deserved credit for 'the destruction of the hierarchical system [of the Catholic church] and for the resurrection of the idea of the church as an inner community of faith'. Additionally, the Reformation had brought 'the inner redemption of religious feeling'; Luther and other Reformists had anchored religious feeling in the experiences of the individual consciousness (1912 II: 89). Even more importantly, 'Luther [had] created less a new *religion than a new ethics*'. According to this proto-Kantian ethics, activity was more important than contemplation, and the concept of duty was primary with respect to the concept of virtue (1912 II: 90–1). Indeed, Wundt held, like Paulsen above, that there was a direct parallel between Luther and Kant. Luther had been the first to 'raise the moral conscience of the individual to the position of a judge over his thinking and doing'. And the final philosophical codification of this 'viewpoint of moral freedom' had come with 'Kantian Idealism' (1920a: 352).

Moreover, Luther deserved credit for having emphasised the importance of the religious community: 'Luther replaced the supremacy of the Pope with the religious community; this was a step which renewed the spirit of community, a spirit that had been the soul of German antiquity and the older German mysticism' (1920a: 329).

Turning from the past of Protestantism to its future, Wundt repeatedly stressed that Protestantism had not yet fulfilled all of its promises. Protestantism represented 'the principle of freedom' in religious matters, but at present it still was guided in too many respects by the Catholic 'principle of authority'. True Protestantism translated into 'protest against any kind of confessional coercion', and thus Protestantism would come into its own only once it became '*a religion of free personal conviction*' (1912 III: 219–21):

> the aim towards which Protestantism is striving, on the basis of its inner principle of faith, is that of being a church without doctrines to which every member has to confess. This is not a contradiction in terms. For what defines the nature of a religious community is the form of worship, not the content of beliefs.

Protestantism had to take on this mission because religious beliefs were not likely to converge. What history showed instead was an ever greater diversity and plurality of beliefs (1923: 538, 550).

Above I quoted Wundt as saying that 'Luther created less a new religion than a new ethics'. Wundt justified this assessment by suggesting that much of Luther's religious ideas did in fact come from German mysticism.[2] This is noteworthy in light of the fact that German mysticism played a key role when Wundt formulated his own religious credo. In his brief statement of his own religious views in his memoirs, Wundt advocated especially the ideas of *Meister* Eckehart. As particularly important, Wundt regarded the idea that 'the human soul is completely one with God – provided it is thought of as completely purified from all those things to which it is tied…due to its embodied state'. Humans could experience this unity with God for short moments, and in these moments they were able to realise that true immortality lay not in a continued supernatural existence of one's soul, but only in one's uniqueness and God's immanence in the human soul. In fact, because there were no proofs of God's existence, the only certainty of God's existence could come from such experiences. God's transcendence could be known only through his immanence in the world and in human souls (1920a: 118–23).

Wundt called 'religious optimists' those believers who were content to forgo hope of individual immortality. 'Religious pessimists' were unable to live without the prospect of an individual eternal life. In line with Protestant voluntarism, Wundt suggested that the decision for religious optimism or pessimism was, in the last instance, a 'decision of the will', and this decision could not be justified or explained (1920a: 122).

Finally, like the Protestant theologians of his time, Wundt did not believe in secularisation. Religion was 'the driving force behind all culture' because the 'ideal of religion' was 'striving beyond the already achieved'. To believe oneself free of religion was to engage in some sort of self-deception (1920a: 223).

Protestant themes in Wundt's metaphysics and psychology

It is one thing to document that Wundt was anti-Catholic and shared some contentions with militant Protestants such as Paulsen; it is quite another to show that Wundt's views on religion and the confessions influenced his psychological theorising, and thus his stance in the controversy over thought psychology. I shall now turn to this latter task.

The clearest link between Wundt's Protestantism and his psychology was undoubtedly his pronounced 'voluntarism' – after all, for both Paulsen and Wundt voluntarism was strongly associated with Protestantism. The same point can also be made *via negationis*. Voluntarism for Wundt was the correct alternative to psychological logicism and intellectualism; and intellectualism was most characteristic of the Brentano school, that is of 'free Neoscholasticism'. As Wundt saw it, the Würzburgers followed Brentano, Stumpf, and Husserl; thus they 'had to' be committed to intellectualism, be opposed to voluntarism, and thus be direct or indirect supporters of religious sentiments that Wundt disliked. No wonder, therefore, that the link between Bühler's non-depictive thought and the scholastics' *actus purus* was so central to Wundt's criticism.

Wundt himself called his psychology 'voluntaristic psychology'. When explaining this title, he typically emphasised that voluntarism did not mean the attempt to reduce all mental events to volitions. Rather, psychological voluntarism was said to contain three assumptions. First, different mental processes, such as representing, feeling, and wanting, were always mere aspects of a unitary event. Second, volition had a 'representative importance' for many other subjective processes, in so far as these other processes could often be most clearly detected when they were part and parcel of a fully fledged intentional and volitional action. And third, the fully fledged intentional and volitional action could serve as something of a paradigm for all psychological processes. This was because in the case of individual actions, it was easy to see that they had the character of unique (i.e. unrepeatable) and dated 'events'. Taking actions as the model of theorising in psychology could thus save psychologists from the temptation of believing that the products of mental acts – for instance, representations – could be exactly reproduced at different times (1908b: 161).

Such cautious programmatic statements did not fully reflect the central role of the will and volitions in Wundt's overall theoretical edifice, however. For Wundt, the will was the key to understanding higher processes of thinking, the essence of the self, and the endpoint of the development of emotional life. The will was important in Wundt's theory of thinking because will and apperception were more or less identical, and because it was apperception that made systematic thinking possible. Recall that in Wundt's account of thinking, associative processes always produced a number of presentations, but that it was the human will, that is apperceptive processes, that chose amongst them (e.g. Wundt 1897: 41). Wundt could therefore write that 'all thinking is inner choosing'. Moreover, Wundt held that the only experienced continuity in inner experience was the 'continuum...of acts of willing'. From this Wundt concluded that 'from the viewpoint of psychology, *the will is the I*' (1914: 121), or, to cite another oft-quoted passage: 'There is nothing outside or inside the human being of which he could say that it is fully his own – except for his will' (1897: 377). And finally, emotions were mere 'preliminary stages' with respect to the development of acts of willing. Volitions and emotions were complexes of elementary feelings, and each one of them could be captured via a characteristic curve through the three dimensions of pleasure versus pain, excitation versus relaxation, and tension versus release. Because volitions were more complex than emotions, the latter were 'preliminary stages' for the former (1903a: 107). None of these views was of course shared by the Würzburgers.

The will also figured centrally in Wundt's metaphysics. According to Wundt, metaphysics after Kant could no longer be a priori speculation. Instead, the metaphysician had to become acquainted with the state of the art in the sciences, and then build theories by speculatively extrapolating from the best scientific theories to date. Not surprisingly, in Wundt's metaphysical treatises

System of Philosophy (*System der Philosophie*, 1897) and *The Perceivable World, and the World Beyond* (*Sinnliche und übersinnliche Welt*, 1914) his own psychology was treated as one such best theory. Now, because psychology taught that the empirical will was identical with the empirically given 'I', metaphysics was licensed to conclude that an individual '*pure will*' was the metaphysical essence of the individual human being. Psychology had to presuppose this pure will 'as the ultimate basis of the unity of all mental processes'. But psychology was not permitted to use this metaphysical entity for explanatory purposes. Put differently, 'pure will' was an 'idea of reason' (*Vernunftidee*), not an observational term (1897: 379).

Such pure will marked an important deviation from traditional conceptions of the soul as a substance. Such received conceptions were intellectualistic and static; they took their starting point from presentation rather than the will, and ended up conceiving of the soul as something impenetrable, isolated, and unchanging. The state of the art in psychology no longer allowed for such a view (1897: 385).

Wundtian metaphysics also extrapolated from the insights of collective psychology. This led first to the assumption of an interaction between pure wills. Here Wundt believed that he could give a metaphysical explanation for how 'presentations' originated; presentations came about once one pure will tried to influence another:

> the presentation is a result of the plurality of wills: On the one hand, it is a product of the interaction between wills; on the other hand, it is an auxiliary tool by means of which the willing elements combine with one another to form higher units of willing.
>
> (1897: 395)

Finally, metaphysics could also make out something about the ultimate building blocks of the world, the ultimate cause of the world, and the history of the world. First, because the world was ever changing, that is active, and because for the metaphysician activity equalled volition, therefore the metaphysical building blocks of the universe had to be elementary pure wills. Second, because cause and effect always had to be similar, the ultimate cause of the world could not be thought of independently of 'the world's content'. Thus, because the content of the world was a plurality of wills, the ultimate cause of the world had to be 'God *qua will of the world*'.

And third, the history of the world was the development and unfolding of the divine will. Wundt wrote 'that the idea of God is here transformed into the idea of a highest will of the world, a will of which the individual wills partake, but which also leaves them with a separate independent sphere of activity' (1897: 434). As these passages make clear enough, Wundt's metaphysics was not only voluntaristic but also monistic. Over and above wills there was nothing.

Wundt also had an answer to the traditional metaphysical question *vis-à-vis* the will: that is, the question of whether or not the will of the human individual was free. As Wundt saw it, he had arrived at a position beyond determinism and indeterminism. The indeterminist thought of the will as being prior to its motives, that is the will as being able to choose between its motives. The determinist, on the other hand, conceived of the will as secondary with respect to its motives, that is as determined by them. Both views were equally mistaken, however, because motives were inseparable from the will; they were neither causes nor effects, they were 'factors of the will' (1918: 355). Put differently, both determinism and indeterminism were wrong in thinking of the relationship between the will and motives as a relationship of physical causality. The correct view was that this relationship was one of mental causality. Neither motives nor the will were Humean causes; instead, they stood in some sort of internal, meaningful relationship to one another (1912 III: 39–40).

Wundt's overall position concerning the freedom of the will was compatibilist. What made a human choice free was not the absence of all determination; what made a human choice free was the presence of the right kind of causality. It was free 'if it happens with *level-headed self-consciousness*....[This means]...a consciousness of one's own personality and of the traits that this personality has acquired in and through the past development of the will' (1912 III: 39). A choice or action was unfree if it happened by chance, and undetermined by the actor's personality (III: 53). Here we need to remember that the concept of personality was one of the key concepts of turn-of-the-century Protestant theology. For the theologians, the further development of culture depended on the emergence of personalities that could combine the production of cultural goods with ethical excellence and religious devotion (Graf 1989). It is striking that Wundt employed the same concept not just in his theory of the freedom of the will, but also in his political philosophy; there he spoke of the individual and of the state as the only two real forms of personality (1912 III: 326–7).

In one important respect Wundt differed from other Protestant philosophers such as Eucken and Paulsen, however. Wundt did not share their admiration for Kant's epistemology. This was made most dramatically clear in Wundt's paper 'What Kant should not be to us' (*Was uns Kant nicht sein soll*, 1892b). Wundt's answer was that Kant 'should no longer be treated as alive by the living'. One could still learn from Kant's mistakes, but one could no longer accept him as an authority (1892b: 2, 6). Elsewhere Wundt even suggested that neo-Kantianism was no better than neo-Thomism; both were forms of '*authority philosophy*' that he regarded as unworthy of thinking philosophers (1909: 266).

Kant's epistemology could no longer be accepted. Psychological research had shown that space and time were not 'originally separate...forms of order for sensations, that is, forms that enclosed the sensations' (1909: 344). Although these forms had a considerable degree of constancy, this constancy was not sufficient reason for treating them as a priori (1897: 140). Kant had failed to tackle the important problem of studying the relations between the forms and various qualities of sensations (1909: 344). He had also overestimated the importance of time. Moreover, Kant's table of the categories of understanding was also obsolete. His way of determining the categories on the basis of the traditional forms of judgements was artificial and unconvincing. Equally unsatisfactory was Kant's way of arriving at the *Ding an sich*:

> It is impossible to understand how the fact that sensations are *given* to us can prove the existence of a reality independent of us. This would be a plausible inference only if we first assumed that these sensations are given to us by something outside of our consciousness. But to make this assumption is to presuppose what is to be proven.
>
> (1909: 345)

Finally, the most implausible aspect of Kant's edifice was his equation of the things-in-themselves (of his epistemology) with the noumenal free will (of his practical philosophy). Wundt suspected that Kant had introduced the whole distinction between a 'world of being' and a 'world of appearance' into his epistemology only because he would later need this opposition in his practical philosophy. In Wundt's view the equation of things-in-themselves with free wills made things-in-themselves 'a mysterious idea...that reminds me of Neoplatonistic notions of emanation' (1909: 345).

In opposition to Kant, Wundt demanded a realistic epistemology. Such epistemology rejected the Kantian opposition between appearance and reality, and it took its lead from the results of physiology and psychology (1909: 419).

The neo-Thomists' criticism of Wundt's philosophy and psychology

We have seen above that F. Paulsen, perhaps *the* most outspoken 'Protestant philosopher' of the time, shared with, or adopted from, Wundt a number of central contentions, such as the principle of actuality, psychophysical parallelism, and voluntarism. Paulsen regarded these very contentions as befitting of a Protestant philosophy. Interestingly enough, Catholic critics of Wundt attacked these very same views as insufficient, atheistic, or materialistic. Reviewing their criticism of Wundt thus strengthens my thesis, according to which Wundt's philosophy and psychology were 'Protestant' by the standards of the time. Introducing the neo-Thomists' criticism of Wundt is also important for my argument for a further reason. We need to take note of the neo-Thomists' main lines of attack against Wundt to be able to recognise later in this chapter that Külpe's and Messer's qualms about Wundt's views were similar to those of the neo-Thomists.

Before turning to the neo-Thomists' critique, we might note that despite all their scepticism with respect to Wundt's edifice, they still valued him much more highly than they valued the neo-Kantians. For instance, in 1914, C. Gutberlet, the most prolific neo-Thomist writer of the time, criticised the neo-Kantian C. Joël because the latter had lamented the low standards of philosophy of the time. Joël had suggested that only a still stronger neo-Kantian movement could provide a remedy. Gutberlet countered by calling Kant 'the real cause of the present philosophical chaos' and chastised Joël for neglecting Wundt's work. While not directly endorsing Wundt's views, Gutberlet insisted that it had much greater 'philosophical significance' and influence than anything the neo-Kantians had produced (1914: 339). Sixteen years earlier, Gutberlet had called Fechner, Lotze, and Wundt 'the most important representatives of non-ecclesiastical philosophy of our time', and continued:

> they are really serious about the truth; they want to gain and offer an explanation of the world that is based upon the current state of science. If only they did not have such enormous

> prejudices against Christian metaphysics; such prejudices are especially visible in Wundt's case. Wundt tries to put down the Scholastics wherever he can.
>
> (Gutberlet 1898: 127)

The first issue on which Gutberlet disagreed with Wundt was the freedom of the will. Gutberlet defended an indeterministic position and censured Wundt for his alleged determinism. Wundt's position was superior to that of traditional determinism in so far as Wundt had 'accepted' the existence of a distinct mental causality alongside mechanical causality. But Wundt had not arrived at the correct viewpoint, according to which 'the adequate cause of a decision is the influence of the *motives*, on the one hand, and the power of the *free will*, on the other hand' (Gutberlet 1907: 261). Wundt's suggestion that to act freely was to act in accordance with one's personality or character was not satisfactory. On this theory Gutberlet wrote:

> I never have the possibility of interfering with my willing: I am the plaything of my character and those motives that happen to be effective. And [on Wundt's premises] I do not have control over the formation of my character; every decision concerning my character is determined by earlier internal and external states. There is no way for this [Wundtian] determinism to escape fatalism.
>
> (1907: 267)

Gutberlet and other neo-Thomists were no less critical of Wundt's actuality theory of the soul, and thus of Wundt's rejection of the idea of a soul-substance. As Gutberlet saw it, one of the major divides in contemporary psychology was that between '*substantialistic*' and '*actualistic*' psychologies. 'Actualists' were 'psychologists without a soul' (ibid.), rejected the notion of a psychological unconscious, were opponents of mind–body interaction, and advocated psychophysical parallelism (1908: 6). This was of course the Wundtian position.

As concerned the 'psychology without a soul', Gutberlet had already claimed in 1898 that its existence was indicative of the low standards of philosophy of the time: 'What philosophical or objective value can one possibly attribute to a "psychology without a soul"? That recently one is even proud of it...is a sign of the deep decay of philosophical knowledge' (1898: 128). A quarter of a century later Gutberlet wrote that 'the denial of the soul leads inevitably into either materialism or Spinoza's pantheism' (1923: 26). The actuality theory was one more unfortunate outgrowth of Kant's philosophy: 'the psychology without a soul, the destruction of all metaphysics, and a general phenomenalism, these are logical developments of Kant's doctrine of the I' (1899: 122). Kant's position was that of 'a psychologism of the "I"' (1899: 131).

For Gutberlet and his neo-Thomist colleagues, 'the psychological *I*' was 'the soul, or the soul-substance' (1899: 105). Gutberlet admitted that the soul-substance was 'unknown' in some sense; but he insisted that in this respect the soul was no different from the ether of natural science. Moreover, one could make reasonable inferences about the nature of the soul on the basis of its 'expressions and activities'. For instance, the soul-substance had to be as spiritual as were its expressions and products (1901: 359).

In arguing for the soul-substance, neo-Thomists often relied on arguments about the unconscious. J. Geyser (1908b) reasoned that conscious events had to have causes, and that such causes were not introspectively accessible. These causes therefore had to be unconscious. But they could not be material: mental and material events were *toto genere* different, and this precluded the possibility that mental events were determined by material events. Thus the causes of mental life had to be unconscious mental causes. Geyser concluded that

> this unconscious is the soul.... Wherever there are activities, there necessarily is an actor. Now we have proven that in us there exist unconscious activities, and that these unconscious activities are of a mental nature. We call *soul* this unitary and persisting real ground of all mental activity, this ground upon which rests individual conscious content in its lived totality.
>
> (1908b: 83)

Geyser also called 'the mental unconscious...the real subject itself, the soul with its innate general and individual dispositions'. He counted 'mental acts' within its realm. Geyser took it to be an argument against the 'monistic-parallelistic theory' that it had not come up with a convincing account of the unconscious (1902: 119).

Other neo-Thomists agreed. E. Grünholz submitted that Wundt had no good arguments against the unconscious. 'The serious and honest thinker Oswald Külpe' had shown this to Grünholz' satisfaction (1913: 317). And Gutberlet wrote that 'no one can, with even a semblance of justification, reject unconscious states and powers'. Such unconscious states and powers were necessary *inter alia* to explain memory (1896: 17).

Elsewhere Geyser added that the actuality theory did not fit phenomenological experience either. Humans experienced an inner unity, a unity of the self, and this experience attested to the existence of a soul-substance. According to Geyser one had to give up not only the actuality theory but also the other Wundtian doctrines that followed from it (1902: 110).

One of these related doctrines was voluntarism. To Grünholz, Wundt's voluntarism constituted something of a category mistake. Rather than adopting a neutral soul-substance as the bearer of all mental activities and states, Wundt identified the bearer with one kind of mental content, that is with volition. This came out clearly when Wundt equated the will with the ego-subject (Grünholz 1913: 313). And, as if to make matters worse, Wundt's voluntarism ruled out epistemological realism. Wundt collapsed presentations and thoughts into volitions, and thereby destroyed the distinction between subject and object. The latter distinction made sense only if presentations and their objects existed independently of the will. And only with the distinction between subject and object could one make sense of realism in epistemology. (1914: 13).

Wundt's treatment of the will contradicted everything Grünholz knew about the soul; he was convinced of 'the existence of an independent, real soul, for which thinking and willing form two completely independent and unique parts of mental life'. And the existence of this 'real soul' in turn lent support to the idea of 'a dualism of the physical and the mental'. Wundt's contrary position was part and parcel of an historical trend that had started with Kant and that was currently ushering in 'a total subjectivism and relativism in all areas' (1914: 20).

Wundt's adherence to psychophysical parallelism, as well as his rejection of psychophysical interactionism, were also unacceptable to his neo-Thomist critics. To Gutberlet it was so obvious that presentations influenced bodily processes and vice versa that Wundt's advocacy of parallelism could be explained only as being due to 'self-deception' (1899: 147) and 'persistent blindness':

> Only with persistent blindness is one able to look upon the mental and the material as the inner and outer sides of one and the same process: The mental and the material are radically different, each so totally unique that no difference of

> viewpoint, no mere process of abstraction can explain their character.
>
> (Gutberlet 1899: 155)

Standard arguments against mind–body interactionism, arguments that had been used by Wundt too, cut no ice with Gutberlet. For instance, it was no good using the law of the conservation of energy against the possibility of interaction. The soul's acting upon the body did not violate the law by creating new energy, and the body's acting upon the soul did not contradict the law by losing energy. Such thinking was flawed because 'the spirit is not governed by the law of conservation of energy'. Moreover, because soul and body formed a substantial union, it was, strictly speaking, wrong to speak of 'interaction' between soul and body anyway (1908: 6–7): 'The only unobjectionable explanation of the causal nexus is offered by the Aristotelian-Scholastic viewpoint, according to which body and soul are linked in a substantial unity' (1899: 147–8).

Gutberlet was also unhappy with Wundt's treatment of Christianity in his *Völkerpsychologie*. Gutberlet complained that Wundt had failed to recognise that Christianity was different from all other religions:

> Christian religion has very prominent, characteristic features that make it incommensurable with all other forms of religions. Already its content...sets it apart from all of the fictions, absurdities, contradictions, superstitious and immoral views, and ceremonies of the heathen; moreover its divine origin is attested to by supernatural criteria, whereas the human or mythical origin of all other religions is obvious to anyone who is willing to see.
>
> (1904: 327)

Gutberlet added: 'After reading Wundt's other writings I would not have thought possible such incompetent product' (1904: 328).

And finally, Gutberlet deeply disliked Wundt's voluntaristic metaphysics. To begin with, it was 'incomprehensible' how presentations could ever result from an interaction between wills. Moreover, Wundt's account of history, as a progress towards ever more cultural heights, was naïve and unfounded. Especially unfounded was the assumption of an infinite progress. Humankind was more likely to move towards 'intellectual shipwreck' (1891: 344–5). Gutberlet found Wundt's theory of God as the will of the world 'nebulous and incomprehensible', and he admitted that he had not been able to make much sense of it all, despite 'multiple rounds of reading and thinking and comparing different passages'. Gutberlet concluded: 'Compared with the philosophical giants of the past, our present-day philosophers appear as a speculatively impotent philosophical race' (1891: 357).

Külpe's criticism of Wundt and Kant

As I shall document in this section, in many key respects, Külpe's criticism of Wundt's thinking was fairly close to the neo-Thomists' fault-finding. This might seem astonishing in light of the fact that Külpe was both Wundt's student and a Protestant. Moreover, until his move to the Catholic Würzburg in 1894, Külpe was strongly influenced by Avenarius and Mach, and thus his views ought to have provoked at least as much resistance amongst neo-Thomists – for instance, in Würzburg – as did Wundt's. After all, Mach and Avenarius rejected all metaphysics, whereas Wundt did not.

The available published and unpublished sources do not allow for a conclusive resolution of this somewhat paradoxical situation. What we do know, however, is that Külpe was not the first choice of the philosophical faculty in Würzburg. A letter by the dean to the senate (of 2 February 1894) listed R. Falckenberg (Erlangen), K. Groos (Giessen), and G. Martius (Bonn) as the top candidates, and regretted that the very best person, T. Lipps, a Catholic, was likely to go to Munich. As further documents in Külpe's personal file indicate, Falckenberg was opposed by the ministry because he was suspected of a Jewish background (letter of the Bavarian ministry of 7 May 1894). Over the next four months, the faculty searched for a more acceptable candidate, and eventually united behind Külpe.[3] Perhaps Külpe's predecessor, J. Volkmann, and Wundt had acted behind the scene to make this possible. Volkmann's professorship became vacant because he was moving to Leipzig. On 1 January 1894, Volkmann reported to Wundt that

> the situation in Würzburg will be difficult. The Catholic Ultramontanists will try to use the occasion to promote Stölzle [a neo-Thomist] to the position of full professor and head of department....But I assume that the faculty and the senate will successfully resist this move.
>
> I have already considered the issue of my successor. Could you tell me more about Dr Külpe whom you have mentioned

> with so much praise? What is needed here is first and foremost a good teacher.
>
> (Wundt Archiv)

Wundt's reply has not survived but in his next letter Volkmann wrote (6 January 1894):

> I thank you very much for the information about Dr Külpe. This information will be most valuable in the negotiation about my successor.
>
> (Wundt Archiv)

Despite the absence of hard evidence, the available source material allows for a number of conjectures. First, to hire the Protestant philosopher and psychologist Külpe in 1894 was not a particularly risky move for the university in Würzburg. Although a Protestant, contemporary sources attest to the fact that Külpe was known to be 'a *religious* Protestant' (Stumpf 1924: 11; my emphasis), as well as 'careful and scrupulously cautious' in theoretical matters (Baeumker 1916: 90). The university in Würzburg was perhaps the most progressive of all Bavarian universities at the time; it had employed the apostates Brentano and Stumpf and was the centre of the German modernist movement. Moreover, after the publication of his *Grundriß*, there could be little doubt that Külpe was one of the leading psychologists in Germany, no doubt the leading one without an *Ordinarius* position.

Second, in light of what the sources tell us about Külpe and the conditions in Würzburg, it does seem possible and probable that Külpe soon adjusted his philosophical views to his Catholic environment. There is no reason to assume that Külpe did not do so for 'good and sound reasons'; my adjustment-thesis does not attribute tactical motives to Külpe. It is conceivable, however, that Külpe was helped on his way towards changing his views by the fact that a philosophy professor in Würzburg needed to attract theology students if he wanted to secure large audiences. In his memoirs, Stumpf reported that he had great difficulties working in Würzburg between 1869 and 1879 for just this reason: Because he was known to be 'a fallen Catholic', theology students stayed away and he was left with few listeners (1924: 11). Be this as it may, if my adjustment-thesis is roughly on the mark then Külpe scholars need no longer suspect that a sudden onset of neo-Kantianism moved Külpe from positivism towards realism (Lindenfeld 1978). In more than one respect this seems an implausible idea.

Third, there can be little doubt that Külpe's thinking was soon regarded – by his contemporaries – as congenial to Catholicism in general, and neo-Thomism in particular. A presentation of the detailed evaluation of Külpe's views by the neo-Thomist M. Grabmann will have to wait until after we have first acquainted ourselves with Külpe's epistemology and metaphysics. But some other evidence can be already adduced here. On the one hand we might note that Külpe was never attacked by, and never attacked, any of the neo-Thomists of the time. On the other hand, it deserves mentioning that of the calls to professorial chairs that Külpe received between 1904 and 1914, most came from universities in predominantly Catholic areas. Amongst these universities were Bonn and Munich – these were calls that Külpe answered in 1909 and 1913 – as well as Münster, Breslau and Vienna. Külpe also negotiated about the possibility of a move to Berlin in 1908, but here his contact was the aforementioned Catholic apostate Stumpf with whom Külpe shared numerous views in psychology, epistemology, and metaphysics (Külpeana, V: 18).

Before turning to those of Külpe's views that he shared with the neo-Thomists, it is worth noting that with respect to two important issues he took up positions quite different from them. Like Wundt he denied that there were any convincing proofs for the existence of God (1910a: 272), and like Wundt he spoke out against indeterminism of the will. Indeterminism was based on a series of 'confusions and misunderstandings' and was no longer compatible with the state of the art in psychology. Freedom of the will made sense only in a rather limited sense. The will was free provided it was not determined by factors outside of the agent, that is provided it was determined only by the agent's own 'task and determining tendencies' (1910a: 251). To the best of my knowledge, none of the major neo-Thomists ever chastised Külpe for holding these views.

What then were Külpe's views that brought him the sympathy of the neo-Thomists? And what was his criticism of Wundt's Protestant traits?

To begin with the divide between 'actualists' and 'substantialists' *vis-à-vis* the soul, we already know (see Chapter 4) that Külpe sided with the latter against the former. Külpe did not believe that the realm of consciousness coincided with the realm of the mental, and he insisted on 'the phenomenality of the inner world'. For Külpe, rejecting Wundt's actuality principle meant adopting psychological realism and an epistemological stance that allowed for the postulating of theoretical entities beyond the given. To this earlier account, we can now add that Külpe also sought to rebut Paulsen's and Wundt's specific arguments against the need for, and plausibility of, a soul-substance. To the argument that the soul-substance could not be observed, Külpe replied that atoms were not observable either. To the objection that the link between the soul-substance and particular mental processes was dubious, Külpe retorted that the link between mental and physical phenomena in Paulsen's and Wundt's psychophysical monism was no less unclear. Moreover, it was unfair to allege, as Paulsen did, that all the substantialists could say about the soul-substance was that its properties were unknown. The substantialists were entitled to say that the soul-substance possessed independence, reality, and unity; and these were not 'negatives'. Furthermore, Wundt was wrong to claim that the assumption of a soul-substance was incompatible with mental change and development. Wundt's argument was plausible only if one assumed an unchanging soul-substance. Finally, Wundt was mistaken also in his assertion that the assumption of a soul-substance was useless for the purposes of psychological explanation. As Külpe saw it, it remained to be seen in the future whether the soul-substance could be used for explanatory purposes. At the same time, Külpe admitted, however, 'that scientific psychology is not yet mature enough for it to be able to make specific assumptions about the nature of the soul' (1910a: 277–83).

A second area of contention in which Külpe sided with the neo-Thomists against Wundt was the issue of 'interactionism versus monistic parallelism'. Four years after his move to Würzburg, Külpe published a paper entitled 'On the Relationships between Bodily and Mental Processes' (*Über die Beziehungen zwischen körperlichen und seelischen Vorgängen,* 1898). In it he defended the coherence of dualism against standard objections by Wundt and others. Interestingly enough, not only did Külpe's paper come four years after taking up his new post in Würzburg, it also came two years after the 'free neo-Thomist' C. Stumpf had made public his views on 'body and soul' (*Leib und Seele,* [1896] 1910). In much greater detail than any of the 'official' neo-Thomists had done, Stumpf defended interactionism and attacked Wundtian and other forms of parallelism. Stumpf's main argument against parallelism was that 'the parallel running of the two worlds [of the mental and the physical, according to psychophysical parallelism] is no less incomprehensible than it is according to the disreputable doctrine of Geulincx and Malebranche' ([1896] 1910: 78). As concerned objections to interactionism, Stumpf was most eager to point out that interactionism could be made to cohere with modern physics, especially with the law of the conservation of energy. For instance, the body's acting upon the mind did not amount to a total loss of a quantum of physical energy because mental events were mere '*side-effects*' of physical events. And these side-effects did 'not absorb physical energy' ([1896] 1910: 83).

In his paper, Külpe followed Stumpf's lead, although he did not accept Stumpf's 'side-effect solution' to the problem of how interactionism could be reconciled with the law of the conservation of energy. Külpe suggested instead that the assumption that mental processes were somehow 'free of energy' be dropped. All problems with the law of the conservation of energy disappeared once one accepted that, in addition to physical and chemical energy, there also was mental energy, and that all three forms of energy could be transformed into one another (1898: 110). Overall, Külpe's advocacy of interactionism in this early paper was rather guarded. He seemed to be more concerned about establishing interactionism as possible rather than advocating it as highly probable or true.

By 1910, Külpe's position had shifted. He now was ready to support interactionistic dualism more strongly. He wrote that dualism fitted 'much better with natural science and psychology than materialism and spiritualism' and went on to explain why:

> It does justice not only to the real differences between body and soul, but also to the empirically known relationships of dependence between body and soul. It harmonises with the aspirations of modern biology to pass beyond the one-sidedly mechanistic interpretation of life processes...and it is able to comprehend the phenomena of consciousness as necessary elements in the development of living beings. It no less agrees with epistemology, for epistemology demonstrates that subject and object are the two phenomenological factors of full experience. And finally, dualism keeps away from the arbitrary extending of the mental, an extending that characterises spiritualism and monism.
>
> (1910a: 203–4)

A third issue on which Külpe and the neo-Thomists saw eye to eye was their assessment of Paulsen's and Wundt's voluntarism. As Külpe judged the case, there were no good reasons to assume that 'the will is the primary function of the soul'. Neither did such good reasons exist now, nor was it likely that such reasons would ever be found. Külpe took on three 'priority' claims that Paulsen and Wundt had made for the will. According to the first, the will was primary in an evolutionary sense; before organisms developed any other mental states, they 'had to' develop a will. This argument did not convince Külpe because there was no empirical evidence for such evolutionary priority. In any case, it was much more likely that various mental functions had emerged from a prior undifferentiated mental life, in which will, feelings, and presentations formed a whole. According to the second priority claim, the will possessed 'psychological priority'; the will determined one's goals in life and thereby shaped one's personality. To this Külpe objected that the will had to choose between something or other, and this 'something or other' had to be other mental entities, such as presentations or thoughts. The third, 'metaphysical', priority claim Külpe attributed solely to Wundt. He referred here to Wundt's metaphysical stance, according to which the essence of the human being is a pure will. Külpe was unconvinced that such '*actus purus*' (Külpe 1902: 96) could exist even according to Wundt's own premises. After all, following Wundt's own psychology, volitions could always be analysed into sensations and feelings. Moreover, Wundt's attempt to 'deduce' metaphysically the existence of presentations from the interaction of pure wills was 'not sufficiently motivated' (1910a: 287).

Although Külpe refused to replace voluntarism by intellectualism (1910a: 287), his warm advocacy of the intellect in some of his writings on the one hand, and his linking of the intellect to scholasticism on the other hand, was telling of his sympathies. In one place he wrote that 'acceptance of the intellect, and emphasis on the spiritual independence of our soul has always given wings to thinking', and he went on to mention, as one of his prime examples, 'the intellectualistic perspective of the culmination period of mediaeval Scholasticism' (1912c: 1107; cf. also 1912a: 33–4).

In some places Külpe categorised Wundt as a 'spiritualistic monist' rather than a 'voluntarist'. Spiritualistic monism claimed that the essence of reality was spirit rather than matter. Külpe submitted that this position was plausible only if one already accepted the actualistic position. Only if one believed that inner perception gave one a reality-as-such, only then could one come to conclude that what one found as *the* reality in inner perception was also the cause of the appearances in outer perception. To Külpe's mind, Wundt's spiritualism was 'a return to errors that have been overcome long ago' (1910a: 194).

Külpe's thought-psychological realism

Külpe's emphasis upon the intellect brings us to the fourth, and final, issue on which Külpe and the neo-Thomists could agree wholeheartedly. I mean the issue of epistemological realism. The link between intellect and realism was provided by non-depictive – Würzburg-style – thinking: proper appreciation of the intellect meant a proper appreciation of non-depictive thinking; and proper appreciation of the nature of thinking translated naturally into an acceptance of epistemological realism. And finally – to many philosophers in turn-of-the-century Germany – to endorse epistemological realism was to side with Aristotle and St Thomas against Kant, German idealism, and the neo-Kantians. Despite Wundt's anti-Kantian stance, Külpe added him to the list of anti-realists. I shall now explain these various links and associations in more detail.

Külpe's main project between around 1900 and his death in 1915 was the development of an epistemological defence of 'critical realism'. Major stages in this project were a book-size criticism of Kant's epistemology (1907a), a paper entitled 'Epistemology and Natural Science' (*Erkenntnistheorie und Naturwissenschaft*, 1910b), a study entitled *On the Doctrine of the Categories* (*Zur Kategorienlehre*, 1915b), and the three-volume work *Realisation* (*Die Realisierung*, 1912a, 1920a, 1923a), of which the second and third volumes were edited posthumously.

According to Külpe, his own 'critical realism' could be distinguished from various other epistemological stances by means of the following four questions. The first question asked whether science was ever justified in positing anything as a real entity, that is as an entity that was not given in original, immediate experience. Realist epistemologies answered 'yes', 'conscientialism' (*Konszientialismus*) answered 'no'. 'Conscientialism', a term invented by Külpe, thus covered various forms of 'immanent' philosophy, positivism, idealism, and – at least within the realm of the psychological – Wundt (1910b: 1027). Indeed, Külpe regarded it as a major inconsistency in Wundt's thinking that Wundt wished to be a realist with respect to natural science but a conscientialist with respect to psychology. Here it was of course Wundt's actuality theory of the soul that committed him to 'psychological conscientialism' (1910a:

191). While Wundtian conscientialism was anti-realist from one perspective – in so far as it denied realities beyond the given – it could also be regarded as a form of 'naïve realism' from a different standpoint: After all, it took the immediately given for complete reality (1912a: 168).

The second key epistemological question applied only to those who had answered 'yes' to the first. It was: 'How is it possible to posit entities as real?' Different realist stances might give different accounts of what, for instance, justified one's belief in an external world (1910b: 1027).

The third question was: 'Can we ever determine the properties of the posited real entities?' This question separated 'general' from 'special' realism. Kantian phenomenalism would answer negatively to this question; although it accepted the positing of real 'things-in-themselves', it insisted that humans could never make out their properties. Kantians thus accepted a general realism, but rejected special realism, that is a positive answer to the third question. Külpe himself advocated special realism (1910b: 1027–8).

Finally, the fourth question differentiated between various brands of special realists. It read: 'How is a determination of the real possible?' To answer this question in a satisfactory fashion demanded that one become a 'critical realist'; that is, that one understood the nature of thought. Real entities were neither mere concepts nor conscious contents of perception. Instead, real entities were 'as Plato put it so well, thought entities [*Gedankendinge*]'. It therefore needed an 'epistemological appreciation of *thinking* as the organ by means of which real entities are posited and determined'. Real entities were not only things like atoms or stars; they also included 'the ideal and fictive objects of which mathematical science makes so frequent use' (1910b: 1028).

As these last statements already indicate clearly enough, for Külpe a sound epistemological realism was inseparable from accepting the results of Würzburg thought psychology. Realism presupposed a human faculty by means of which knowers could free themselves from being confined to their intra-mental world of sensations, presentations, and feelings. As long as knowers remained on the level of sensations, presentations, and feelings, they never had reason to posit anything 'beyond', anything 'real'. Obviously, the needed faculty was thinking. But not just any account of thinking would do. If thinking itself was nothing but a sequence of sensations, feelings, and presentations, then nothing was gained. Only a thinking with its own independent laws, only a thinking for which sensations, feelings, or presentations were accidental vehicles at best, only such thinking could justify the assumption of a mind-independent, real world. At the same time, however, thinking had to be informed by sensations, presentations, and feelings, otherwise its constructions and assumptions would be without any checks. In Külpe's words:

> experience *and* thinking must cooperate in order to posit and determine real entities. To the question of whether or not there are such entities, experience holds the key; but only thinking can abstract from all of the subjectivity that so richly attaches to our mental experiences....For the natural scientist, neither the totality of experience nor any of its parts is tantamount to the realm of the real....Sensations do not fall, do not attract one another, and do not repulse one another....We can understand the realism of natural science only as a product of both experience and thought, of both the reality of consciousness and reason-based consideration, and of both sensations and the activity of the understanding.
>
> (1910b: 1029)

Külpe did not find this 'thought-psychological' realism in Kant, the neo-Kantians, or Wundt. I have dealt with Külpe's criticism of Wundt's psychological conscientialism already, but some aspects of his comments on Kant and the neo-Kantians are worth noting. Külpe's Kant book of 1907 criticised Kant's epistemology in great detail. With respect to Kant's doctrine of space and time, Külpe argued that Kant had failed to provide convincing arguments for his phenomenalistic conclusions. Kant had not shown that the a priori character of time and space justified regarding time and space as subjective. Nor had he established the further claim that time and space somehow stood between the knower and the known (1907a: 49–79).

More importantly for present concerns, Külpe claimed that Kant's account of thinking was badly flawed. Kant was wrong to assume that all thinking was tied to his twelve categories. If thinking were tied to Kantian categories, Külpe maintained, then thinking could concern only well-ordered materials. But this was not the case: thinking could be about anything, even 'the orderless, chaotic material of sensations, a material that Kant assumes as the raw material of sensory perception'. Moreover, Kant's theory failed also to explain how thinking could be reflective. Kant had no answer to the question: 'By means of which forms of thought do we think about the categories?' (1907a: 80).

Külpe therefore insisted, against Kant, that there existed no 'subjective forms of thinking that are operative in the understanding once and for all'. The very opposite was true: 'Whatever determines thinking comes *only from its objects*'. This was not to say that thinking had no inner direction or determination; but such direction and determination came from the *tasks* to which subjects could commit themselves. And whether or not a task could be solved was again determined by reality, not by the subject (1907a: 80).

Elsewhere Külpe located his own realistic epistemology historically. Basically, there were three ways of thinking about categories: the Aristotelian, the Lockean, and the Kantian. For Aristotle, categories were objective: 'predicates that apply to beings'. For Locke, categories were subjective: 'mere givens of consciousness, formed according to psychological laws'. Finally, for Kant, categories were both objective and subjective. They were objective in so far as they were not formed by the individual, and subjective in so far as they were not true of things-in-themselves. Külpe sided with Aristotle (1915a: 4, 52–58).

The neo-Thomists on Külpe

I have now explained the link between Würzburg thought psychology and Külpe's realism. We have also seen how thought psychology informed Külpe's attack on Wundt's epistemology. It remains for me to link up Külpe's critical realism with Catholicism and neo-Thomism. This connection was belaboured in great detail by the neoscholastic philosopher M. Grabmann in two studies in 1913 and 1916. Lest his positive assessment of Külpe is suspected of being an isolated case, some other approving comments by other neo-Thomists are perhaps worth alluding to, at least briefly. I have already mentioned that Grünholz used Külpe's criticism of Wundt's epistemology and psychology in his own attacks upon Wundt (1913: 13; 1914: 317). C. Schreiber used Külpe's critique of Mach's phenomenalism as a confirmation of his thesis that the best epistemologists were returning to the epistemology of St Thomas. Schreiber's paper was entitled 'The Epistemology of St. Thomas and Modern *Erkenntniskritik*' ('*Die Erkenntnislehre des hl. Thomas und die moderne Erkenntniskritik*', 1914). And the leading German neoscholastic journal, the *Philosophisches Jahrbuch der Görres-*

Gesellschaft, translated and published a twenty-page review of the first volume of Külpe's *Realisation*, written by the leading Italian neo-Thomist A. Gemelli. Gemelli spoke of 'the high genius, the depth of thought, and the well-known scientific seriousness of the famous professor of psychology at the university in Bonn' (1913: 361). Gemelli's review was mostly a faithful and sympathetic summary. The fairly tame critical section of the paper confined itself to wondering 'whether Külpe had really destroyed *all* enemies of realism'; Gemelli thought that this was not the case, but was confident that Külpe would get closer to this goal in the second and third volumes of his epistemological *opus magnum* (1913: 378–9).

Grabmann's booklet *The Contemporary Value of Historical Studies of Mediaeval Philosophy* (*Der Gegenwartswert der geschichtlichen Erforschung der mittelalterlichen Philosophie*, 1913) was the text of his inaugural lecture in Vienna in 1913. As the title of the lecture indicates, Grabmann was keen to show that scholastic philosophy was still relevant to the philosophy of the twentieth century. He tried to make his case by pointing out that even the most important non-ecclesiastical philosophers of the day either used Scholastic insights, or else had reinvented central scholastic doctrines. His cases in point were Husserl and the Würzburg school. Both were credited with having paved the way from 'logical psychologism back to logical objectivism'. And objectivism was, according to Grabmann, the heritage of scholasticism (1913: 49).

In passing we might note here that Husserl's phenomenology, especially his *Logical Investigations*, was held in high regard by official neo-Thomists; indeed, contemporary sources suggested that Husserl's influence upon the neoscholastics was second only to that of St Thomas and Aristotle (Ueberweg and Oesterreich 1951: 631). Of course, the fact that both the Würzburgers and the official neo-Thomists valued Husserl's work – whereas both Wundt and the neo-Kantians were more sceptical – strengthens my case for an intellectual proximity of Würzburgers and neo-Thomists.

Grabmann's 1916 study was devoted exclusively to an investigation of Külpe's critical realism in its relation to the standpoint of Aristotelian–scholastic philosophy ('*Der kritische Realismus Oswald Külpe's und der Standpunkt der aristotelisch-scholastischen Philosophie*', 1916). Grabmann sought to document in great detail that 'Külpe's critical realism brings him very close to Aristotelian-Scholastic philosophy' (1916: 335). To make his case, Grabmann picked out various passages from Külpe's œuvre and showed that either St Thomas or else his modern-day followers had said pretty similar things. He also used Messer as an additional witness. In his obituary for Külpe, Messer had written that his former colleague had wanted 'to prove that the Kantian epistemological revolution was without justification'; Külpe therefore had attempted to return to 'the *pre*-Kantian conception of knowledge, a conception central to Aristotle and the Scholastic' (quoted in Grabmann 1916: 356).

All in all, Grabmann found seven 'points of contact' – three with respect to method, four regarding doctrinal content – between Külpe and the medieval schoolmen. First, Külpe's use of 'immanent criticism' reminded Grabmann of Aristotle and Thomas Aquinas. Second, both Külpe and the scholastics took their starting point from the empirical sciences. Third, and especially noteworthy, Külpe's advocacy of the method of self-observation would have delighted the likes of Aristotle, Augustine, and St Thomas. In this context Grabmann commented more generally on the Würzburg thought psychology. He regretted that it could still appear as if there was a difference between the Würzburg position and St Thomas's way of linking thinking to 'phantasmata'. At the same time, Grabmann expressed his hope that more detailed historical study 'of the older school of Thomists' might bring the two positions closer together, and eventually even unite them. Grabmann applauded the fact that a number of neo-Thomists had trained in thought psychology, some even in Külpe's institutes. This additionally proved that neoscholasticism and thought psychology were not at odds (1916: 357–60).

The fourth parallel between Külpe and the scholastics was that both took a keen interest in the problem of realism – the scholastics had done so under the title of 'the problems of universals'. Fifth, St Thomas agreed with Külpe that the object of knowledge was extramental. Sixth, both men had emphasised the importance of the concept of 'intention', and seventh, and finally, both had spoken of the 'dignity and task of thinking'; this task was 'a deep and complete knowledge of reality' (Grabmann 1916: 361–7).

Bühler, Marbe, Messer

For most of this chapter I have concentrated on the opposition between Wundt's Protestantism and Külpe's Catholicism. Even if little could be said about other key Würzburgers, my case for a link between Würzburger thought psychology and Catholicism would not collapse. After all, as the head of the research school, as supervisor, colleague, and experimental subject, Külpe did influence the work done in his laboratory. Nevertheless, at least three other key Würzburg thought psychologists can be linked to Catholicism.

Bühler, Marbe, and Messer were all reared in the Catholic religion. Interestingly enough, as young students they all were in contact with E. Krebs, a priest in Freiburg who nowadays is remembered primarily because of his friendship with the young Heidegger almost two decades later. After Heidegger had informed Krebs of his intention to leave the Catholic church in 1919, Krebs entered into his diary the following comment: 'Having qualified to lecture on Catholic philosophy, Heidegger will get himself into a lot of trouble for now changing sides. He is growing away from Catholic thinking, going the same way I saw Bühler going.' To the apostate Bühler, Krebs added a list of further names of one-time Catholic philosophers, amongst them Marbe and Messer (quoted in Ott 1993: 109).

Little can be said of Marbe in this respect. He himself reported having received 'a strict religious (Catholic) education', but added that already from his gymnasium days he 'had not used it any more' (1945: 49). Bühler did his MD at the University of Freiburg in 1903; this presumably was the time of his close contact with Krebs. The following year Bühler went to Strassburg to do a PhD with the neo-Thomist philosopher C. Baeumker. His PhD was a somewhat unambitious summary of 'the system' of the eighteenth-century Scottish philosopher Henry Homes (Bühler 1905). From Strassburg Bühler moved on to Berlin, where he acted as Stumpf's assistant in the academic year 1905–6. And from working with this 'free neo-Thomist', Bühler then went on to Würzburg to begin his work on his *Habilitationsschrift*. Some sources report that evidence that Bühler was a '"fallen" Catholic' did not emerge until 1916, when Karl Bühler married Charlotte, who was from an assimilated Jewish family and who had been reared a Protestant. The Bühlers married in a Protestant ceremony (Gardner and Stevens 1992: 141).

Much more information is available about August Messer, and first and foremost from his own pen. Messer's relationship with the Catholic church was complicated. On the one hand, he always insisted on his 'high regard and affection for this religion' (1924: 121). On the other hand, he found unacceptable the church's position that religious doubt was a sin, and that independent thinking in religious matters was acceptable only if it ended up confirming official church doctrines (Friedwalt [alias Messer] 1905; Messer 1921: 172; 1922: 15; 1924: 128; 1927: 3).

Messer had strong sympathies for the leading German

modernist, the Würzburg theology professor H. Schell. Schell became famous, or notorious, when in 1897 he published the pamphlet *Catholicism as the Principle of Progress* (*Der Katholizismus als Prinzip des Fortschritts*, 1897). His book was soon put on the index because he demanded that the Catholic church open itself to modern science and philosophy, and that individual church members should have the right to think freely and independently (1897: 10, 56, 77). Messer's sympathies for Schell were obvious from the fact that he included a chapter of Schell in his *Contemporary German Philosophy* (*Die Philosophie der Gegenwart in Deutschland*, 1927, §4), that he sought a dialogue with Schell when he worked in Würzburg during the academic year 1904–5, and that he left the Catholic church over Pius X's condemnation of modernism (Messer 1922: 15).

As far as Messer's epistemological and metaphysical views were concerned, these largely coincided with Külpe's. He rejected Wundtian psychophysical parallelism and argued for the assumption of a soul-substance, (1914c: 370; 1920: 66: 1924: 96), and he opted for critical realism. Messer was more outspoken, however, on the relationship between his (and Külpe's) epistemological views and neo-Thomism:

> Neothomist philosophy does not deserve the contemptuous treatment to which it is usually subjected....Neothomism is right to base knowledge of reality upon the combined activity of (outer and inner) experience on the one hand, and the interpretation and explanation of the material of experience by rational thinking, on the other hand.
>
> (1927: 7)

> My fundamental epistemological convictions...coincide with those views that, through the efforts of Thomas Aquinas, have become the dominant views within Catholic philosophy. Clemens Baeumker is a case in point.
>
> (1922: 26)

In one respect, Messer moved closer to neo-Thomism than Külpe had done. He opposed determinism *vis-à-vis* the human will, and labelled Wundt and Paulsen 'determinists' (1925: 133). As Messer saw it, 'indeterminism best fits immediate experience' (1920: 156); and it remained a credible option as long as no one was able to prove that 'all of reality can...be explained according to natural-scientific method. The claim of indeterminism is therefore acceptable: Decisions of the will do not fall under this mode of explanation because they cannot be deduced completely from prior events' (1911: 101).

Summary

In this chapter I have tried to make plausible the idea that the religious conflict between Protestants and Catholics influenced the debate over thought psychology. First, neo-Thomists took a keen interest in psychology in general, and Würzburg thought psychology in particular. Psychology in general was important to them because it potentially threatened traditional Catholic articles of faith, such as the belief in an immortal soul, or the freedom of the will. Because Wundt (alongside others) spoke out against such articles of faith, he was attacked by neo-Thomists. Würzburg-style thought psychology, on the other hand, seemed to provide a confirmation for some scholastic and neo-Thomistic views of the mind, of thought, and of the knower's relation to the world.

Second, Wundt was a Protestant who made no secret of his distaste for Catholicism and neoscholasticism. Moreover, Wundt's psychological and metaphysical voluntarism was – by the standards of the time – a Protestant philosophical position. As we have seen in earlier chapters, the Würzburgers rejected Wundt's psychological voluntarism; for instance, they denied that thinking could be best thought of as an 'inner choosing'. We can now see that such rejection threatened more than just a specific thesis regarding thought processes. Because Wundt's metaphysics was 'inductive' and as such based upon his psychology, any assault upon psychological voluntarism also threatened metaphysical voluntarism. And finally, because metaphysical voluntarism was a 'Protestant' viewpoint, any denunciation against psychological voluntarism was a criticism of the religion of Wundt's childhood.

Third, although Külpe was not a Catholic himself, he increasingly sided with philosophical stances that – again by the standards of the time – were looked upon as Catholic. His opposition to voluntarism, psychophysical parallelism, the actuality theory, and epistemological conscientialism were all rather telling in this respect. Thought psychology was tied into Külpe's criticism of these doctrines, and it also provided the main pillar of his anti-Wundtian epistemological realism. Because Külpe's attacks on Wundt's metaphysics and epistemology happened at the same time as the Würzburger thought psychologists' criticism of Wundt's psychology, it would be anything but surprising if Wundt did not see the connection between the two sorts of criticism.

Finally we have seen that Marbe, Messer, and Bühler all had some connection to Catholic philosophy. Little can be documented about the cases of Marbe and Bühler, although it is known that both sought to become Catholic philosophers at one stage in their careers. It is worth registering, however, that neither man seemed to be particularly concerned by Wundt's insinuation that they had reinvented a scholastic wheel. In his reply, Bühler wrote that such coincidence would be interesting and would motivate himself and others to acquaint themselves further with scholastic doctrines (1908a: 112). Messer did largely side with Külpe, and he made the parallels between their critical realism and the epistemology of the neoscholastics explicit.

7 Conclusions

Over the last six chapters, I have provided a social history of the controversy over thought psychology in Germany at the beginning of the twentieth century. It remains for me to summarise my main argument, to draw some philosophical and methodological conclusions, and to relate my study to other work in sociology, psychology, and the history of science.

Summary

I have given a fairly detailed narrative summary of my study in the Introduction. I shall not repeat it here. Figures 7.1 and 7.2 may serve to capture some key elements of the above chapters from a slightly different angle.[1]

Figure 7.1 represents how Wundt's opposition to the Würzburgers' thought psychology was linked to his 'recluse' model of leadership, his purism, his collectivism, and his Protestantism. *Mutatis mutandis,* Figure 7.2 does the same for the Würzburgers. The linguistic elements within the pictures (e.g. 'metaphysical voluntarism', 'recluse model of leadership') refer to beliefs and actions of the historical actors. The lines and arrows of the pictures can mean that one element 'suggested' another element to the historical actor; that belief in one element 'caused' belief in another; or that one element was said, by the historical actor, to 'imply' another.

These figures bring out one aspect that I have dealt with only in passing in Chapters 3–6: the interconnections between Wundt's and the Würzburgers' respective positions in the four social dimensions (recluse versus interlocutor, purist versus promiscuist, collectivist versus individualist, and Protestant versus Catholic). Some of these links were stressed by the historical actors themselves, as when

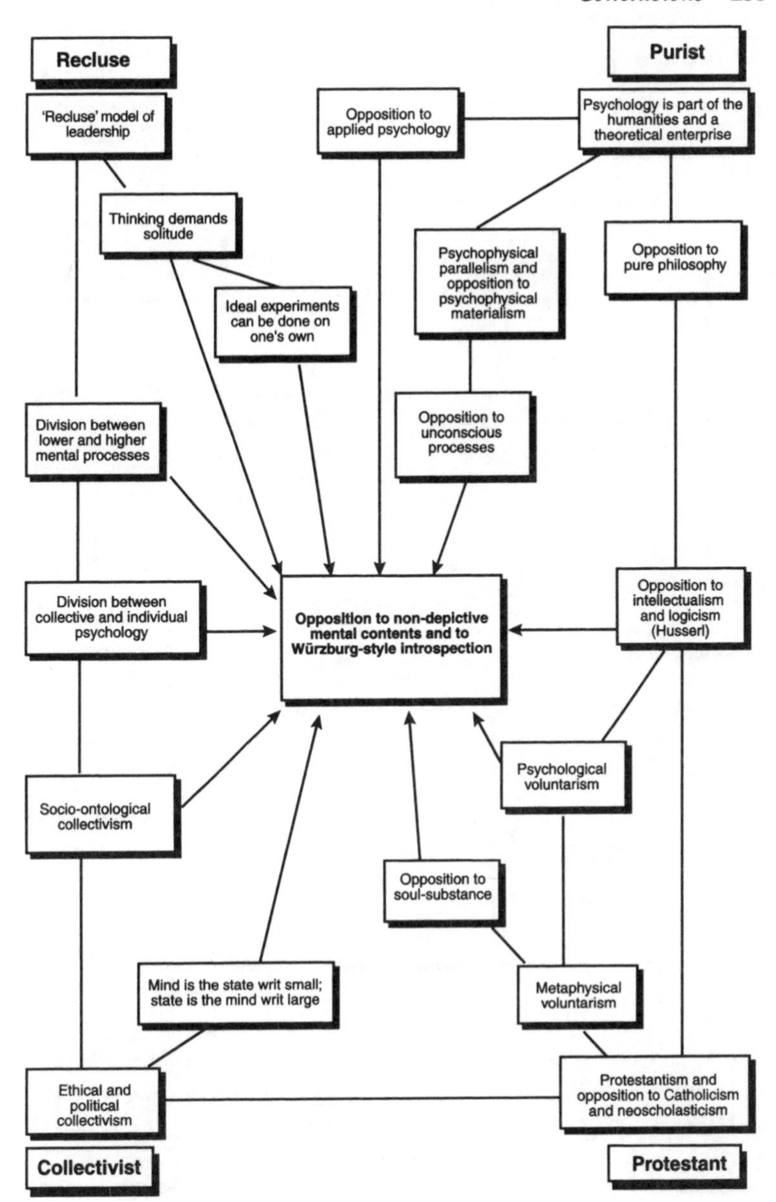

Figure 7.1 Summary of the analysis: Wundt

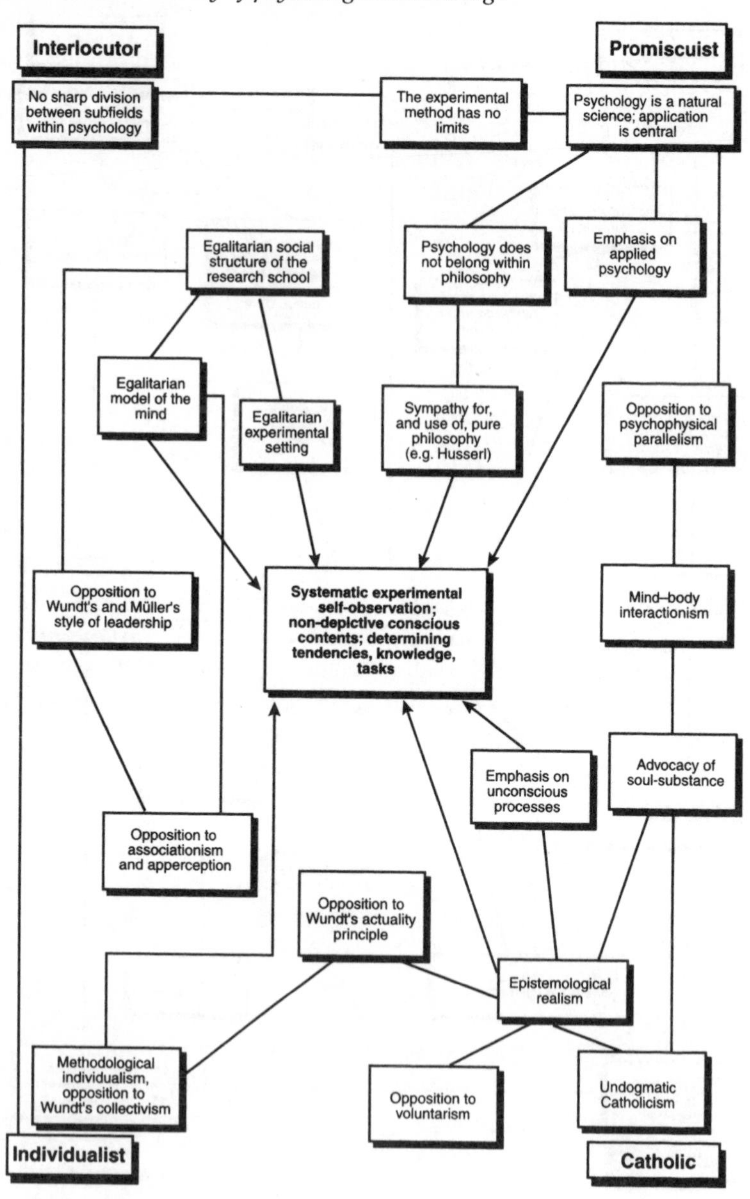

Figure 7.2 Summary of the analysis: the Würzburgers

Wundt praised Luther's emphasis on the community. Other links were more indirect and more tenuous. I shall not here pursue this issue further. One point, however, is worth making. The links between Wundt's or the Würzburgers' respective positions in the four dimensions were historically contingent; they were not compelling for writers at all times; and they were not even accepted by everyone during the first two decades of the twentieth century. Indeed, at the time Protestantism was sometimes linked to individualism and subjectivism, and Catholicism to collectivism (e.g. Switalski 1925).

Institutions

I started this book with two general claims. The first was that scientific-psychological theories (and other bodies of psychological knowledge) are social institutions. According to the second, to analyse a given body of beliefs and actions as a social institution is to do the following: to identify a self-fulfilling or self-referential structure of beliefs; to describe its relations to human individuals; to depict its characteristic actions and artefacts; and to pinpoint its links to other entities of the same structure. I now want to return to these general claims and relate them explicitly to the historical narrative.

Self-reference

Of the four aspects of analysis of social institutions, it is perhaps the aspect of self-reference that deserves, at this point of the argument, a slightly longer explanation than that given in the Introduction.[2] Remember again the example of money. What makes a metal disc a coin, that is an instance of money? It is nothing more nor less than that a collective treats the metal disc as money, that it talks of it as money, that it uses it as money, and that it sanctions and enforces this talk and this use. Money, we might say, is what we collectively take to be money. Or take the social institution of a group. What makes a given person a member of a group? Nothing more nor less than that the group members treat that person as a member, and that they believe him or her to be a member. Or, finally, take the person possessing social authority. What gives a person social authority is that others take that person to have authority.

We can render these informal observations more precise by saying that, when taken collectively, institution-sustaining beliefs (and actions) are self-referring. If persons A, B, and C believe that C is the authority in their group, then C's authority is constituted by the mutual belief of A, B, and C. If they all change their minds, then C's authority evaporates into thin air. C's authority is nothing but this mutual belief in C's authority. The mutual belief is thus a belief about this mutual belief; that is, it is self-referring. By the same token, it is also self-validating: The mutual belief of A, B, and C that C has authority is what makes this very same belief true.

Obvious cases of social institutions

Some of the social entities that play a role in my case study are clearly social institutions in both the common understanding and in the sense just defined. To be a member of the Catholic church is of course like being a member of a group; you are a member if you are taken to be a member. And to be a church or confession is to be taken to be a church or confession. The analysis for the state or political parties is also fairly straightforward and will occasion little resistance: Something is a state if it is taken to be a state by a sufficient number of its citizens, and something is a party if it is taken to be party by members (and perhaps non-members). It is not fully in accordance with ordinary language to call a scientific discipline (and its internal order) or a university department (and its social order) a social institution. But these cases too are easily brought under the definition. After all, to be a scientific discipline is to have a social status, and a social status can only be had if then there exists a self-referring belief that creates it. The same obviously holds for the university department.

Experiments as social institutions

Can we find self-referentiality in the case of psychological experiments? Were the Würzburgers' thought experiments social institutions in the sense that they were created by a self-referential and self-validating belief? The answer is not so straightforward. After all, there were many people, such as Wundt and Müller, who did not accept the Würzburgers' experiments as 'proper' experiments, and there also were disagreements within the Würzburg school itself. Nevertheless, I want to suggest that the answer to the above questions is 'yes'.

Imagine, first of all, that someone is sitting across the table from you, with a stopwatch in his or her hand, asking you for a reaction to 'Eucken's world-historical apperception' or some aphorism by Nietzsche. Is this a Würzburg-style psychological experiment, a quiz, a psychotherapeutic session, or a school exam? Obviously, this will depend on what you and your opposite take this situation to be. It will be a thought-psychological experiment only if you collectively take it to be a thought-psychological experiment. Just like a piece of metal becomes a coin only through the self-referential collective belief that it is a coin, so also – *mutatis mutandis* – for a sequence of verbal behaviour and the experiment.

Second, for something to be a social institution, it is sufficient that *some* collective has a self-referential belief about it. Moreover, it is not necessary that this collective be extensive. Thus there would be no contradiction in saying that systematic experimental self-observation was a social institution in the Würzburg Psychological Institute (and a few other German psychology departments), but not in Leipzig, or Göttingen. Indeed, the Würzburgers were of course trying to enlarge the collective for which systematic experimental self-observation was a social institution, whereas Wundt and Müller did their best to make this collective shrink to but two or three psychologists in Würzburg.

Third, to deal with the school–internal disagreements, I need to draw attention to one further crucial characteristic of social institutions. It is important to recognise the *performative* or *finitist* aspect of social institutions.[3] Social institutions are performed, and their path is not predetermined by rules and norms. Recall again our mini-society of A, B, and C, which has made C their leader. Moving from context to context, A, B, and C have to make decisions regarding C's authority: Is this a situation in which C is to be granted authority? Is this a case in which C wishes to risk disobedience? Is this the point at which A will decide to opt out, and so forth? To put it another way, members of social institutions have to make judgements of similarity and dissimilarity and the only criteria they can go by are their experience with a finite number of earlier occasions and earlier judgements.

It is clear how this idea can be applied to the case of the Würzburg experiments. The experiments of Marbe, Ach, Messer, Bühler, and the many other Würzburg psychologists were not following a set of prescriptions on what constituted a Würzburg thought experiment. To be a Würzburg thought experiment was not to follow explicitly formulated rules; it was to have a design similar to the series of experiments that had been done before, and it was to achieve results that could be construed as extending or modifying the outcomes of earlier experiments. But this left plenty of room for deviation and for internal criticism. We might say that different Würzburg psychologists tried to develop 'their institution' in different directions. Alternatively, one could perhaps also say that some of them were seeking to establish new institutions. The former way of putting things will seem more adequate if one wishes to emphasise that, despite their disagreements, the Würzburgers all agreed on three important points: that thought-psychological experiments were possible; that they were best done as reaction experiments involving both an experimenter and an experimental subject; and that the key to thought processes was retrospective reports coming from the experimental subject.

Fourth, before leaving the topic of psychological experiments *qua* social institutions, I want to highlight an interesting historical development with respect to psychological experiments. This historical development comes out clearly in my case study, but it is perhaps worth formulating explicitly. I mean the social order within different experimental designs. For Wundt, the perfect experiment was done on one's own. Let us call this the 'recluse model' of the psychological experiment. The Würzburgers first favoured an egalitarian two-person setting in which experimenter and experimental subject were each other's equals in terms of social and cognitive authority. Above I suggested the name 'interlocutor model' for this scenario. Later they adopted a form of experimentation that was similar to the 'drillmaster model' of Müller's Göttingen Institute. This was a model in which the experimenter has social and cognitive authority over the subject. Schematically, we might distinguish these scenarios as follows:

Recluse	*Interlocutor*	*Drillmaster or interrogator*
S	S|E	$\frac{E}{S}$

S here stands for the experimental subject, E for experimenter; an E above an S (separated by a horizontal line) indicates that E has social and cognitive authority over S; S on the same level as E (separated by a vertical line) means that S and E are equals in terms of their social and cognitive authority.

The schematic representation has the advantage of directing our attention to further possible forms of social organisation in a psychological experiment. Thus we can have a single E, or an S over an E. And furthermore, we might want a way to indicate within our little formalism that according to some writers on psychological experiments, certain participants in the psychological experiment might be inessential, or that their properties as humans or 'as subjects' are actually irrelevant to the experimental situation. (I shall indicate this possibility by putting brackets around these inessential elements.) Thus, for instance, Wundt thought that second best to doing an experiment on one's own was to do it with an experimenter who could be replaced by a machine. Adding these possibilities to our earlier list, we get something like the following:

Recluse I	*Recluse II*	*Stimulated observer*	*Interlocutor*	*Interrogator or drill-master*	*Behaviourist*	*Simulator*
S	$\frac{S}{(E)}$	$\frac{S}{E}$	S|E	$\frac{E}{S}$	$\frac{E}{(S)}$	E

Of what use is this classification? Here are some rough and ready suggestions, theses, and questions. First, the history of psychological experimentation has, by and large, moved from the left-hand side to the right-hand side. Second, recent years have seen the emergence of a new model of psychological experimentation, the model that I call 'simulator'. At least the psychologists that I found around me while writing this book (at the Centre for Adaptive Behaviour and Cognition of the Max-Planck Institute for Psychological Research in Munich) were taking great strides towards making the experimental subject (even as the social and cognitive underdog) less and less crucial to psychological experiments. Computer simulations were increasingly replacing both human and animal subjects. Third, the taxonomy might be useful for identifying important junctures in the history of experimental psychology. The emergence of, or the return to, specific models of cognitive and social order in psychological experimentation might well be key moments in the history of psychology. Finally, and moving on from theses to questions. What is the genealogy of these various social models? How does the history of psychological instruments relate to the history of these social models? To what extent did these experimental models define the theoretical models of psychological theory? And to what extent is it possible to show that social order of experiment is similar to social order of research schools? Some of these questions I have addressed in this book; the others I hope to tackle elsewhere.

Introspection as a social institution

Someone might agree that the Würzburgers' two-person experiments were institutions of sorts but contest that there was, or is, anything social about introspection itself. This is a natural intuition. But it is wrong nevertheless. To see why, we have to make a small detour through the sociology of experimentation.

One important conception in the sociology of scientific knowledge has been H. M. Collins' theory of 'the Experimenters' Regress' (Collins 1985). Imagine that two groups of scientists, say the As and the Bs, argue over the existence of some particular form of gravitational waves, called 'high-flux gravitational waves'. The As claim that such waves cannot possibly exist. The Bs insist that gravitational waves do exist because they appear as peaks on the screen of their newly developed high-flux gravitational-wave detector. Assume further that the As wish to challenge this claim not only on theoretical but also on experimental grounds, and thus build a device similar to that of the Bs. The As' device does not find any gravitational waves, however. Will the two sides be able to come together and agree on the superiority of one of the two detectors? Not as things stand. The As will be certain that the Bs' 'detector' is a bad piece of equipment because it detects something that does not exist. The Bs will be certain that the As' device is a bad detector because it fails to detect waves that 'obviously' are there. Here than is the regress: the two sides cannot agree on the detectors because they do not agree on the existence claims; and they cannot agree on the existence claims because they cannot agree on the detectors.

This case illustrates nicely both what happens when an instrument *is* a social institution, and what happens when it *is not*. Amongst the Bs, their device is a social institution in so far as it has the social status of being a high-flux gravitational-wave detector. And *to be* a high-flux gravitational-wave detector is *to be taken to be* a high-flux gravitational-wave detector. Or put differently, the peaks on the screen are, or have the theoretical status of, high-flux gravitational waves because they are taken to be high-flux gravitational waves. For the As, however, the Bs' so-called 'high-flux gravitational-wave detector' is really just a device producing lines and peaks on a screen, and whatever it detects, it surely is not high-flux gravitational waves. The Bs' device could only convince the As if the As adopted the Bs' instrument as a social institution.[4]

Now we can return to the case of introspection. The cases of introspection we are concerned with here are not the introspections of some total isolates, but of psychologists participating in psychological debates over the contents of consciousness. These debates often had a structure strongly reminiscent of the experimenters' regress. To honour the special case at hand, we might speak of the introspectionists' regress. Two or more groups of introspectionists disagreed over the contents of consciousness (for instance, are there non-depictive conscious contents; is there a judgement quale?) and they used their own introspections trying to convince the other side. For the one side, anyone who *did not* find non-depictive conscious contents in their introspections was the victim of their theoretical, sensualistic biases (Bühler). For the other side, anyone who *did* find non-depictive conscious contents in their introspections was a careless observer (Müller, Wundt). The two sides could not come to agree on their theories because they could not agree on who was a competent introspectionist; and they could not agree on who was a competent introspectionist because they could not come to agree on their theories. Bühler tried to solve the problem of the competent introspectionist by giving overwhelming prominence to Külpe's introspective reports in his first paper on thought psychology (1907b); Külpe, after all, was respected by both Müller and Wundt. But even that move did not succeed. While Külpe was one of the most important 'non-depictive thought detectors' within the Würzburg school, he was just a victim of suggestion for psychologists in Leipzig and Göttingen. The moral should be clear: introspection is social because being a reliable introspectionist is a social status. Introspection can be a source of knowledge for a community only if the producer of that knowledge, that is the introspectionist, is taken to be a producer of such knowledge.

The fact that the 'introspectionists' regress' is but a variant of the 'experimenters' regress' has an interesting implication for evaluating the ultimate demise of introspectionist psychology. A naïve observer might be tempted to think that the disappearance of introspectionist psychology was somehow inevitable and rational. And the observer might justify this view on the ground that introspectionist psychologists with different theories were unable to reach agreements on their data and instruments. And here our naïve observer might contrast this sad state of introspectionist psychology with that of experimental physics (cf. Watson 1913). But in the light of the above, this whole line of reasoning is flawed. As Collins' work on gravitational waves shows so well, experimental physicists are plagued by the very circularity that our naïve observer saw as a good reason for the demise of introspectionist psychology.

Theories as social institutions

I have already shown that Wundt's, Müller's, and the Würzburgers' theories of the mind fulfilled the following criteria of social institutions: they were fought over by individuals with divergent social, political, scientific, and confessional interests; and they were linked to institutions such as experiments, introspectors, social orders of research schools, universities, the states, and the confessions. It remains for me to make explicit the self-referential aspect of psychological theories, on the one hand, and the ways in which individuals – their actions, behaviour, and mental states – can be shaped by psychological theories, on the other hand.

To bring out the self-referential character of psychological theories, we need first to remind ourselves that borders and boundaries are social entities; they are social institutions or parts of social institutions, or dependent upon social institutions. The border between two countries, for example, say between France and Germany, is what is collectively taken to be this border. Moreover, the social institution of a border comes with rights and obligations, with duties and restrictions. Not everyone is allowed to cross a given border without, say, a visa; one has the duty to present a valid passport, to declare goods above a certain value, or to leave the country again after a certain period. Furthermore, a border between countries comes with specific roles and types of actions. There are guards who secure the border and who check travellers' papers; there are customs officers and customs investigators; and there are judges that pass sentences on illegal immigrants.

Second, theories, categorisations, arguments, and reasons are about drawing conceptual boundaries. Categorisations and theories set boundaries between things of different kinds, and arguments and reasons are arguments and reasons for or against drawing boundaries in some given way or other. Most importantly, theories and webs of arguments draw boundaries on the rational–conceptual level and – if successful – on the social, institutional level. Put differently, whatever the kinds of entities around which, and within which, a given theory draws boundaries, it always – and by the same token – draws boundaries around, and within, a group of humans.

Let me demonstrate these abstract claims by means of an example. One of the main pillars of Wundt's theory of the mind was the following argument (I):

I 1 The two main methods of psychological enquiry are those of collective and experimental psychology.
2 The experimental method is successful with respect to simple processes.
3 The methods of collective psychology are successful with respect to complicated processes.
4 Thought processes are complicated processes.

---

5 Thought processes are to be studied by collective psychology.

Clearly, Wundt's argument set a boundary between experimental and collective psychology. It is also easy to see that if we accept this argument, certain types of institutional arrangements suggest themselves. Assuming that we want to find out about both simple and complicated processes, it will naturally follow that we accept the existence of two separate subfields within psychology, experimental and collective psychology, perhaps eventually with their own professorships, institutes, and forms of training. Indeed, in the Leipzig Institute under Wundt's leadership, the border between collective and experimental psychology was most important. It marked a crucial social-cum-cognitive divide. Why was it *the* central divide in the theory and practice of the institute? It was the central divide because Wundt and his students and colleagues took the above argument to be one of *the* central theoretical pillars of *their* work in psychology. It defined their work as distinct from that of psychologists elsewhere. In brief, in the Leipzig Institute, the above argument or theory was a social institution.

Of course it was an institution in Leipzig only, and certainly not an institution in Würzburg. Bühler's and Külpe's alternative argument or theory went roughly like this (II):

II 1 The experimental method has no principled limitations, and it is the most important method of psychology.

2 The experimental method is particularly successful when the processes to be studied are relatively simple.
3 Thought processes are not complicated processes.

---

4 Thought processes can be studied successfully by means of experiments.

Theory II undermined the borderline so central to Wundt's social and cognitive enterprise. And it suggested the doing away with any fundamental border between a psychology studying thought, and a psychology studying other mental processes. Again, II was both a theoretical move *and* a social institution: It was taken to be the argument defining the Würzburg position *vis-à-vis* Wundt, and thus one of the key arguments that constituted the Würzburgers as a distinct scientific-psychological collective.

Beyond the dualism of the rational and the social

Recognising that theories and experiments are social institutions has an important implication. It allows us to avoid the dualism of the rational and the social. On a traditional construal of the distinction, one could ask of my historical case study, above, whether or not it seeks to 'reduce' rational arguments to 'mere' social factors, or whether it denies that rational arguments and theories can ever have any persuasive force. But once we come to see that theories and arguments are social institutions (or parts of social institutions), we can abandon this dualism for a form of monism that I suggest calling 'sociologism'. Sociologism is the claim that so-called 'rational' factors, that is theories, arguments, and reasons, are in fact social factors.

I do not wish to rule out a priori any one type of causal relationship between social institutions, individuals' beliefs, or individuals' goals or interests. I do not rule out the possibility that one's scientific-psychological theory comes first and causes one to pick a specific religion or political programme. All I deny is that this is a case of 'the rational' being prior to 'the social'. I also permit – at least as a logical possibility – that one might form views and opinions about scientific matters without the causal influences of any of the grand old institutions of the church, the class, or the state, or even of the new institutions of the research school or the discipline. But even in this latter case, the process of forming these views and opinions is still a social process in that it involves the social institutions called theories.

Having said that I do not rule out the possibility that one chooses, say, one's confession to fit one's scientific psychology, I hasten to add that on the basis of what is known about humans and their decision making, such order is rare and unlikely. It is also implausible in the case of the main antagonists of my historical case study. It does not seem a credible account of Wundt, for instance, to suggest that he chose his religion, politics, or model of leadership on the basis of his psychological insights. Wundt's religious and political views predate his work on psychology by decades, and his reclusive model of leadership fitted with what Wundt had experienced as Helmholtz' assistant in Heidelberg (Turner 1993, 1994), and with his own character (Diamond 1980). Nor is it likely that Wundt's psychological thinking developed largely independently of his political or religious views, or that the links between the psychological and the political were only a *post hoc* rationalisation. Such independence seems improbable for a number of reasons. First, the social order – of his institute, or of the state – provided important models for Wundt's psychological theorising. Second, Wundt's psychological, metaphysical, ethical, and political-philosophical writings were all written or revised in parallel, and overlapped considerably in content. And third, Wundt's psychological texts themselves contained numerous statements on institutions such as psychology as a discipline, the educational system, the state, or the confessions.

The case of the Würzburgers is a bit more complex because here we are dealing with a group of people from different backgrounds. But here too it seems unlikely, in the light of my historical case study, that the Würzburgers picked a social organisation for their research school in light of their theory of the mind, rather than vice versa.

Kinds of kinds

To call scientific theories social institutions might easily arouse the suspicion that I deny that scientific theories can be *about* an independent reality; that is, that I advocate some sort of sociological idealism. Not so. To claim that a theory is a social institution is to say that it centrally involves a self-referential component, but it is not to say that this self-referential component exhausts the reference of the theory. The point is easiest to make in terms of 'kinds' rather than full theories, and by distinguishing between three kinds of kinds. These three kinds are *natural kinds*, such as 'tiger', *social kinds*, such as 'authority',[5] and *artificial kinds*, such as 'typewriter'[6].

All three kinds are social institutions in the sense that they have a self-referential component. But the self-referential component works differently in the three cases. Take first a social kind such as 'authority': 'Authority-talk' (and 'authority-action') ultimately refers to other 'authority-talk' (and 'authority-action') and it refers to nothing else. The referent is created by the talk (and the action): ultimately, it *is* the talk (and the action) *itself*. The reference here is, as it were, 'exhausted' by the self-reference. The case of natural kind terms such as 'mountain' is different. 'Mountain' has what we might call an '*alter*-reference'; that is, the term refers away from itself towards individuals in the physical world, individuals that exist independently of the reference. But 'mountain' also possesses a self-referential component. This self-referential component consists of the criteria for classifying individuals as mountains, that is of models, paradigms, and prototypes. It is easy to see why models and paradigms are self-referential components: nothing is a model for anything in and of itself; a model or paradigm is what is collectively taken to be a model or paradigm.

Artificial kinds stand between these two cases: 'typewriter' does not just refer to talk about typewriters, and thus it differs from 'authority'. It has *alter*-reference, and there are models, paradigms, and prototypes of what a typewriter ought to look like and how it ought to work. Yet 'typewriter' differs from 'mountain' in that the individuals classified as typewriters do not exist wholly independently of the classifying activity. Instead, the classifying activity is part and parcel of a social process – indeed, a social institution – that essentially involves physical actions bringing the classified individuals into existence. In the case of artificial kinds, human action makes it so that the individuals referred to fit the prototypes, rather than vice versa.

What kind of a kind was a Bühlerian thought?

The issue that arises on the basis of the last section is of course the position of psychological kinds with respect to my classification of kinds. Naturally, the correct account must emphasise diversity. Some psychological kinds are no doubt natural kinds, while others, say those investigated by social psychologists, belong more within the category of social kinds.

More important for my concerns here is to address the specific question of whether at least some scientific-psychological kinds are artificial kinds. In proposing an affirmative answer, I also make good the promise of explaining how psychological theories can sometimes be self-fulfilling promises, and shape the very mental life that they are meant to describe and explain.

It is best to focus the discussion of this point on *Bühlerian thoughts,* or *Gedanken* for brevity and clarity. *Gedanken* were non-depictive conscious experiences of immediate knowledge, usually unaccompanied by sensations, presentations, or feelings. As we have seen, Müller, Wundt, and others contested the existence of such *Gedanken.* These critics claimed that at the typical points in the stream of consciousness where Bühler's subjects had observed *Gedanken,* they (the critics) noticed feelings or hazy presentations.

Were *Gedanken* a natural kind, and was '*Gedanke*' a natural kind term? For many of the Würzburgers the answer is easy to provide. *Gedanken* clearly formed a natural kind for those Würzburgers who followed Bühler, adopted his theory, and trusted the self-observations of his introspectors.

Interestingly enough, some later-day interpreters also treat *Gedanken* as a natural kind. C. Burt, for instance, submits that both the Würzburgers and their critics were right: there really were entities with *Gedanken* properties in the Würzburgers' streams of consciousness, and there really were no such entities in, say, Wundt's or Müller's. It was all due to individual differences (1962: 230). *Gedanken* thus were natural kind terms in Würzburg, but not in Leipzig or Göttingen. A different account is offered by A. Brock who seems to think that the Würzburgers and their critics had competing ways of classifying the same description-independent streams of consciousness (Brock 1991). In this case too, *Gedanken* were at least natural kind terms.

For the Würzburgers' critics, however, *Gedanken* were something on the order of a social kind. There was nothing to which *Gedanken* could refer other than to more talk about them. Of course, this was not the exact way in which the critics put their point, but it is likely that they would have accepted this formulation.

It seems to me that neither of these alternatives is very appealing. Should we really say that either the Würzburgers or their critics were unable to notice whether or not they had imagery or other depictive contents in their minds when solving a problem? That seems like a rather uncharitable interpretation. Burt's proposal, on the other hand, seems odd as it stands. Would it not be somewhat miraculous that by some process of self-selection people without depictive contents turned up in Würzburg whereas those who did have such contents studied in Leipzig, Göttingen, or Cornell?

My own proposal is to think of *Gedanken* as an artificial kind rather than a natural kind or a social kind. This is not a particularly radical claim. It is little more than a new formulation of the sometimes-heard claim that our psychological classifications are constitutive of our mental states and events. Our psychological vocabulary does not classify mental states and events that exist wholly independently of the vocabulary. Instead, having *this* vocabulary rather than *that* vocabulary causes us to have – or be more likely to have – one kind of mental experience rather than another. Now if *Gedanken* were an artificial kind, if being trained in one psychological theory of consciousness causes one to have mental experiences corresponding to this theory, then obviously these kinds were more than a mere social kind. They were *real* – *real as artefacts*.

Research schools

Having now completed my plea for the claim that scientific-psychological theories are social institutions, it remains for me to connect my case study – and the underlying methodological and philosophical assumptions – to work of at least some other investigators in science studies: to the historians' study of 'research schools'; to the psychologists' research into heuristics of discovery; and to the sociologists' ways of analysing the relation between the rational and the social. I begin with the historians.

In his 1981 article on 'Scientific Change, Emerging Specialties, and Research Schools', G. L. Geison tried to bridge what he perceived as the gulf between the interests of sociologists of science and those of historians of science. As Geison saw it, sociologists leaned towards the study of 'emerging specialties' – for example, the emergence of experimental psychology in the 1890s or of radio astronomy in the 1940s and 1950s – while historians tended to focus more on 'research schools', that is on 'small groups of mature scientists pursuing a reasonably coherent program of research side by side with advanced students in the same institutional context and engaging in direct, continuous social and intellectual interaction'. Geison's paper tried to persuade sociologists of science to join historians in the study of research schools. To this end he discussed, at some length, D. Edge and M. Mulkay's book *Astronomy Transformed* (Geison 1981; Edge and Mulkay 1976). He found much to praise in this first book-length case study in the sociology of scientific knowledge, particularly its interest in different leadership styles and their effects on scientific knowledge.

Looking back over the past decade, one notices that, by and large, Geison's invitation has not been taken up by what he called the 'cognitive sociology of science', and what nowadays is more typically referred to as the 'sociology of scientific knowledge'. The sociology of scientific knowledge has not yet turned to the study of research schools, nor has it built on Edge and Mulkay's sensitivity to the effects of leadership styles. For instance, a recent 800-page *Handbook of Science and Technology Studies* contains no references to historical studies on research schools, and the term *research school* does not appear in its index (Jasanoff *et al.* 1994).

Historians of research schools have also paid little attention to work done in the sociology of scientific knowledge. A recent anthology of historical scholarship on research schools, co-edited by Geison, contains a few passing references to sociologists' work on tacit knowledge but makes use neither of their studies of laboratory life nor of their more general concerns with the interrelations between social order and natural order (Geison and Holmes 1993).[7]

My historical case study above constitutes an attempt to remedy this situation of non-communication. It seeks to follow Geison's 1981 suggestion and study research schools from 'cognitive-sociological' perspectives. It is motivated by the belief that the sociology of scientific knowledge and the historical study of research schools can fruitfully learn from one another. The sociology of scientific knowledge can benefit from historians' emphasis on leadership styles, teacher–student relationships, and training.[8] The historical study of research schools, on the other hand, can profit from sociologists' focus on social variables of scientific knowledge in general and from work on how scientific knowledge is used to justify social order in particular.

I have tried to make these abstract claims plausible by means of a case study of German psychological laboratories during the first two decades of the twentieth century. German psychological institutes of the time have been cited as paradigm examples of 'research schools', and thus they ought to qualify as a test case for my attempt to combine the concerns of the historian of research schools with those of the sociologist of scientific knowledge (Olesko 1993: 22). It is up to the reader to decide whether I have done so successfully.

Heuristics

There also are some interesting points of contact between my case study and some work done in cognitive psychology. G. Gigerenzer

(1991), amongst others, has suggested that cognitive psychologists turn to the study of the largely implicit heuristics by means of which scientists make their discoveries. Gigerenzer is particularly interested in a heuristic he calls the 'tools-to-theories heuristic' (1991). This heuristic can be seen at work when scientists model their subject matter on the tools by means of which they study this very same subject matter. Gigerenzer's examples come from psychology itself: psychologists first used statistical tools to understand the mind, and then came to construe the mind as an 'intuitive statistician'. Or, cognitive scientists first employed computers for processing data about the mind, only then to start thinking about the mind as a computer itself (Gigerenzer and Goldstein, 1996).

I am unable to find any examples of the 'tools-to-theories heuristic' in the work of Wundt, Müller, or the Würzburgers. But there is something still better that my case study provides for the cognitive psychologist in search of discovery heuristics. A good part of my story about the links between various social orders – of the laboratory, of the experiment, of the discipline, or of the state – on the one hand, and various 'psychological orders' of theories of the mind, on the other hand, suggests the existence of a *social-order-to-theory heuristic,* at least within psychology. In Wundt's psychology, the mind had the same kind of hierarchical structure that we found in the Leipzig Institute; in Müller's psychology, persevering presentations stayed on top for conspicuously similar reasons to those supporting Müller's position; and Bühler gave the mind the same sort of egalitarian organisation that he found in Külpe's research school. Further examples not covered in this study are easy to think of. Instances of the *social-order-to-theory heuristic* can be found from antiquity (Plato) to the present (Dennett 1991a; Minsky 1985). Both Dennett and Minsky construe the mind/brain on the model of a bureaucracy.

This is not the place to pursue this topic in any detail. I shall be satisfied if at least some psychologists, historians, and sociologists agree that the study of the *social-order-to-theory heuristic* might be an interesting meeting point of psychologists, historians, and sociologists. But at least two somewhat speculative comments are hard to suppress.

First, what makes the numerous social metaphors in mainstream mind-or-brain sciences – that is, psychology, philosophy of mind, cognitive science, neuroscience – so intriguing is that they occur within an overall individualistic framework. These fields like to focus on those aspects of human behaviour and 'information processing' that they assume to be universal, and thus hope to be 'unpolluted' by diverging social environments. The irony is unmistakable: The social realm that is being excluded a priori by individualistic commitment returns *a tergo* in the form of social metaphors. The collective that is not allowed to exist outside the mind–brain returns as its inner structure.

Second, in my historical case study we have not only encountered examples of the *social-to-psychological-order heuristic*, we have also met with the reverse phenomenon, the *psychological-to-social-order heuristic*. Wundt's concepts for understanding the collectivity were all drawn from individual psychology. Of course Wundt has not been the only theoretician using this heuristic. Durkheim might well be another case in point. The fascinating point here is that the *psychological-to-social-order heuristic* seems to be at work most strongly when a theoretician wishes to emphasise that collective phenomena cannot be reduced to individuals' actions and interactions, that is when a theoretician seeks to combat the intrusion of individual psychology into social theory. The irony *vis-à-vis* some sociological theories is thus the reverse of the irony in the case of individualistic mind-or-brain sciences and the social metaphors: the psychological that is excluded a priori by collectivistic commitments returns *a tergo* in the form of individual-psychological metaphors.

Versions of sociologism

Finally, I wish to situate myself with respect to some other types of work within sociological studies of science in general, and other types of sociologism in particular. I suggest distinguishing between three versions of sociologism: 'eliminative', 'reductive', and 'anomalous' sociologism. All three views are united in believing that arguments, reasons, and theories are social entities, but they differ in how they elaborate on this common insight.

Eliminative sociologism is the position that many philosophers ascribe to the sociology of knowledge in general (Brown 1989; Frede 1987, 1988; Gracia 1992; Laudan 1990; Newton-Smith 1981; Normore 1990). It amounts to saying that reasons, arguments, and theories are not what they claim to be, and that our usual ways of looking at them constitute a radically mistaken 'folk theory'.[9] It is difficult, however, to identify anyone who actually advocates this view. I find resonances of it, for example, in M. Foucault's paper 'Nietzsche, Genealogy, History', although I hesitate to attribute it to either Foucault or Nietzsche himself (Foucault 1984). Be this as it may, Foucault here presents Nietzschean genealogy as concerned with showing that arguments are no more than 'a mask'. What they hide are 'the passions of scholars, their reciprocal hatred, their fanatical and unending discussions, and their spirit of competition...aspects of the will to knowledge: instinct, passion, the inquisitor's persistence, cruel subtlety, and malice' (1984: 78, 95). I find it inviting to interpret Foucault's reading of Nietzsche as follows. Our ordinary understanding of the rational is fundamentally flawed; it is a folk theory that misconstrues the factors that move us and others. This folk theory is upheld by those in power to hinder us from empowering ourselves.

It is difficult to see how one could argue with an advocate of this view. No doubt the eliminativists would reject any criticism of their stand as yet another power move. But one can address an argument to the interested bystander who is still trying to make up his or her mind. One might argue – along Davidsonian lines – that we cannot make sense of the idea of being wrong about almost everything (Davidson 1984). Or one might question the status of the eliminativist theory itself. And one might analyse eliminative sociologism as deriving from an instance of the fallacy of equivocation: It is true that some arguments are social in so far as they are nothing but masks for vices. It is also true that all arguments are social in the sense of being social institutions. But to infer eliminative materialism from these two premises is to overlook the different meaning of 'the social' within them.

Turning from eliminative to reductive sociologism, I take the latter to hold that types of rational entities (arguments, theories) are numerically identical with types of social institutions. Sociorational laws state such type–type identities. The latter are, for instance, of the form 'arguments of type A are identical with what is credible in a social structure of type S'. Reductive sociologists will make claims such as the following: radical philosophical scepticism will be widespread amongst people who have a privileged position but lack influence in an arbitrary powerful political system (Douglas 1986); or 'primitive exception barring' is the typical intellectual reaction to novelty in a society characterised by internal fragmentation and the absence of a common enemy (Bloor 1983: 140–5).

Bloor's and Douglas's work informs pretty much all of what I myself have written on the sociology of knowledge. Nevertheless, I feel that reductive sociologism is probably mistaken. The point is this: the type–type reduction aimed for by the reductive sociologist is no more than an ideal. Although we would all perhaps like to have strict laws reducing the rational to the social, all we have so far are the vaguest of generalisations. All of the proposed laws are acceptable only with open-ended escape clauses such as 'other things being equal'.

Furthermore, I am doubtful about whether socio-rational laws can ever be more than such rough-and-ready generalisations, and thus doubtful about whether reductive sociologism is helpful even in the limited role of a regulative ideal. To see this, we need to distinguish between two ways in which the analyst might choose to identify types of the rational. These types are either actors' categories, that is categories used by the historical actors themselves, or else categories of the analyst, that is categories introduced and defined by the sociologist of knowledge. In the first case, the reductive enterprise is doomed for reasons that closely approximate to the failure of reductive materialism in the mind–body arena (Davidson 1994). Holism stops both reductive materialism and reductive sociologism in their respective tracks. In other words, scientific and philosophical arguments and theories come in webs and networks, and the character of any given argument or theory depends on endless other arguments or theories. Turn-of-the-century associationism was linked to a web of other psychological and philosophical positions, and these links were essential to what turn-of-the-century associationism was all about. The same holds for other occurrences of associationism at other times. And thus a type–type reduction is impossible: all tokens of associationism will – owing to their essential links to various other positions – be realised differently in the social realm. And they will be realised differently because they are part of a theoretical whole that is realised in the social realm holistically rather than atomistically. In the second case, the case where the analyst's categories provide the types, the enterprise stands in danger of turning from reductive sociologism into eliminative sociologism. It will be eliminative at least in so far as the analyst's categories of the rational will transcend those of the actors. And if these analyst's categories are chosen on the basis of a prior theory of the social, then the enterprise stands in danger of delivering social laws *simpliciter* rather than ratio-social laws. If theories of social institutions determine how the rational realm is carved up into types then the conceptual independence of the rational is quickly eroded.

In light of these considerations the more promising alternative to eliminativism seems to me to be 'anomalous sociologism'. This view is of course modelled on Davidson's stance with respect to the mind–body problem. Davidson's anomalous materialism holds that mental events are identical with physical events, denies the possibility of strict psychophysical (type–type) laws, rejects epiphenomenalism, and conceptualises the mental as supervening on the physical (e.g. Davidson 1994). Likewise, anomalous sociologism identifies historically situated ('dated') arguments, reasons, and theories with historically situated ('dated') social institutions, denies the possibility of strict ratio-social (type–type) laws, rejects sociological epiphenomenalism, and conceptualises the rational as supervening on the social.

In saying that the rational supervenes on the social I wish to emphasise the notion that types of arguments, reasons, and theories are multiply realisable. Associationism in turn-of-the-century Germany and associationism in present-day cognitive science departments in, say, the United Kingdom are both social institutions of sorts, but as social institutions they are rather different. Aquinas's theory of the mind was often regarded as similar to the Würzburgers' thought psychology, but it is highly questionable whether the two instances of the same type of psychology had any common features as social institutions. However, multiple realisability must not be mistaken for ontological independence. That the rational supervenes on the social rules out the following possibility: it cannot happen that a scientific culture shifts from, say, associationism to Würzburg-type thought psychology without a change in social institutions, social arrangements, or social interests. It should not be difficult to see why this is so. If thought psychology replaces associationism, there will be new authorities, new favoured methods of psychological work and argument, new relations between subfields of psychology, new curricula, and new standard textbooks.

Note that supervenience does not amount to epiphenomenalism. I am not denying that arguments, theories, or reasons can have causal powers. It would be wrong to say that arguments or theories are like the causally inefficacious steam whistle that accompanies the working of a locomotive engine. Their historically situated ('dated') instances have causal powers precisely because they are identical with social and psychological events. There is nothing wrong with saying that Bühler's arguments were part of the cause that ultimately led to a marginalisation of the Wundtians within German psychology. It is quite proper to say that, even though this will not be the full story of the Wundtians' defeat.

Interlude

The performative theory of social institutions

Introduction

In this 'Interlude' I shall provide a more fine-grained account of the 'performative theory of social institutions' that underlies the arguments of both Part I and II of this book. Nearly all of the key ingredients of this theory were developed by two writers in the sociology of scientific knowledge: B. Barnes and D. Bloor (Barnes 1983, 1988, 1995; Bloor 1995a, 1995b, 1996a, 1996b, 1997a, 1997b). A number of other authors have presented similar ideas independently (Anscombe 1976; Balzer 1993; Haugeland 1990; Itkonen 1978, 1983; Lagerspetz 1995; Searle 1995; Tuomela 1995).

Most important for the argument of Part II will be Bloor's insistence on the distinction between 'pattern matching' and 'concept application'. At least some forms of pattern matching can occur in the absence of social institutions, but concept application is dependent on the existence of social institutions. To put the central point in a nutshell: it is an essential feature of concepts that they can be applied correctly or incorrectly, that is that they 'ought to' be applied in some ways but not in others. 'Correctly' and 'incorrectly', however, are *normative* notions, and normativity is a social phenomenon. 'Oughts' exist only for members of groups.

Bloor's argument will be crucial in Part II. There I shall apply this general philosophical argument about concept application to the special case of folk-psychological concepts. I shall argue that the normativity of folk psychology can only be captured by treating folk psychology itself as a social institution.

In the first half of this chapter, I shall give special prominence to Barnes' paper 'Social Life as Bootstrapped Induction' (1983). This difficult paper laid the foundation, and provided the conceptual framework, for Barnes' and Bloor's subsequent work.

Two stereotypes of concept application (Barnes 1983: 525–6)

Barnes' starting point is the distinction between two stereotypes of concept application and reference. The first of these two stereotypes concerns the application of *natural kind terms*, such as 'tree' or 'man', the second the application of *social kind terms*, such as 'husband' or 'convict'. According to the stereotype for natural kind terms, concept application is based on *pattern recognition*. Individuals learn and remember patterns, and these patterns are tied to labels. When these individuals encounter worldly entities that fit the patterns, they attach the respective labels. The empirical properties of the encountered entities determine whether or not they fit the pattern; the empirical properties thus determine how the entities will be labelled or classified.

According to the second stereotype, the stereotype for social kind terms, neither pattern recognition nor empirical properties play any role in concept application. There cannot be patterns of empirical properties in this case because social kinds cannot be discriminated by means of our senses: a husband need look no different from a bachelor, and a prisoner no different from a free man or woman. Concept application therefore works differently in this case. Think, for example, of how the concept 'husband' is applied by a priest during a marriage ceremony. The priest might say something like 'I hereby declare you two husband and wife'. This 'performative speech act', as Austin called it, *creates* a social reality by being uttered.

One important difference between these two stereotypical cases is this. Concept application is self-referential in the second, but not in the first, case: if I am the priest declaring Otto to be a husband – by saying 'You, Otto, are now a husband' – then my referring to Otto as 'husband' is correct for no other reason than that I in fact refer to him in this way. 'Husband' therefore refers to my referring, and thus its use is self-referential. By the same token, its use is of course also self-validating: Otto is a husband because I say that he is a husband. Nothing of the sort occurs in the case of the idealised natural kind terms. According to the stereotype for natural kind terms, nothing becomes a tree just in virtue of my saying so. Whether or not the concept 'tree' can be applied to a given entity is determined by the internally stored pattern, and by the empirical properties of the entity. It is determined by nothing else. Referring to this entity is free of all self-reference.

N-devices (Barnes 1983: 527–9)

To bring the distinction between these two kinds of concept application into still sharper relief, Barnes suggests the following 'reification' of the two cases. This reification amounts to thinking about classification in terms of so-called 'designation devices'. Designation devices model 'heavily routinised classificatory activities, activities we carry out unthinkingly and automatically, as a matter of habit and custom' (1983: 527).

Let us start from the reification for the application of natural kinds terms; Barnes calls the designation devices needed here '*N* designation devices', or '*N*-devices' for the sake of brevity. An *N*-device is a simple machine. It takes physical entities (= *X*) as inputs, modifies some of them, and then releases all – the earlier modified ones in their modified form – as its outputs. An *N*-device consists of two subsystems that work on the inputs in a fixed sequence. The first subsystem is a 'pattern recognition system'. It contains a stored pattern (= *P*) and matches incoming *X* against this *P*. Those *X* that do not fit *P* are immediately released, unchanged, back into the environment. Those *X* that do fit *P*, however, are moved on to a second subsystem, the 'pattern attachment system'. As the name of this system indicates, the second subsystem attaches a permanent pattern to those *X* that are passed on to it from the pattern recognition system. Call this attached permanent pattern '*N*'. Think, for instance, of the following way in which a shepherd might mark his rams (say for recognising them more easily from a distance): if the sheep (i.e. *X*) have horns (i.e. *P*), then he sprays some blue colour (i.e. *N*) on their backs; if they do not have horns, they are released unchanged.

Next, consider the possibility that particulars might be put through more than one *N*-device, or that they may be put through the same *N*-device more than once. Obviously this opens up the possibility of confirming and refuting inductive inferences. Imagine that our shepherd is being followed around by a scientist who wishes to study the migrations of the shepherd's rams by attaching small electronic tagging devices to their ears. Unlike the farmer, however, the scientist attaches patterns, the electronic devices, based upon whether or not the encountered sheep have male sexual organs. This invites the inductive inference – valid as long as we stay on the shepherd's land – that sheep with blue colour patches on their backs are sheep with electronic devices tagged to their ears. We can also imagine that the same particulars are passed through the same *N*-device on successive days: This licenses, for example, the inference that sheep with blue colour patches on Monday are sheep with blue colour patches on Tuesday.

The *N*-devices considered thus far capture, in more precise terms, the intuitions underlying the idealised stereotype for the application of natural kind terms. First, we have here a more precise rendering of the idea that concept application in the case of natural kind terms is based on pattern matching. Second, the model embodies the intuition that natural kind terms give rise to inductive inferences. Third, the model affirms the belief that things to be classified must exist independently of the classifying activity, and that whether or not a concept is applicable is largely determined 'by the world' rather than 'by us'. This is captured by the stipulation that the pattern recognised by the pattern recognition system is always different from the pattern attached in the pattern attachment system. Not only must this hold in the case of a single *N*-device, it also must apply across the pattern recognition systems and pattern attachment systems of different *N*-devices. From the time they have passed through the shepherd's *N*-device, those *X* that were judged to be similar to *P* can be discriminated on the basis of two patterns: on the one hand, the shepherd's *N*-device can recognise their similarity to the 'initial pattern' *P*. On the other hand, the 'attached pattern' *N* might itself be stored in the pattern recognition system of another designation device, say a designation device belonging to the scientist. For instance, our scientist might use the shepherd's blue dots as the basis for tagging. Here the pattern recognition system of the scientist's designation device checks for *N*. Could the scientist's tagging still be construed as similar to the application of natural kind terms? Not according to the intuitions that underlie the stereotype of concept application in the case of natural kind terms. A natural kind term must be applied on the basis of 'initial' or 'original' patterns, that is patterns that have not been previously attached by others. Only in this way can all forms of self-reference be excluded.

S-devices (Barnes 1983: 529–30)

Barnes also introduces a designation device (or rather a family of similar devices) to model the application of social kind terms. He calls these '*S*-devices'. Like the *N*-devices above, an *S*-device is an input–output system with two subsystems: a pattern recognition system and a pattern attachment system. An *S*-device takes physical entities (= *X*) as input, modifies some of them, and then releases all – the earlier-modified ones in their modified form – as its output. The pattern recognition system contains a stored pattern (= *S*), and matches incoming *X* against this *S*. Those *X* that are not similar to *S* are immediately released, unchanged, back into the environment. Those *X* that are similar to *S* are moved on to the second subsystem, the pattern attachment system. The pattern attachment system attaches a permanent pattern to those *X* that it receives from the pattern recognition system. This attached pattern is – and this distinguishes *S*-devices from *N*-devices – the very same pattern *S* in terms of which *X* was recognised in the first place.

The functioning of an *S*-device would not be described correctly by saying that it attached another token of the *initial* pattern *S*. The *S*-device does not add another set of horns to those sheep it finds to have horns. Horns (i.e. *P*) are initial, natural, or original patterns; initial patterns are patterns that can be found in pattern *recognition* systems but *never* in pattern *attachment* systems. On the other hand, no pattern can occur in the pattern recognition system of an *S*-device unless it also occurs in the pattern attachment system of the same, or another, *S*-device.

This difference between *S*-devices and *N*-devices has an interesting corollary. In the case of *N*-devices, the pattern recognition system 'dominates', as it were, the pattern attachment system: The pattern recognition system discriminates whereas the pattern attachment system has the less demanding job of attaching labels to whatever is passed on to it. In the case of *S*-devices this hierarchy is collapsed or reversed: the pattern recognition system can get to work only once the pattern attachment system has attached the relevant patterns. It is this difference that makes it natural to say that someone possessing an *N*-device can *learn* (by means of their inductive inferences), whereas someone possessing an *S*-device (and no other device) can only *stipulate*.

Because in an *S*-device both subsystems use the same pattern, and presuppose one another – no pattern recognition without prior pattern attachment, and no pattern attachment without pattern recognition – it is difficult to see how an isolated *S*-device could ever get started, how it could, as Barnes says, ever get 'primed'. Possible solutions to this priming problem include the following.

First, we might assume that the *S*-device follows the lead of an *N*-device. The *N*-device attaches a pattern (on the basis of a prior recognition of an initial pattern); the *S*-device recognises the attached pattern and then adds the very same type of pattern. We might well imagine our shepherd having such an *S*-device built into him: because colours fade, he periodically might want to redo the spraying job; here he is assumed to pick out the sheep to be resprayed on the basis of their already having blue colour patches on their backs.

Second, an *S*-device might also get primed by a new kind of designation device, an 'arbitrary designation device' (an '*A*-device'): instead of a pattern recognition system, the *A*-device possesses a randomising system; the decision of which *X* to leave unchanged, and to which *X* to attach a pattern, is made at random. For instance, a shepherd might divide his flock in two by randomly spraying green colour on every second of his sheep.

Finally, note the important point that an *S*-device has the structure of a tautology: '*X* is *S* (pattern *S* is attached to *X*) because *X* is *S* (*X* fits recognised pattern *S*)'. This marks yet another sharp contrast with *N*-devices in which such tautology was principally ruled out by the stipulation that the two subsystems must contain different patterns.

A system of *S*-devices (Barnes 1983: 531–4)

Consider a multitude of *S*-devices and their interconnections. Assume that the output of one *S*-device can become the input for another *S*-device. Ignoring for the moment the problem of priming, once set up, the system obviously functions as follows. Some *X* in the environment have *S*-patterns attached to them, other *X* have not. Once an *S*-device recognises one of the *X* to have an *S*-pattern (i.e. once one of the *X* fits the stored *S*-pattern of the *S*-device), the *S*-device attaches another *S*-pattern to that very same *X*. Over time, the number of *S*-patterns attached to any given '*S*-patterned' *X* will thus increase.

This increase of *S*-patterns attached to different *X* allows us to model the emergence of social referents and social institutions. To do this, we only need to assume that different *S*-devices are sensitive to the *number* of *S* attached to given *X*. If the number is above a certain threshold *t*, *S*-devices attach a further *S*; if the number is below the threshold, *S*-devices leave the *X* unchanged. Different *S*-devices are assumed to have *different* such thresholds (starting with threshold $t = 1$).

Of course, this still leaves open the question of how the first *S* ever gets attached to a particular *X*; but we can leave this issue aside here. At this point it is more important to note that if we run the same set of *X* repeatedly through this egalitarian system of *S*-devices, eventually all *S*-devices will attach *S*-patterns; eventually all thresholds will be passed. We might say that the system has reached its *saturation* point at this stage.

Assume that we have a group of shepherds in the process of moving from a system of barter to using a medium of exchange. Imagine that they are moving to using the bones of dead sheep as currency. Initially only some shepherds will accept bones as a medium of exchange, and different shepherds will have different thresholds for accepting it. Some will regard this 'currency' as trustworthy only if it is involved in almost all exchanges. Others will be more trusting or ready to take risks: they might start storing sheep bones and accepting sheep bones in exchange for live sheep or grain, even though these bones are not accepted as currency by all shepherds and for all transactions. Every new transaction carried out in sheep bones, however, will give every shepherd more reason to accept sheep bones as currency. Saturation is reached once all shepherds accept sheep bones as a medium for all forms of exchange.

Intriguingly, the system process *prior to* saturation point is a process in which a referent is increasingly being *created* by the egalitarian system. The more *S*-devices attach *S*-patterns to a given *X* (because their threshold for attaching *S* has been crossed), the more *S*-devices identify or recognise this *X* as an *S*. Past saturation point the egalitarian system *maintains* the referents. In our example, initially there was no referent 'money'; and once the system starts working, money is being increasingly created as a referent.

Induction and tautology in an egalitarian system of S-devices

Note that an egalitarian system as a whole can be thought of as one big *S*-device: it takes *S*-patterned *X* as inputs and produces *S*-patterned *X* as outputs. Like the isolated *S*-device considered above, it too has a fully tautological structure: *X* is *S* for the array of *S*-devices because *X* is *S* for the array of *S*-devices. Sheep bones are money because the farming community takes them to be money. Or formulated from the inside, as it were: 'What we take to be currency is currency'.

The egalitarian system of *S*-devices *taken as a whole* 'celebrates a tautology', but individual *S*-devices that are part of an egalitarian system do not. (I shall call such *S*-devices 'embedded *S*-devices'.) An embedded *S*-device is able to learn, and thus able to confirm inductive inferences. For instance, with respect to the temporal inductive inference, the embedded *S*-device can confirm that 'All *X* with *i* number of *S*-patterns (with $i \geq t$, i.e. my threshold) to which I attach another *S*-pattern today will be *X* to which I attach another *S*-pattern tomorrow.' Or, more briefly: '*S* today, *S* tomorrow'. Why is this a case of learning and not a case of stipulation? Because the stipulation is now the action of the egalitarian system of *S*-devices as a whole, not the exclusive action of the single embedded *S*-device.

True, each one of the embedded *S*-devices carries out a part of the egalitarian system's overall action of stipulation. And thus learning about *S*-patterns is never quite like the case of learning about *N*-patterns. But the former case approximates the latter asymptotically as the number of *S*-devices in the egalitarian system increases. The less each of the *S*-devices contributes to the overall stipulation, the more it can learn about that stipulation. (At least in parentheses it is worth noting that a certain 'social fallacy' or 'social illusion' may arise with respect to this collective stipulation. Individuals note that they do not – individually – stipulate the existence of the social institution. They also recognise that none of their fellow subjects – individually – carry out such stipulation. They conclude mistakenly that no stipulation has happened. In this way, social institutions get misinterpreted as 'abstract, Platonic entities', as entities of some mysterious 'third realm'.)

To put it in terms of shepherds and sheep bones: the shepherds collectively impose the status 'currency' upon the sheep bones. The bigger the group of shepherds, however, the less each individual shepherd matters, or contributes. And the bigger the group of shepherds, the less each shepherd is able to look upon the creation and maintenance of the currency as his own doing. And thus he can *learn* that members of his group, including himself, will accept as money tomorrow what they accept as money today. Ultimately, it is of course the sum of such acts of learning that maintains the whole system. Because each shepherd individually draws the inductive inference that, in his group, money today is money tomorrow, therefore each one has a reason to accept money today (for using it tomorrow). And this alone guarantees that the referent, that is the currency of bones, is maintained, or recreated, from day to day.

Self-reference in the egalitarian system

Finally, it remains to analyse the full extent to which an egalitarian system of *S*-devices is self-referential. Barnes puts the central point as follows: 'taken as a whole, the referring activity which is the operation of the array is the same as what is being referred to as the array operates' (1983: 533).

Fleshed out for the case of money, this amounts to saying that what individual shepherds, in particular acts of referring, refer to as 'money' is not the bones themselves *as bones* but the sum of such particular acts of referring to bones as money. What makes bones money is that they are referred to as money; and thus to refer to bones as money in a given situation is to refer to that which makes them money, that is the referring activities.

The point is difficult to make fully transparent in terms of Barnes' *S*-devices in the egalitarian system. Somehow the language of attaching is ill-suited to bringing out what is going on. It seems odd to say that when a given *S*-device attaches an *S*-pattern it is attaching the other *S*-devices' acts of attaching *S*-patterns. We can capture the phenomenon better by shifting from the language of attaching patterns to the language of emitting and receiving signals.

Assume that the *S*-pattern-attachment systems (of our *S*-devices) are replaced by *S-signal-emission systems,* and that the *S*-pattern-recognition system is replaced by, first, an *S-signal-detection system,* second, a *memory,* and third, a *tracking device* that follows and records the movements of physically identical particulars X. All *S*-devices, as well as the *X*, are spatially near to one another so that each *S*-device is able to detect signals emitted by all the others, and close enough so that all *X* can be continuously tracked by all *S*-devices. The *X* move about at random, but they regularly bump into one or another of the *S*-devices (they then move on, before colliding with another *S*-device). When such a collision with a particular X occurs, *S*-devices either emit or do not emit *S*-signals on the basis of the following rule: they emit an *S*-signal if and only if their memory tells them that all other *S*-devices have previously emitted an *S*-signal when colliding with the same particular *X*. They do not emit an *S*-signal when their memory tells them that not all *S*-devices emitted an *S*-signal under this circumstance. The self-referential nature of acts of signalling can then be put by saying that each *S*-signalling (with respect to a given *X*) signals the sum of all of the prior *S*-signallings (with respect to the same *X*) by the same and all the other *S*-devices.

Intermediate summary

Let us take stock before moving on. I introduced social institutions earlier as self-referential systems of belief: money is what we collectively take to be money, or marriage is what we collectively take to be marriage. Having taken the difficult route through Barnes' paper, we can now replace this brief and intuitive starting point with a more detailed picture. Here I assume of course that Barnes' egalitarian systems of *S*-devices model social institutions of sorts.

Barnes' paper contributes the following ideas to our understanding of social institutions:

1 It makes the emergence of a social institution problematic (the problem of 'priming') and suggests simple mechanisms to solve this problem (the interplay of *S*-devices with *N*- and *A*-devices; the difference of thresholds of different *S*-devices, etc.).
2 It suggests a measure for distinguishing stages in the priming process (pre- and post-saturation).
3 It highlights the ways in which a social institution creates and recreates (or maintains) reference.
4 It identifies the basic modes of concept application that form the precondition for all reference to social entities.
5 It makes tractable various forms of, and interrelations between, the collective action and individual actions (individual and collective referring, learning, acting on the basis of inductive inferences, stipulating, signalling to others, tracking objects, and storing and calling up information about others' actions).

In at least these respects (1–5), Barnes' paper goes beyond other approaches in the literature. For all their sophistication, these alternative theories either fail to move much beyond the initial insight that social institutions are constituted by mutual beliefs (everyone believes that everyone else believes that so and so) (Lagerspetz 1995; Tuomela 1995), or else get side-tracked into subordinate problems: on what kinds of *X* can one impose what kind of *S*-status; if language is an institution, does its introduction presuppose another language; or how many types of collectively imposed statuses are there (Searle 1995)?

Up to this point, I have concentrated on interpreting, clarifying, and reconstructing Barnes' paper 'Social Life as Bootstrapped Induction'. Needless to say, this paper does not exhaust all that needs to be said about social institutions – indeed, it does not even include all that Barnes himself has said about the topic. In what follows I shall introduce five further key aspects of social institutions and concept application into the framework developed above, that is the framework of designation devices and their interaction. These five aspects are (1) the interaction of *N*- and *S*-devices, (2) concept possession and normativity, (3) the idea of local consensus and individual differences, (4) finitism, and (5) the function of social institutions.

The interdependence of *N*- and *S*-devices

Barnes' paper highlights the differences between *N*-devices and *S*-devices to bring the nature of the latter into the sharpest possible relief. Having followed Barnes along this road, we can now turn around and emphasise that the two kinds of devices – and the two mental processes that they model and reify – presuppose one another. (I shall use the language of 'devices' for both the mechanical models and the mental processes.)

Note, first of all, a point already made in Barnes' paper (1983: 538). In both everyday life and scientific practice, the patterns that we recognise often are not initial patterns; instead they are patterns that were attached by ourselves and others at an earlier point in time. When we buy food, medicine, or paint, we usually trust the labels on the products, and we do not perform a chemical analysis to determine the content of the can, tube, or pot. Scientists behave in the same way: the biologist, too, mostly trusts the labels on bottles and test tubes.

Second, in many cases in which we employ our *N*-devices (and thus work on the basis of initial patterns), we additionally check for attached (or signalled) patterns, patterns attached (or signalled) by others. Even when we try to determine, for instance, the ingredients of a food product (on the basis of initial patterns), we still usually pay some attention to the label on the product, for example to confirm or to challenge it. Or else, while engaging in *N*-pattern matching ourselves, we might also be attentive to the *N*-pattern matching performed by others around us. We might do so, for example, on the grounds that we all occasionally make mistakes. It thus seems reasonable to compare our results with those of others, and to rerun our pattern-matching device whenever, on a given occasion, its outputs differ from those of the majority of other devices around us.

Third, the kinds of self-referential and self-validating structures that we encountered in the case of single *S*-devices and arrays of *S*-devices also occur in certain arrays of *N*-devices. Take, for instance, the relationship between recognised and attached patterns in an array of *N*-devices, all of which attach *N* on the basis of the initial pattern *P*. Leaving aside cases of innately determined pattern recognition and attachment, it is usually a question of stipulation (i.e. convention) which patterns are regarded as relevant, or best suited, for picking out a certain group of *X*. It is equally conventional which pattern is chosen for labelling these *X*.

For instance, a community of shepherds interested in picking out the rams from the ewes might use any one of the following initial patterns: presence or absence of horns, shape of primary and secondary sexual organs, height, weight, or hormonal structure. None of these criteria is fully equivalent to the others. This phenomenon of collective stipulation returns us to the logic of *S*-devices. Whichever criterion the farming community selects, the relationship between the criterion and the concept will be a social institution: it will be marked by self-reference and self-validation. The criterion will *be* the criterion because the collective *takes* it *to be* the criterion. And, as before, each individual member of the collective will draw his or her inductive inferences about the collective as a whole, thereby participating in maintaining the institution itself. The farming community might say that 'horns are the decisive criterion for being a ram because we collectively take them to be the decisive criterion'.

The same type of analysis applies (trivially) also to the selection of the label. For instance, the farming community might use blue, green, red, or some other colour dots to mark out the rams from the ewes. Whichever colour is chosen, it will be *the* label for ram-hood because it is collectively taken to be the label for ram-hood.

Fourth, *S*-devices presuppose *N*-devices. Social statuses and functions are imposed upon something or other, and ultimately this something or other must be identifiable on the basis of its spatio-temporal features. A scientist might identify the rams on the basis of the blue dots, and the shepherd might identify the rams on the basis of some other earlier-attached pattern. But this chain of attached patterns must come to an end somewhere. To put it in J. Searle's terminology (Searle 1995), 'institutional facts' presuppose 'brute facts'. Social statuses are 'imposed' upon spatio-temporal entities; if we were unable to identify the latter then we would be incapable of identifying the former.

Normativity, proto-normative systems, and the possession of concepts

Fifth, and most importantly, we need to highlight properly the differences between pattern matching (e.g. on the model of *N*-devices) and the application, and possession of, a meaning or concept. What distinguishes an isolated *N*-device from a human that possesses the concept '*N*'? The crucial difference, analysed in considerable detail by Bloor (1995a, 1995b, 1996a, 1996b, 1997a, 1997b), is this. Possession of a concept differs from simple pattern matching in that the former is inseparable from the ability to distinguish between correct and incorrect, right and wrong, acts of classification or matching. The simple pattern matching of an isolated *N*-device does not allow for any such distinction. The *N*-device acts in accordance with its internal mechanics, and, at least from within the perspective of that mechanics, there is no way of distinguishing between right and wrong pattern matching. Should the internally stored pattern change over time, then this will be invisible to the device itself. All of its pattern matchings will be on an equal footing; none of them are more correct than others. And thus the whole distinction does not come into play.

What then has to be added to a simple *N*-device for there to arise the possibility of right and wrong, for there to arise normativity? The answer to be developed here is based on the idea – familiar from Durkheim, Hume, and Wittgenstein – that normativity is a social phenomenon; that 'right' and 'wrong', 'correct' and 'incorrect', mark the various distances that might exist between an individual's action, on the one hand, and a group consensus about what such a type of action ought to be like, on the other.

Assume then that we have an aggregate of individuals such that each one of them possesses at least one *N*-device. The question of how to explain or understand normativity can then be put in this way: what further devices must these individuals possess for them to have norms? And how must these individuals be related to one another? My answer is close to the suggestions made in Bloor (1996a) and Haugeland (1990). Such individuals need to be equipped with the following three further kinds of devices:

1 An *imitation device* that makes individuals attach patterns in the same ways as (most) others do; for instance, that makes individuals select the same initial patterns, use the same attached patterns, and so forth.

2 A *sanctioning device* that makes individuals (a) direct negative sanctions at those individuals who produce categorisations that contradict the collective or majority view; and (b) direct negative sanctions at those individuals who fail to do (a).
3 An *adjustment* or *self-correction device*: This device registers 'incoming' negative sanctions and corrects accordingly the workings of the *N*-device involved, the functioning of the imitation system, the operation of the sanctioning system, or the doings of the adjustment system itself.

I shall call such a system of individuals – where thus each individual possesses at least one *N*-device, plus the three devices mentioned under 1, 2, and 3 – a 'proto-normative system'.

Bloor and Haugeland suggest that such a proto-normative system provides the non-intentional, meaning-free underpinning for all social institutions, and thus also for social institutions such as norms or meanings. Only where we have a proto-normative system can we have 'right' and 'wrong', 'correct' and 'incorrect' applications of concepts. Of course social institutions as we know them are much richer and much more complicated than the proto-normative system. Our institutions involve language, norms, intentionality, concepts, and meaning. But each of these of course is in turn a social institution, and we need an account of what makes all social institutions ultimately possible. The proto-normative system provides the answer.

The *proto*-normative system is not a *pre*-normative system. It is not *pre*-normative because the system can be thought of as embodying a *de facto* norm (Bloor 1996a: 13; Haugeland 1990: 405). First, the proto-normative system brings it about that the operations of the *N*-devices will become more and more similar. Second, the class of similar operations of *N*-devices is itself the standard that determines which operations of individual *N*-devices are going to be sanctioned and adjusted. Thus, third, the class of similar operations is the *de facto* norm. Of course it is a *de facto* norm only; it needs language and verbal interaction to make the norm explicit.

It is important to understand in detail that the proto-normative system is similar to, but not identical with, the earlier-considered egalitarian system of *S*-devices. The crucial difference is that the egalitarian system of *S*-devices worked only with *one* pattern: *S*. *S* was *both* the recognised pattern *and* the attached pattern. In the present system, the difference between recognised patterns and attached patterns is preserved in that each of the individuals operates his or her *N*-device: that is, the individual attaches a pattern (or signals) *N* if and only if an encountered *X* fits the internally stored pattern *P*.

Another difference between the egalitarian system of *S*-devices and the currently considered proto-normative system is the absence of sanctioning in the former. Nevertheless, it is precisely the presence of sanctioning in the latter that brings out the similarity between the two cases most clearly. To see this connection, remember, first of all, that an individual in the proto-normative system engages in four kinds of acts: an individual imitates others; operates his or her *N*-device, that is attaches *N*-patterns; sanctions other individuals; and adjusts his or her own *N*-device in light of sanctions coming from others.

Moreover, the way in which each individual *operates*, or – as we might also say – 'tunes' *his or her own N-device* depends on how all the other individuals have tuned theirs. This dependence is due to imitation, sanctions, and sanction adjustment.

Furthermore, the way in which each individual *sanctions* others with respect to their operation of their *N*-devices is dependent upon how other individuals have sanctioned in the past. This might not be immediately obvious on the basis of what was said above. There I defined the sanctioning device as being cued to the majority (or collective) view of the *N*-devices. But obviously the second cue (majority or collective view concerning *N*) is equivalent to the first cue (majority or collective view concerning sanctioning). This equivalence holds because whatever is in line with the collective view will not be sanctioned; and whatever will not be sanctioned is in line with the collective view.

We can therefore say that what is deserving of a sanction – or what is *wrong* – is determined by a collective stipulation and is self-referential. (And likewise what is not deserving of a sanction, or what is *right*.) Each individual carries out only a small part of that collective stipulation and thus can learn, and draw inductive inferences, about the working of the system as a whole. Each individual sanctions – signals a sanction – on the basis of how the others have sanctioned or signalled; thus we can also say that to signal a sanction *vis-à-vis* a certain operation of an *N*-device is to signal the sanction signalling of other individuals (*vis-à-vis* a certain operation of a given *N*-device). This is how self-reference appears in the proto-normative system.

This kind of collective stipulation and self-reference is of course precisely what egalitarian systems of *S*-devices are all about. What makes the present case more complicated, and therefore also more interesting and more realistic, is that the self-referential circle both is longer and maintains a reference to the world. The circle is longer and contains more elements because it is made up of more, and different, kinds of actions than just 'recognising pattern *S*′ and 'attaching pattern *S*′. It is made up of the more complex actions of tuning one's own *N*-device to bring it in line with the operation of others; and of sanctioning in light of the collective consensus. And the circle maintains a causal link to the world because the *N*-devices involved in the system continue to work as *N*-devices: their pattern recognition system scans *X* on the basis of an initial pattern *P*, and the pattern, *N*, attached in their pattern attachment system is different from the initial pattern.

In such a system, the collective consensus and the process of sanctioning constantly shape and determine one another. The consensus sets the standard for the sanctioning, and the sanctioning protects and recreates the consensus. Neither phenomenon can be reduced to the other without distorting the overall process. If one gives insufficient attention to sanctioning, it becomes difficult to understand how a collective consensus can stay in place and determine individual actions. And if one focuses one-sidedly only on sanctioning, that is on determining correctness, one easily loses sight of the origins of the standard. It is then easy to slide into believing that the standard is somehow provided 'by the world'.

After these admittedly somewhat laborious explorations I am now in a position to summarise the difference between simple pattern matching on the one hand, and the possession and application of a concept on the other hand. Pattern matching is something any isolated machine or individual is able to do. It differs from the possession and application of a concept in that it has no space for the distinction between correct and incorrect. Concepts are 'possessed' primarily by normative systems of individuals, and they are possessed by individuals only in so far as they are parts of such systems. A concept is a community-wide set of similar dispositions with respect to how a given pattern should be matched and a label applied; these similar dispositions are themselves the outcome of sanctioning, self-correction, and imitation.

Individual variation and local consensus

In the last two sections I have developed the framework of Barnes (1983) in two ways: by allowing for interaction between *N*- and *S*-

devices, and by introducing normativity and concepts. Both moves were meant to narrow – in a controlled and tractable way – the gap between Barnes' abstract model and the social reality we know. In the present section I want to pinpoint, and in part bridge, a third such gap.

As it stands, the theory of social institutions expounded in this chapter does not yet explicitly incorporate two important truths about social life: that individuals differ, and that consensus is often a local, rather than community-wide, phenomenon. Social institutions do not turn individuals into carbon copies of one another; even socially shaped dispositions do not fully coincide; and members of social institutions can start trends that eventually change or destroy the social institution itself. Moreover, in the case of larger institutions, none of their members are able to observe or sanction the doings of but a small number of others. This shows that it is something of an idealisation to suppose that individuals can 'know' what the collective consensus actually amounts to.

Again it might be useful to have a 'reification' of these phenomena. Assume that we have a simple aggregate of *N*-devices – nothing as fancy as our earlier proto-normative system. Moreover, each *N*-device now has a certain degree of individuality: let us say that the recognised pattern *P* of each of the devices has its own characteristic tendency to go out of shape; in different devices this happens at different speeds, and also the direction of distortion is different. Over time, our devices will end up differing more and more concerning the *X* to which they attach labels. We thus have a 'mechanical' model for individuality.

Of course, as long as we let these devices act independently of one another we have no analogy of social life, or social institutions. Social institutions have the function of reducing divergence, and of increasing similarity of output. How can we model the function of social institutions in the present case? More than one way of reducing the divergence can be imagined. I shall mention three.

The 'way of the single authority' is to have all *N*-devices linked up to one authoritative *N*-device. This master device regularly 'resets' the shapes of all *P*-patterns of all the *N*-devices to whatever happens to be its own shape. (Like a master clock might send an electronic signal to all other clocks, thereby setting them all to its time.)

The 'way of the single consensus' is to have all *N*-devices linked up to one consensus-forming mechanism. At regular intervals, this mechanism simultaneously receives information about the shapes of all *P*-patterns of all the *N*-devices. The mechanism calculates the average of all of the readings it receives and resets the shapes of all *P*-patterns of all the *N*-devices to this value. (Like someone with two clocks, one showing two minutes to twelve, and one two minutes past twelve, might reset them both to twelve o'clock.)

Finally, according to the 'way of the multiple, local consensus', we allow our *N*-devices to move about freely, without being linked up to any kind of master device. But the space in which they move about is limited, and thus they randomly collide. Whenever two (or more) of them collide, they are able to perform the following operation (they thus need a little more machinery for doing this). They observe the shapes of each others', and their own, *P*-patterns, calculate the average of these two (or more) readings, and reset their shape to this value. They then continue their journey until they collide with another (or the same) *N*-device.

It is of course this third case that answers the challenge formulated early on in this section. The 'third way' models more adequately the scenario of a social institution in which no member has access to the action of all, and where some degree of conformity co-exists with divergence. Indeed, only under very special circumstances will all devices in the third scenario momentarily ever have *P*-patterns of the exact same form.

Moreover, the simple model can also capture phenomena of drift and change. Imagine, for instance, that from the above aggregate, we pick out a subset of devices and isolate them from the rest. Depending on what kinds of individuals we have chosen, that is depending on the direction and speed of the bending of their *P*-patterns, we might well end up with two communities that increasingly drift apart. Such drift might be measured by our calculating the average of all of the devices in each group. (Think, for instance, of how Canadian French has come to differ from the French spoken in France.)

This is not the place to develop a more detailed mathematical treatment of this intriguing third case. Suffice it here to emphasise only that the difference between the three cases is of great theoretical significance. Only too often is the social realm misunderstood as the realm of total conformity, or as determined by some mysterious mechanism of consensus formation. The third case shows that there are very simple models to dispel these illusions.

Finitism

Of course individuals do not just differ with respect to how quickly, and in what directions, their patterns change their shape. Much more important is the fact that the learning histories of different individuals differ from one another. Each individual has been trained to count certain *X* as being relevantly similar to one another in specific respects; such training leads to the formation of patterns used in the pattern recognition systems. The important point here is that individuals learn about such similarities on the basis of a limited, a *finite*, number of examples. And this finite number of examples is all they are able to go on. Different individuals will typically have been trained on slightly different sets of particulars. And thus it is likely that their spontaneous pattern matching will – in the absence of consensus-forming mechanisms – lead to different results. Even if we throw our individuals into a proto-normative system (of the multiple local consensus variety) such divergence will not be eradicated. Just as different finite sets of examples make for some divergence in pattern matching, so also do different finite sets of sanctions.

Even more important is another consequence of the fact that learning happens on the basis of a finite set of examples. I mean the consequence that Barnes and Bloor treat under the title of 'finitism' (e.g. Barnes *et al.* 1996). Because learning happens on the basis of a finite set of examples, because all the learner can go on is his or her knowledge of a finite set of cases, every new classification, every new pattern matching, tacitly or explicitly involves a judgement of similarity. The learner has been taught, say, that particulars *X*1, *X*3, *X*7, and *X*9 all belong together, that they all fit the same pattern *P*. But what about a newly encountered *X*8? It was not part of the learning process, but does it fit *P*? Is it sufficiently similar to the elements of the training set? This is the issue that the learner has to decide. Applying a concept is thus more like extending an analogy than acting under 'conceptual compulsion'. Past acts of pattern matching, past acts of classification, do not predetermine the next case.

This is not to deny that the application to the 'the next case' is usually done in a routine and automatic fashion – on the basis of one's learning history. The point is rather that what appears as the 'natural' matching to the possessor of one history of learning need not strike another individual as 'natural' at all. Judgements of similarity have a fuzzy logic; more can always be said than the simple 'match' or 'no match'. Various interests of individuals and groups can lead them to favour one similarity over another, to regard two entities, actions, or sanctionings as either highly similar or highly dissimilar.

This means, *inter alia,* that no individual can ever be absolutely sure that his or her acts of classification will not be challenged, criticised, and sanctioned by others. It also means that concepts, norms, and other social institutions can never have a final, ultimate content or structure. No concept can ever have a final, ultimate content because to have such content would mean to be no longer subject to the logic of fuzziness or the logic of analogy. And this can be had only at the cost of not being applied at all.

The same holds for every social institution. The participants all are acquainted with only a finite set of past situations, that is of examples, of what it is to act 'correctly' within this institution. Treating new situations in light of these past examples is again to extend an analogy creatively.

Institutions as collective goods

Throughout the past two sections, I have emphasised the fact that the existence of social institutions is compatible with – nay, implies – divergence and variation in a number of dimensions. In conclusion, it is worth stressing that social institutions nevertheless have the important function of containing and constraining variation. It is this that makes them collective goods.

Social institutions make humans more predictable to one another. Although humans differ in their dispositions to act (e.g. in their dispositions to classify, and to sanction) in at least the ways encountered above, social institutions make it so that such dispositions become and remain similar. For instance, we all have our individual idiolects, our regional dialects, and our styles of speaking and writing English. We have been taught at different times, perhaps in different countries, by different teachers, and to different standards. But for all that we are usually still able to understand and predict the grammatical structures of each others' sentences.

Different groups of people with different interests and goals will of course differ with respect to the question of in which areas of life increased predictability is desirable. It is thus not surprising that we find very different social institutions in different parts of the world, in different natural environments, and in different classes or strata of society.

On the other hand, it is equally obvious that *qua* human beings, *qua* biological creatures of a certain kind, all humans share a strong inclination to increase predictability in certain specific areas: we all have to eat and drink; we have to co-operate to defend ourselves against some predators and natural forces; we have to find mates and organise the raising of offspring. We also have to figure out what the humans around us want, what they are trying to convey to us, and what they are trying to do. Accordingly, we find in all cultures social institutions such as 'the meal', 'marriage', 'education', or folk psychology and language.

Our biology thus 'suggests' the emergence of certain types of social institutions. As a species, we would not have done quite so well if we had not solved certain problems of co-ordination by means of norms, sanctions, and conformity, that is by means of social institutions. Moreover, our biology does not only enter as a factor that defines species-wide co-ordination problems. Our evolutionary history also provides us with the natural capacities that enable us (to learn how) to conform to norms, to sanction, and to detect deviation.

Summary

In this chapter, I have reconstructed and extended Barnes' and Bloor's work on social institutions. This theoretical work will be my central resource in arguing the thesis that folk psychology is a social institution. I conclude with an analytical summary of the main claims and ideas set out above:

1. Social institutions are self-referential systems of actions and dispositions.
2. Social institutions are systems of sanctioning, imitation, and self-adjustment.
3. Social institutions are collective stipulations.
4. Individual members of social institutions contribute to the collective stipulation. The more members a social institution has, the smaller is each individual's share in the overall, collective stipulation.
5. Individual members of social institutions learn about the collective stipulation.
6. The condition of the possibility of all social institutions is the proto-normative system.

7 The existence of social institutions presupposes the existence of an independent natural world.
8 Social institutions are subject to the 'logic' of finitism. That is, social institutions are 'applied' or 'recreated' from situation to situation, and this application is not predetermined by the history of the institutions.
9 Concepts are social institutions.
10 Social institutions both imply and constrain individual differences.
11 Social institutions are collective goods: They facilitate action co-ordination.
12 The existence of social institutions in certain areas of life is contingent upon the interest or need to increase the predictability of human behaviour in these areas. Some of these interests are species wide.
13 The existence of a social institution is compatible with members' limited knowledge of other members' doings.

Part II

The sociophilosophy of folk psychology

Introduction to Part II

Defending the thesis that bodies of psychological knowledge are social institutions is a twofold task. One needs to argue the case *both* for *scientific- and* for *folk-* psychological knowledge. Above, I have tried to make plausible the view that *scientific*-psychological knowledge is best thought of as a social institution. I now turn to presenting my evidence for the case of *folk* psychology.

The easiest way to establish the thesis of this book with respect to folk psychology is to present philosophical rather than historical arguments. These philosophical arguments take issue with the currently dominant philosophical and psychological theories concerning the nature of folk psychology. 'Taking issue with' these theories means exposing their tacit *individualism*. By individualism I mean the view that the primary subject of knowledge – the primary 'knower' – is the *individual* human being. 'Individualism' contrasts with 'collectivism'. Collectivism holds that the primary subject of knowledge is irreducibly plural in nature; that is, that the primary subject of knowledge is the *collective*. To say that bodies of knowledge are social institutions is but an alternative formulation of collectivism. Therefore I can also describe my project in this second part as developing a collectivistic alternative to the dominant individualistic theories of folk psychology. I propose 'sociophilosophy of folk psychology' as a shorthand label for such a collectivistic interpretation of folk psychology.

One could argue that for at least two reasons the nature of folk-psychological knowledge must be a central topic in the more general debate between individualism and collectivism. First, folk-psychological knowledge is the most frequently employed and the most ubiquitous form of knowledge in human societies. It is involved in every single human interaction, and in every act of self-reflection. If collectivism turns out to be fundamentally right about this central type of knowledge then it seems unlikely that it could be principally wrong about other kinds of knowledge. Of course the same holds, *mutatis mutandis,* if individualism proves adequate for understanding folk psychology. To win the argument, collectivism must show that folk psychological knowledge is collectively sustained, conventional, and historically variable. Individualists will gain the upper hand if it turns out that the central features of this knowledge can be explained without invoking social processes and entities.

Second, both individualism and collectivism draw their initial plausibility from certain basic intuitions. For instance, individualism is supported by intuitions like the following: that 'the mind is its own place' (cf. Shapin 1990); that no one can know my mind better than I know it myself; that rationality is a property of the individual mind; and that the individual's rationality depends in part on his or her ability to shut out the corrupting influence of society. Faced with these individualistic intuitions, what can the advocate of collectivism do? Clearly, just pitting his or her own intuitions against those of an opponent is not enough. To take the argument further, the collectivist must somehow weaken the individualistic intuitions. One way this might be achieved involves folk psychology. Assume the collectivist can show that the aforementioned intuitions are just the kind one would expect to arise if folk psychology had the character of a social institution. Imagine the collectivist could make plausible not merely that our folk psychology is the 'intuition pump' informing individualism, but also, and more importantly, that the central individualistic intuitions actually spring from prima-facie plausible, but mistaken, renderings of the 'grammar' of folk psychology. Surely such dialectic ought to move the individualist.

I shall attempt such a Hegelian twist towards the end of my essay. Because the route taking us there will be long and complex, I had best begin with a brief overview. I will liken folk psychology to social institutions such as money. In both cases, we find something of a circular and self-referring structure: money is what we collectively take to be money, and psychologically understandable, or rational, behaviour is what we collectively take to be psychologically understandable, or rational, behaviour. Moreover, in both cases we find obligations, duties, and rights: for instance, the obligation to pay for the goods we want, or the obligation to make psychological sense to others. Furthermore, both institutions involve the creation of artefacts: in the case of money, coins and bills; in the case of folk psychology, mental states such as doubts, desires, beliefs, and emotions. I make the latter claim because I hold that the concepts we use to express or describe our mental experiences are partly constitutive of these very experiences themselves; part of what it is to be in doubt is to have the concept of 'doubt'. And finally, just as using the social institution of money does not imply a full understanding of the workings of that institution, so also being trained in folk psychology does not lead us to understand correctly the way folk psychology functions. In both cases, actors are prone to mystify or misunderstand the phenomena: 'to be money' is easily thought of as being an intrinsic, non-social property of certain metal discs, and 'to have a desire' is easily thought of as being an intrinsic, non-social property of certain entities called selves or minds. Sociophilosophy and sociology make it their business to overcome such mystifications.

8 The folk psychology debate

Introduction

In this chapter I shall prepare the ground for the 'sociophilosophy of folk psychology' that will be developed in the next chapter. I shall begin by providing a summary of the current debate over the nature of folk psychology. This debate concerns a number of different issues that can conveniently be grouped around two main questions. First, is our ability to understand others and ourselves based upon the possession of a folk theory of human mental life? And second, assuming that the answer to the first question is 'yes', is it conceivable that this folk theory is empirically false, and thus deserving of 'elimination' in favour of some *scientific* theory of the mind–brain? I shall report the various answers to these two questions separately. Subsequently, that is in the last section of this chapter, I shall identify what I take to be the central individualistic biases of the key positions in the debate.

The theory theory

In everyday life we explain, understand, predict, and express others', and our own, actions and mental states. For instance (if you were able to see me right now), you could explain my *action* of drinking from a glass of water by attributing to me both the *desire* for liquid and the *belief* that drinking from the glass will fulfil my desire. What is involved in this ability to make sense of one another and ourselves? According to the most widely held view in the recent philosophy of psychology, the answer lies in our possession of a 'folk theory' of human psychology. The basic theoretical terms of this theory are 'belief', 'action', and 'desire', and its laws are platitudes such as 'someone who is thirsty will usually try to get something to drink'. This explanation of our basic psychological abilities is often referred to as the 'theory theory' (Morton 1970). Although it might look innocent enough on first sight, almost everything about it is controversial. Philosophers and psychologists disagree not only over the question of which form the 'theory theory' ought to take, but also over whether any version of the theory theory is true. Equally contested are claims concerning the implications of the theory theory. To start us off, I shall introduce some prominent versions of the theory theory; some of these versions were first proposed by professional philosophers, others were first formulated by developmental psychologists.

David Lewis's and Stephen Stich's theory theory (1983)

The starting point of this version of the theory theory is David Lewis's work on the 'implicit functional definition' of theoretical terms (Lewis 1983 [1970]). Lewis suggests that theories introduce and define new terms by specifying causal roles. For instance, a new theory of the Kennedy assassination might introduce some person – call him or her *X* – who allegedly coerced Oswald into shooting the president. The theory might go on to claim that *X* held a grudge against Kennedy because of the latter's political success, and that *X* later rose to fame and power in the oil industry. According to Lewis, *X* is a *theoretical term*: it is a term introduced by the theory; a term that specifies a causal role; and a term that has meaning only in and through the theory. In these respects *X* differs from other terms used in this new assassination theory – for instance, terms such as 'grudge', 'Kennedy', or 'Oswald'.

Lewis proposes that folk psychology can be thought of as such a term-introducing theory. Common-sense mental concepts such as 'belief', 'desire', 'feel', 'think', and so forth acquire their meanings in the same manner as did '*X*' in the new assassination theory (Lewis 1972). Here is how Stephen Stich puts the idea:

> They are 'theoretical terms' embedded in a folk theory which provides an explanation of people's behaviour. The folk theory hypothesises the existence of a number of mental states and specifies some of the causal relations into which mental states enter – relations with other mental states, with environmental stimuli and with behaviour.
>
> (Stich 1983: 19–20)

According to Lewis (1972), we can assemble the central ingredients of our folk-psychological theory by collecting platitudes about causal relations between mental states, sensory stimuli, and motor responses. Stich tries to make this proposal more precise by sketching a formal characterisation of a fragment of folk-psychological theory (1983: 20). Let *M1, M2,…,M6* stand for a half-dozen (monadic) mental predicates, like 'desires ice-cream', or 'is thinking of Plato'; let *B1, B2,…,B6* stand for a half-dozen (monadic) predicates characterising behaviour; and let *S1, S2,…,S6* stand for predicates characterising environmental stimuli. Writing '->' for 'typically causes', we can then distinguish between the following types of folk-psychological generalisations.

First, there are generalisations that concern the interrelations between mental states, for example:

$$\text{for all } x \;\; \text{M1}x \;\&\; \text{M2}x \rightarrow \text{M3}x \;\&\; \text{M4}x$$

Second, there are platitudes that specify causal links between environmental stimuli and mental states, for example:

$$\text{for all } x \;\; \text{S1}x \;\&\; \text{S2}x \rightarrow \text{M1}x \;\&\; \text{M4}x$$

Third, other platitudes cover causal relations between mental states and behaviour, for example:

$$\text{for all } x \;\; \text{M1}x \;\&\; \text{M2}x \rightarrow \text{B3}x \;\&\; \text{B4}x$$

And fourth, still other generalisations concern more complex causal links between, say, combinations of stimuli and mental states on the one hand, and combinations of mental states and behaviour, on the other. For instance:

$$\text{for all } x \;\; \text{M1}x \;\&\; \text{S2}x \rightarrow \text{M3}x \;\&\; \text{B1}x$$

As Stich sees it, 'the whole of our folk psychological theory might be represented as a conjunction of principles' like these (1983: 20). We use these principles when we try to understand and predict the actions of others and ourselves. Finally, concerning the origins of folk psychology, Stich believes it to be the product of speculations of the early humans, that is of 'ancient shepherds and camel drivers' (1983: 229).

Paul Churchland's theory theory

A crucial inspiration for all theory theories is an early paper by Wilfred Sellars (1956). Sellars' work is particularly important to Paul

Churchland (see e.g. Churchland 1994). Sellars' paper puts forward something of an 'origin myth' for folk-psychological theory. He hypothesises an early stage of human development in which people's language lacked all vocabulary for inner mental states and processes. Humans' only resources for explaining and predicting each others' actions were some dispositional terms – more precisely, dispositional terms that could be operationally defined in terms of observable circumstances. Under these conditions, humans were severely limited in their ability to make sense of one another. Things changed dramatically, however, once 'Jones' appeared on the stage of history. Jones postulated the existence of unobservable internal mental states and events, taking overt phenomena as his models. Having observed that many actions were preceded by, and based upon, spoken-aloud reasoning, Jones theorised that such reasoning was there even when it was not uttered aloud. Actions could be the result of a hidden sequence of speech-like deliberations. Similar postulations introduced sensations, beliefs, intentions, and desires; that is, all of the central entities that are now part of the ontology of our folk psychology. Sellars supposes that Jones' contemporaries learnt Jones' theory, and that they eventually even started to use it spontaneously for first-person ascriptions.

According to Churchland, the lesson of Sellars' origin myth is clear:

> Our modern folk psychology has precisely the same epistemological status, logical functions, and modelling ancestry as the framework postulated by Jones.…It is, in short, an empirical theory.
>
> (1994: 309)

In his early work – that is, up until the late 1980s – Churchland believes that this theory is but a list of platitudes. Churchland's *Scientific Realism and the Plasticity of the Mind* (1979) cites the following laws, amongst others:

1. Persons tend to feel pain at points of recent bodily damage.
2. Persons denied fluids tend to feel thirst.
3. Persons who engage in vigorous activity tend to feel fatigue.
4. Given normal attention and background conditions, persons tend to perceive the observable features (i.e., the normal colours, shapes, textures, smells, sounds, and configurations) of their immediate environment.

5 Persons in pain tend to want to relieve that pain.
6 Persons who feel thirst tend to desire potable fluids.
7 Persons who are angry tend to be impatient.
8 Persons who believe that *P*, where *P* elementarily entails that *Q*, tend to believe that *Q*.

(1979: 92–3)

Laws such as these are needed for psychological explanations because such explanations are deductive nomological. For instance, John's thirst at a given time and place is explained by bringing his case under the 'covering law' (2).

Churchland strongly denies that these platitudes are analytic. They are not analytic because they constantly figure in causal explanations and because people continuously modify and qualify them. Nevertheless, it is true that laws such as the ones listed above do at first *appear* to be analytic. This is due to the fact that they are of great practical importance to us; we all do our best to protect their meaning (1979: 93–4).

One important consequence of the theory theory, Churchland claims, is that it provides a straightforward solution to the old sceptical problem of the existence of other minds. The assumption that other minds exist is not based on an inductive analogy with one's own case, nor is it deduced from observations of others' behaviour. Instead, this assumption is an explanatory hypothesis that – together with the laws of folk psychology – is crucially involved in explaining, understanding, and predicting others' behaviour. Given the general success of this hypothesis, it is reasonable to stick to it and regard it as true (1989 [1981]: 3).

A further important consequence – clearly visible already in Sellars' origin myth – is that the theory theory undermines the notion of first-person privileged access (Churchland 1979: 96–9). Mastery of mental vocabulary does not start from one's learning to identify mental states in one's own case. There is no principled difference between how one knows one's own, and how one knows another's mind. Both cases are structurally identical in that theoretical terms, hypotheses, and laws must always be involved. Here Churchland praises Kant for 'insisting that knowledge of oneself is entirely on a par with knowledge of the world external to oneself' (1979: 99).

Finally, it needs to be mentioned that in his writings from the late 1980s onwards, Churchland revises his account of the theory theory (1989, 1989 [1988]). Churchland now deems it a mistaken 'empiricist assumption' to think that theories are sets of sentences or propositions (1994: 309). In light of the results of neuroscience, and of Kuhn's emphasis on tacit knowledge, it is more appropriate to interpret theories as configurations of synaptic weights in the individual human brain:

> To learn a theoretical framework is to configure one's synaptic connections in such a fashion as to partition the space of possible neuronal activation patterns into a system or hierarchy of prototypes. And to achieve explanatory understanding of an event is to have activated an appropriate prototype vector from the waiting hierarchy.
>
> (1994: 315)

According to Churchland, these new conceptions of theory and explanation do not undermine the original claims of the theory theory. They strengthen them. Most importantly, the new interpretation not only eliminates the need to postulate thousands upon thousands of laws and platitudes, it also manages to escape from the many good criticisms of the deductive-nomological model of scientific explanation (1989 [1988]: 121).

Jerry Fodor's theory theory

As we shall see later in this chapter, Churchland's and Fodor's views of folk psychology are very far apart. Both philosophers agree, however, in taking folk psychology to be a theory of sorts.

Fodor often reminds his reader how easy it is to overlook our constant reliance upon 'common-sense psychology'. Usually, folk psychology works so smoothly that we hardly notice its presence. And yet it is involved in every interaction with other humans; for instance, in agreeing to meet someone at an agreed time and place (1987: 3).

Fodor is particularly concerned with defending folk-psychological generalisations against the charge of vacuity. Someone might object that all folk-psychological generalisations contain *ceteris paribus* clauses, and that these clauses make it impossible ever to disconfirm these generalisations. For instance, the objector might say that the generalisation 'all else being equal, persons who are angry are impatient', comes down to just this: 'persons who are angry are impatient – unless they are not' (cf. Fodor 1987: 4). Fodor rejects this objection by drawing on a parallel between folk psychology and geology. No one denies the scientific credentials of geological laws, and yet these laws always have *ceteris paribus* clauses. The 'modest truth of geology: A meandering river erodes its outside bank' also contains an implicit *ceteris paribus* clause; the law is true only if we exclude the possibilities of frozen rivers, the end of the world, and much else besides. Certainly, 'all else being equal, a meandering river erodes its outside bank' does not mean 'a meandering river erodes its outside bank – unless it doesn't'. It rather means something like 'A meandering river erodes its outside bank in any nomologically possible world where the operative idealisations of geology are satisfied'. And this is far from being vacuously true (1987: 5). The same holds for folk-psychological generalisations.

Fodor is also eager to emphasise the 'depth' of folk-psychological theory. This means, first of all, that the important folk-psychological generalisations are not folksy platitudes like 'money can't buy happiness' or 'all the world loves a lover' (1987: 6). Folk-psychological explanations do not subsume given situations under such folk wisdoms. Most folk-psychological explanations are much closer to 'the "deductive structure" that is so characteristic of explanation in real science'. That is to say: 'The theory's underlying generalisations are defined over unobservables, and they lead to its predictions by iterating and interacting rather than by being directly instantiated' (1987: 7). Usually, when we are trying to understand another person's action, we need to rely on something as complicated as an implicit decision theory: We need to invoke a bundle of interrelated generalisations concerning the interactions between beliefs, preferences, and behaviours (ibid.).

Last but not least, Fodor has a specific proposal on the mode of transmission, and the truth, of folk psychology:

> Here is what I would have done if I had been faced with this problem in designing *homo sapiens*. I would have made a knowledge of common-sense *homo sapiens* psychology *innate*; that way nobody would have to spend time learning it. And I would have made this innately apprehended common-sense psychology (at least approximately) *true*....The empirical evidence that God did it the way I would have isn't, in fact, unimpressive.
>
> (1987: 132)

The theory theory in developmental psychology: Allison Gopnik, Josef Perner, Henry Wellman

Over the past decade, the theory theory has become increasingly important in *developmental psychology*. Child psychologists now study how children's ability to conform to, and abide by, 'adult folk psychology' develops and increases over time. A number of developmental psychologists take (adult) folk psychology to be a theory, and they suppose that every (normal) child eventually comes to reinvent this theory. More precisely, the child hits upon this theory after having first tried out less successful mentalistic theories or generalisations.

The theory theorists in psychology are particularly eager to draw parallels between scientific theories and children's theories; between theory change in science and in childhood; and between scientist and child in general (e.g. Gopnik 1993, 1996; Gopnik and Wellman 1994; Perner 1991; Wellman 1990). To provide a flavour of the approach, I shall here summarise the paper 'The Theory Theory' by Allison Gopnik and Henry M. Wellman (1994).

The authors begin by acknowledging that there are important differences between child and scientist. For instance, theory change in science is a social–collective process, whereas cognitive change takes place in the individual (1994: 258). Nevertheless, there exist 'deep similarities' between the child's and the scientist's cognitive activities. Both are trying to learn about the (social and natural) worlds around them, and both have to organise, produce, and store knowledge. Thus it seems natural to 'use the scientific notions as a touchstone for characterising an important sort of cognitive development' (1994: 259).

To make the case for this parallel in more detail, Gopnik and Wellman begin with a brief reminder of some key features of scientific theories. Experience can be organised in two ways: either by 'empirical typologies and generalisations' or else by 'theories'. Empirical typologies and generalisations order and partition the evidence of experience. In so doing, however, they do not go beyond the vocabulary in which the evidence is presented. Moreover, empirical typologies and generalisations do not postulate unobservable entities to explain the evidence. (Think, for example, of a plant taxonomy based on visible features of the plants.) Theories, on the other hand, go beyond the observable evidence; their vocabulary is usually different from the vocabulary in which the evidence is presented, and they postulate lawfully interrelated explanatory unobservables. Examples to which Gopnik and Wellman refer include Darwin's theory of evolution and Kepler's theory of planet movements (1994: 260).

While both empirical generalisations and theories can be the starting point of explanations and predictions, explanations and predictions made on the basis of theories are deeper and more far reaching. A plant taxonomy (based on visible characteristics) is restricted to explanations like 'this plant is stemless because it belongs to the class of stemless plants'. A Darwinian theory, however, can refer to adaptations and environments, and to principles of natural selection. As far as predictions are concerned, theories often allow for predictions about novel cases, that is cases that were not known when the theory was first formulated (1994: 261).

Finally, theories differ from empirical typologies and generalisations also in that they call for interpretation of the evidence. This much is clear already from the fact that the language of the theory is different from the language of the evidence. Because of the gap between the two vocabularies, theorists must translate, or interpret, the evidence. Such interpretation sometimes amounts to a rejection of data as mere noise (1994: 262).

Given Gopnik and Wellman's concern with the question of development, it is not surprising that they also remind their readers of some key features of theory change in science. Three features are particularly important for them. First, advocates of a theory often initially deny the importance of accumulating counterevidence. Second, at a later stage, counterexamples are dealt with by means of *ad hoc* hypotheses, that is by hypotheses that are formulated to deal with particular cases only. And third, a new, alternative theory usually emerges only very slowly. It often has its humble beginnings as a minor conceptual structure within the earlier theory. Only eventually is it realised that this minor conceptual structure also can successfully deal with the central evidence and counterevidence (1994: 263).

With this account of theories in place, Gopnik and Wellman go on to present the child's developing understanding of the mind as a succession of theories. The main theory shift is from the theory of the 2-year-old to that of the 4-year-old. The 3-year-olds constitute an intermediate phase.

The theory of mind of the 2-year-old is structured around the concepts of perception and desire. Both concepts are understood in non-representational terms. Thus the child assumes that the viewer will perceive whatever is within the viewer's line of sight; or that whoever desires the candy on the table will inevitably reach for it (1994: 265).

From roughly the third birthday on, the child starts to extend its mental ontology. The child starts to talk about beliefs, and begins to have an understanding of pretences or dreams as 'representational but fictional mental states'. The 3-year-old does not yet have an understanding of belief as representational, and thus as potentially misrepresenting. Instead, the child treats beliefs as if they were direct reflections of the world, that is as if they were 'copies' of reality. This tendency of the 3-year-old comes out clearly in the child's performance in various 'false-belief tasks'. In one such task, the experimenter shows the child a candy box with pencils inside. The experimenter then asks the child what another child, currently not present, would think the candy box contained, candy or pencils. The 3-year-old usually answer 'pencils'. The 3-year-olds' concept of belief does not allow someone to believe that p when they (the 3-year-olds) know that not-p (1994: 266).

The main difference between the theory of the 3-year-old and the theory of the 5-year-old is that the latter contains a 'representational model of mind'. The child now realises that 'almost all psychological functioning is mediated by representations'. Mental states involve representations of the world, rather than the world itself. This new insight enables the child to conceptualise misrepresentation; and thus the child now also passes false-belief tasks with ease (1994: 267). This conceptual 'revolution' brings about other changes as well: The child comes to see the mind as an active processor rather than a mere repository of beliefs; comes to distinguish between thinking and believing; and learns to understand the complex interplay of thinking, perception, sensation, emotion, desire, intention, and action (Wellman 1990: 108–9). Later, this theory does get elaborated still further, mainly through the introduction of psychological traits, such as, for instance, intelligence, vanity, selfishness, or introversion (Wellman 1990: 115).

But why conceptualise the child's psychological knowledge as a theory? Gopnik and Wellman advance several replies. First of all, they point out that children's explanations of human actions are theory informed. The 2-year-olds construct their explanations mainly around perceptions and desires; older children also use the category of belief (1994: 268). Second, in their predictions too, children of different ages rely on the characteristic resources of their respective theories. Thus 2-year-olds are quite successful in predicting that fulfilled desire leads to happiness, or that people will perceive an object in their vicinity. Moreover, the influence of their theories is particularly telling where children make incorrect predictions. The paradigm case here is of course the false prediction in the above false-belief task; that is, the 3-year-old child's prediction that a newly arriving child will think that the candy box contains pencils (1994: 268–70). Third, interestingly enough, children also interpret evidence so as to make it fit their theories. In a variation on the original false-belief task, children were asked 'When you first saw the box, before we opened it [i.e. and before we showed you that it contained pencils rather than candy], what did you think was inside?' The 3-year-old children consistently replied 'pencils', thus misreporting, that is misinterpreting their own beliefs (1994: 271). Fourth, and finally, Gopnik and Wellman also find 'transitional phenomena' in the shift from the non-representational theory of the 2-year-old to the fully representational theory of the 4-year-old. The period of transition is that of the 3-year-old. The youngsters of that age group begin to handle mental representations successfully, but only for 'representational but fictional mental states' such as pretence or dream (1994: 272).

Theories and modules

Much of Gopnik and Wellman's detailed comparison between child and scientist is meant to combat an alternative view on the child's development and ability. According to this alternative conception, the child's theories of mind are 'embodied' into one, or several, innate modules (Leslie 1991; Baron-Cohen 1995; Segal 1996). Advocates of the modularity thesis see themselves as theory theorists, too, albeit that the theory is hardwired into the brain, or genetically programmed to mature over time. Modularity theorists advance a number of arguments for their position. To begin with, they maintain that folk psychology is much too difficult for children to (re)invent. Children need to be 'pre-programmed' to hit upon it (Segal 1996: 152). Moreover, only the modularity theory can explain the uniformity of folk psychology within and between cultures; a Gopnik-and-Wellman-style account is faced with the problem of the underdetermination of theory by the data (1996: 153). Furthermore, children's adoption of new theories differs from scientists' theory choice in that only scientists engage in 'metatheoretical reflection'. When faced with counterevidence, scientists seek out alternatives, recheck their data, and wonder what went wrong. Children do nothing of the sort (1996: 154). And finally, the genetic disorder of Williams syndrome also supports modularity. Children suffering from this disease have very low intelligence (IQ of about 50), and are quite incapable of acquiring theoretical, explanatory knowledge. And yet, these very same children have excellent social skills, skills that presuppose mastery of an elaborate theory of mind (1996: 154).

Gopnik and Wellman deny that modularity theory is a variant of theory theory. They agree that the just-born child possesses 'an initial state, a first theory' (1994: 281). But only this first theory is innate. The further theories of the developmental sequence are not innate structures that eventually 'come on line', and they are not the mere result of a progressive parameter fixing with respect to the original theory. Because the child's theorising is similar to that of the scientist, Gopnik and Wellman attribute the development of successive theories to the interplay of three factors: the innate first theory, data from the environment, and general theory-forming capacities. The universality of the data and the innateness of the capacities suffice to explain why folk psychology is ubiquitous; the modularity thesis is not needed to account for this result:

> Children the world over may develop the same understandings because they are specified innately, or they may converge on the same conceptions because the crucial evidence is universally the same, and so are theory-formation capacities. All people do, in fact, have minds, and so it is not too surprising that all normal children reach this conclusion about them.
>
> (Gopnik and Wellman 1994: 282)

Criticisms of, and alternatives to, the theory theory

The theory theory has been attacked by a considerable number of authors, both in philosophy and in developmental psychology. In this section, I shall summarise the main criticisms and alternatives.

Peter Hacker's and Kathleen Wilkes's Wittgensteinian criticism

The criticisms of Peter Hacker and Kathleen Wilkes are fairly similar. Both seem influenced by Wittgenstein: either by Wittgenstein's insistence on the diversity of language games (Wilkes), or by Wittgenstein's thesis concerning the primarily expressive nature of psychological concepts (Hacker).

Wilkes (1984, 1991) has repeatedly challenged the main assumption underlying the theory theory: to wit, that science and common sense work along similar lines. According to Wilkes, this view is 'deeply wrong' (1984: 340).

Wilkes warns of the dangers of overextending the scope of 'theory'. The concept loses all its content once we allow that any assumption, or any network of concepts, falls under it (1984: 342–3). Accordingly, Wilkes seeks to identify a number of respects in which folk-psychological activities differ from scientific practices.

To begin with, although 'other things being equal' clauses appear in both scientific laws and in folk-psychological platitudes, only in the case of the former can the 'other things' be at least roughly specified (1984: 344–5). Moreover, whereas scientists seek to restrict the number of intersections between different theories, lay psychology knows of no such limits. Folk psychology is marked by 'massive intersection' with ' "theories" about the weather, food, health, politics, economics, transport, theology, the family, sociology, anthropology, geography, history…etc.' (1984: 345–6).

In addition, generalisations play little to no role in folk-psychological explanations. Folk-psychological explanations are usually based upon the particularities of the given individual case (1984: 346). Lay psychologists simply are not interested in generalisations or in natural kinds. They are only concerned with understanding the behaviour of specific individuals. And 'we usually neither want nor expect our explanations to generalise' (1984: 350–1). In the same vein, Wilkes also notes that lay psychologists are more preoccupied with exceptions than with the rule (1984: 353).

Furthermore, it is most natural to think of folk psychology as a 'knowing how' rather than a 'knowing that': folk-psychological knowledge contains few explicit generalisations, and it is usually employed in an unthinking and automatic fashion (1984: 347).

In addition, folk psychology has many more functions than explanation, description, and prediction. It is also needed to

> blame and praise; to warn, threaten, prohibit, deter, urge, and cajole; to sneer, hint, insinuate, imply, and suggest; to assess and judge; to plead and to excuse; to promise and contract; to joke and pun; to write poems and plays.

No scientific theory is ever used for such diverse purposes (1984: 347–8). And as if this was not enough, many of these purposes presuppose that the vocabulary is vague, highly flexible, and ambiguous (1984: 349).

Folk psychology differs from science also in that it is based on a normative assumption: to wit, that people act by and large rationally. The existence of this assumption explains why folk psychology provides little help when it comes to understanding problems such as akrasia or self-deception. Folk psychology simply is not meant, nor equipped, to deal with such problems (1984: 356).

Finally, Wilkes cites a list of features that make for a good theory:

> good theories are: (a) predictively accurate (within certain limits), (b) internally coherent, (c) externally consistent, (d) have unifying power, (e) are fertile of new hypotheses, allowing for extension to new domains, and (f) are simple and economical.
>
> (Wilkes 1984: 357)

Lay psychology satisfies (a) to some limited degree, but it can offer nothing to answer the demands (b) to (f). To Wilkes's mind, this failing does not suggest that lay psychology is a bad theory; instead, it indicates to her that folk psychology is no theory at all (1984: 358).

Hacker (1996) presents six arguments against the theory theory. First, Hacker thinks it a mistake to equate being concept laden with being theoretical. It is true that humans use concepts in articulating their observations concerning others' behaviours and attitudes. But it is false to think that these concepts amount to anything like a theory:[1]

> There must be a contrast between what is theoretical and what is non-theoretical, if the expression 'theoretical' is to have any content. A scientist's description of photographs of particle decay in a cloud-chamber will indeed be theory-laden; a description of my room as containing a desk, a sofa and two armchairs, all standing on a red carpet, etc. is not.
>
> (1996: 528)

Second, as Hacker sees it, much of the theory theory of folk psychology is influenced by Chomsky's claim that linguistic competence is based upon the possession of a grammatical theory. But Chomsky is wrong. To learn a language is to learn a skill rather than a theory. If a theory were involved then it should make sense for someone to disbelieve, or be sceptical, about it. But this is clearly impossible: 'One cannot disbelieve or be sceptical about English' (1996: 529).

Third, although our psychological expressions do indeed form an interconnected 'network', they do not, in virtue of this interconnection, constitute an empirical theory: 'The propositions of the network are *grammatical* propositions, not theoretical ones' (1996: 530). In other words, these propositions are rules on how the psychological concepts are to be used. They specify what are meaningful and meaningless ways of using these concepts.

Fourth, it is a grammatical mistake to say that we use concepts such as 'believe' or 'want' to ascribe mental states to others and ourselves. To begin with, beliefs and wants are not mental states at all. This can be seen from the fact that the question 'What mental state are you in today?' cannot be answered meaningfully by 'I am in a state of believing that Richard III did not murder his nephews' (1996: 531). Furthermore, even those psychological concepts that can sometimes be used to ascribe mental states – that is, concepts such as 'expectation', 'hope', or 'fear' – also have other, more typical uses where no such ascription is made: '"The Director is expecting you in his office at 10:00" does not mean that the Director is in a state of expectation' (1996: 534). And finally, as pointed out by Wittgenstein, the primary use of psychological concepts is in first-person present-tense sentences that *express* rather than describe mental phenomena (1996: 527, 535).

Fifth, it is not the case that – led by folk-psychological theories – we form hypotheses concerning the existence of unobservable mental states. Again it is the first-person use of psychological vocabulary that provides one of the decisive arguments: 'To groan or cry out in pain is not to make any hypotheses about one's pain'. But even disregarding the case of the first person, there is no direct parallel between mesons or genes, on the one hand, and beliefs or desires, on the other. After all, 'one can hear a person's thoughts, if he tells them to us' (1996: 536).

Sixth, and finally, 'explanation of human action...is radically unlike scientific, causal explanation' (1996: 538). To say that one picked up a friend at the airport *because* one had promised to do so, is not to give a causal explanation at all. Instead it is to provide a *reason*. It is to invoke a 'grammatical' rule, that is a rule that defines what a promise is, and thus, a rule that is partly constitutive of the concept of promise (1996: 537).

Craft and rationality: Daniel Dennett's normative alternative

Daniel Dennett is far less hostile towards the theory theory than are Wilkes and Hacker. Nevertheless, Dennett rejects the idea that our folk-psychological capacities are due to our employment of an empirical-descriptive, realist and causal psychological theory.

Instead of a 'theory' of folk psychology, Dennett prefers speaking of a 'craft'. In doing so, Dennett feels vindicated by Churchland's claim that theories are best thought of as configurations of synaptic weights rather than as sentential structures:

> Churchland...now wants to say that folk psychology is a theory; but theories do not have to be formulated the way they are formulated in books. I think that is a good reason for not calling it a theory, since it does not consist of any explicit theorems or laws.
>
> (Dennett 1991b: 135)

It is thus a craft – partly innate, partly learnt, mostly tacit – that allows us to predict the actions of others. Nevertheless, in addition to the craft, there is also some sort of folk-psychological theory. Here by 'theory' Dennett means 'ideology', that is the folks' and philosophers' speculations about the functioning of the craft, and the structure of the mind. Dennett suspects that the majority view in the realm of ideology is still some form of Cartesian dualism (1991b: 137).

How then does the craft work? As Dennett sees it, the core of the craft is a specific 'stance' *vis-à-vis* other humans, animals, or even machines. When we adopt this stance, we treat these entities (i.e. the humans, animals, and machines) *as if* they had beliefs and desires, *as if* their beliefs and desires were fully rational, and *as if* these beliefs and desires led, in rational ways, to rational actions. Folk psychology is thus both normative and abstract: It is normative because it allows us to calculate what agents *ought to do*, and it is abstract (or instrumentalist) in that the attributed beliefs and desires

> are not – or need not be – presumed to be intervening distinguishable states of an internal behaviour-causing system....The role of the concept of belief is like the role of the concept of centre of gravity, and the calculations that yield the predictions are more like the calculations that one performs with a parallelogram of forces than like the calculations one performs with a blueprint of internal levers and cogs.
>
> (Dennett 1987: 52)

Dennett does not deny that we also find *some* empirical generalisations within folk psychology; generalisations such as that young children believe in Santa Claus, or that people can be influenced by propaganda. But such generalisations are 'laid over' the essentially normative framework (1987: 54).

Dennett's view seems to commit him to saying that folk-psychological explanations are rationalising rather than causal. That is to say, he must deny that the 'because' in sentences such as 'She quit smoking *because* she believed it was affecting her health' has the same meaning as the 'because' in sentences such as 'The fire started *because* the electrical system was overloaded'. Dennett's critics are not convinced that this move is plausible (Egan 1995: 187).

Ambiguities of the theory theory: Stephen Stich and Ian Ravenscroft

One of the foremost early advocates of the theory theory, Stephen Stich, has recently deplored its alleged ambiguities. Together with his co-author, Ian Ravenscroft, Stich has suggested a number of different interpretations of what the theory theory might be, and he has invited his colleagues to make explicit just which of these views they wish to advocate (Stich and Ravenscroft 1996 [1994]).

A first crucial distinction is that between an 'internal' and an 'external' account of folk-psychological theory. One gives an internal account of folk psychology if one assumes folk-psychological theory to be an 'internally represented "knowledge structure" used by the cognitive mechanism underlying our folk-psychological capacities' (1996 [1994]: 132). Such internal account can take at least four different forms depending on how one assumes the information is stored. It can be stored

> [1] in a mental model or a connectionist network that maps onto a set of propositions in a unique and well motivated way; [2] in a system consisting of rules and sentence-like principles or generalisations; [3] in a system consisting only of rules; [and 4] in a mental model or connectionist network that *does not* map onto a set of propositions in a unique and well motivated way.
>
> (1996 [1994]: 132)

One adheres to an external account if one takes folk-psychological theory to be the work of the scientific psychologist who studies the folk-psychological intuitions of lay subjects. The psychologist's theory is either the set of platitudes to which the subjects assent, or else a systematisation of this set (1996 [1994]: 132).

One important upshot of this classification is the observation that only under *some* of these interpretations does it make sense to speak of the truth or falsity of folk-psychological theory. One cannot regard this theory as either true or false if it only consists of rules (internal [c]) or if it is stored as a network or mental model that does not map onto propositions at all (internal [d]).

The simulation theory: Alvin Goldman, Robert Gordon, Paul Harris, Jane Heal

Of the various alternatives to the theory theory, the so-called 'simulation theory' has recently won the most attention.[2] In part this might be due to the fact that – in marked contrast to other alternatives – advocates of the simulation theory have offered detailed new explanations of the developmental–psychological data. Simulation theory might also appeal to many because – at least in some of its versions – it re-establishes a difference between the (allegedly) 'direct' way in which I know of my own mental states, and the (allegedly) 'indirect' way in which I know of the mental states of others.

The simulation theory has a long past but a short history: it has important ancestors in the Scottish Enlightenment's theories of sympathy, and in German Romanticism's notions of *Einfühlung* (empathy). Nevertheless, the currently debated variants did not emerge until the mid-1980s (Goldman 1989; Gordon 1986; Harris 1991; Heal 1986).

The starting point of Robert Gordon (1986) is the observation that we often are able to make certain kinds of predictions about our own future actions: these predictions concern our actions in the immediate actual future, actions about which we have already made up our mind. For instance, I (MK) can, as I am writing these words, currently predict with great confidence the following: I shall now finish this paragraph before making myself a cup of coffee; and I shall now reread this paragraph.

The important point about such predictions is that they do not seem to be based on any laws or theories. The predictions made in the last paragraph have no deductive-nomological or inductive-nomological basis. Indeed, if they had such basis, we would not be equally confident about them (Gordon 1995 [1986]: 61).

Predictions such as the ones cited above are usually the product of *practical reasoning*, that is the product of reasoning that leads us to do something. Thus my decision to reread a just-written paragraph might be based on the following kind of reasoning: I must make my train of thought clear; I do tend to overuse certain (Germanic) grammatical constructions; I often know what I am really trying to say only after I have read a first insufficient version; and so on. According to Gordon, 'the important point is that declarations of the form "I shall now do X" offer a bridge between such practical reasoning and prediction' (1995 [1986]: 62).

Having this bridge enables us to predict how we would act even in counterfactual situations: all we need to do is *pretend* that we are in these counterfactual situations, apply our practical reasoning, and then see what we decide to do. Put differently, we can use *simulated* practical reasoning to predict our (possible) actions. Note here that in moving on from the original case (of predictions about our actions in the immediate actual future) to the current case (of predictions about counterfactual situations), we have not moved from practical to nomological reasoning. We still operate on the basis of the former, not of the latter (1995 [1986]: 62.).

It is just one more step along the same line to extend finally the idea of such a simulation from one's own case to that of another human being. There are two ways of developing this idea, proposed by Goldman and Gordon, respectively. According to Goldman, to understand and predict the actions of another human being, we take ourselves as the model. Assume that I predict that my friend Barbara will read on once she has reached this point in this chapter. How do I predict that? I run a first-person simulation on myself: in this simulation I am deciding whether or not *I* would read on in a chapter of a friend's book, a chapter in which I have been personally named. I would. I then assume that Barbara is a lot like me (at least in this respect) and predict her action accordingly (Goldman 1995 [1989]). Goldman's proposal thus presupposes two manoeuvres: I first *introspect* my own decision making, and then *project* – via an inference of analogy – my decision onto my friend.

Gordon's suggestion differs from Goldman's in that he seeks to avoid all inference of analogy (Gordon 1995 [1992]). This is because such inference seems to let the theory theory back into the game: it is hard to see how I could conclude that Barbara and I are alike (in the relevant respects) without relying on some theory about similarity in psychological make-up. This then is Gordon's way of dealing with the Barbara case. I place myself in Barbara's position and adopt as many of her circumstances and attitudes as I see fit. I then run a simulation concerning her practical reasoning. Then, as this 'pretence-Barbara', I make a decision.

Goldman and Gordon also differ over how self-ascriptions of mental states are to be understood. Goldman submits that we self-ascribe mental states on the basis of *qualia*: all mental states have their characteristic qualitative 'feel'; this feel is recognised introspectively, and thus provides the criteria for deciding in which mental state we are (Goldman 1993). Gordon rejects this neo-Cartesian solution. His position is that self-ascription is parasitic upon self-expression combined with simulation of others' mental states. As a first step, the child learns to *express* his or her beliefs verbally. Thus the child becomes able to express the belief that p by uttering 'p', or 'I believe that p'. Subsequently, the child learns to attribute beliefs – for instance, the belief that p – to others. This the child does by means of simulation. Only once the child has mastered the simulation of mental states of others can he or she 'turn around', as it were, and attribute beliefs to him- or herself. In so doing, the child in fact applies the simulation technique to him- or herself. As a result, the child knows that whenever he or she expresses the belief that p, this belief can also be ascribed to him- or herself (Gordon 1995 [1992], 1996; Carruthers 1996).

All of the major advocates of the simulation theory are quite willing to grant that we do not rely exclusively on simulation; empirical generalisations too play a limited role. But the use of such generalisation is determined and defined by simulations. Assume for instance that I had had my office for years right next to Barbara's, and that I had observed, time and time again, that she never stops reading in the middle of a chapter. I therefore might be in possession of the 'law': '*ceteris paribus*, Barbara never quits in the middle of a chapter'. When it comes to deciding whether Barbara will finish reading *this book, my book*, however, I will still have to simulate the circumstances of this case: whether she will be annoyed about being mentioned here, whether she thinks my book should be the exception that proves the rule, and so forth (Gordon 1995 [1992]: 106; Goldman 1995 [1989]: 83).

Having the broad philosophical framework in place, we can turn briefly to the application of the simulation theory to developmental psychology. The main point can be stated briefly: the child's increasing competence in correctly ascribing mental states to others in general, and the child's ability to pass the various false-belief tasks in particular, is due to an increase in the child's ability to simulate. Take for instance the above-mentioned candy-task. The 3-

year-old is not yet able to simulate another mind that holds (what from the youngster's point of view are) false beliefs. In 'transforming' herself into the other, the 3-year-old cannot suspend his or her own beliefs, but must attribute them to the other as well. The 5-year old, on the other hand, has no such difficulties (see e.g. Harris 1995 [1992]).

Interestingly enough, simulation theorists too insist on deep parallels between the child and the scientist. Simulation theorist and theory theorist locate these parallels in different places, however. Thus the simulation theorist, Paul Harris, argues that the analogies attended to by theory theorists are 'false' ones, and that only simulation theory can capture the 'real similarities'. The most important of the false analogies concerns theory change. Whereas in science theory change is a social process, a process of negotiation and persuasion, changes in the child's views and competencies are due to processes of maturation. The first is a problem for sociology, the second for psychology (Harris 1994: 306). The most important 'real' similarity between child and scientist is this: neither the child's attribution of mental states to others, nor the 'normal' scientists' puzzle solving, involve well-structured and explicit theories. Harris's picture of the 'normal' scientist goes of course back to Thomas Kuhn (1962). Harris writes:

> The child assimilates a new situation to a previously encountered instance or to a prototypical model of those instances, and then uses the instance or prototype [in simulations] to extrapolate forward to likely outcomes in the new situation. Similarly, the scientist assimilates the outstanding puzzles of normal science back to the paradigmatic exemplar.…The history of science suggests that a great deal of day-to-day thinking in the laboratory is guided not by an explicitly stated theory but by concrete procedures, specific puzzle solutions, that embody potential solutions to outstanding problem.… The conclusion…is that children and scientists think alike because in neither case does theory guide their thinking.
>
> (Harris 1994: 303)

Eliminativism

I now turn to the second contested issue with respect to folk psychology. This is the issue of whether or not folk psychology will 'smoothly' reduce to *scientific* theories of the mind or brain. In other words, the question is whether the types of entities posited by folk psychology will map onto types of entities posited by cognitive science or neuroscience. If such mapping proves possible, then folk psychology will have been vindicated. A number of influential authors doubt, however, that this prospect will come true. In their view, already existing philosophical arguments and scientific results suggest a very different future for folk psychology: Folk-psychological theory will turn out to be an empirically false theory. It follows that its posited entities – for instance, beliefs, desires, hopes, or memories – will prove not to exist at all. In brief, the ontology of folk psychology will be 'eliminated' in favour of a scientific ontology. 'Eliminativists' base their pessimistic evaluation upon the following considerations:

1 A first argument draws attention to the fact that folk theories in general have had a pretty bad track record. Only rarely have they been vindicated by scientific research. Folk astronomy was false astronomy, folk biology was false biology, folk chemistry was false chemistry, and folk physics was false physics. This invites the inductive inference that folk psychology will prove to be bad psychology (Stich 1983: 229–30).
2 Some advocates of eliminativism claim that folk-psychological theory 'suffers explanatory failures on an epic scale' (Churchland 1989 [1981]: 9). Folk-psychological theory is unable to explain mental illness, creative imagination, intelligence differences between individuals, the functions of sleep, motor co-ordination, perception, perceptual illusions, memory, or learning. According to Paul Churchland, such failures suggest that folk-psychological theory 'is *at best* a highly superficial theory' (1989 [1981]: 7).
3 By no stretch of the imagination can folk psychology be judged to be a progressive research programme. Its history is one of 'retreat, infertility, and decadence' (1989 [1981]: 7). For instance, over the last few centuries, natural science has expelled folk psychology from the study of nature. It is no longer acceptable to interpret natural processes in intentional terms. One only needs to imagine this retreat continuing to predict that eventually folk psychology will be replaced by science in all domains.
4 According to Sellars' origin myth, Jones posited covert mental states – thoughts or beliefs – in analogy to overt speech.[3] Many philosophers of mind agree that folk psychology takes inner mental episodes to be quasi-linguistic. Eliminativists argue,

however, that folk psychology is wrong in doing so. On the one hand, it just does not seem plausible that toddlers and dogs – beings to whom we attribute beliefs – are in possession of quasi-linguistic media of representation. On the other hand, neuroscience has not felt any need to assume sentence-like structures in the brain (Stich 1996: 19–20).

5 Folk psychology seems to construct beliefs and other propositional attitudes as 'propositionally modular'. This means that propositional attitudes are taken (by folk psychologists) to be 'semantically interpretable states that can be causally implicated in some cognitive episodes and causally inert in others' (Stich 1996: 21). Take for instance the belief that dogs have tails. As folk psychologists we are inclined to think that this belief will be causally involved whenever we try to determine whether Dalmatians have tails. We would not suppose, however, that this same belief has any causal role to play when we are considering whether all cats purr. Are we folk psychologists right about this? Stich (1983) and Ramsey *et al.* (1996 [1991]) think not. The propositional modularity thesis is wrong if the human brain functions in a way similar to certain kinds of connectionist networks. These connectionist networks store information *holistically* rather than atomistically. In other words, the decisions of these networks are causally determined by *all* of the information stored within them. If our brain is anything like such a network, then even our belief that 2 + 2 = 4 is causally implicated in determining whether all cats purr. The eliminativist concludes that mental states as posited by our folk psychology – mental states characterised by propositional modularity – simply do not exist.

6 Some researchers in social psychology hypothesise that humans have two largely separate cognitive subsystems (of beliefs). The first older and non-verbal subsystem is largely unconscious; the second younger and verbal subsystem is largely conscious. The first system controls much of actual behaviour; the second produces (or invents) explanations of the actions of the first system, using socially shared theories about the self and the situation. Our folk psychology, however, does not posit such bifurcation in our beliefs. Thus, if the social psychologists are right, then folk psychology must be wrong (Stich 1983: 230–7; 1996: 21–2).

7 Folk psychology seems committed to a 'wide' account of content identity. Recall Hilary Putnam's 'Twin Earth' example

(Putnam 1975): imagine that somewhere in the universe there exists another planet that is almost identical to our Earth. Call this other planet 'Twin Earth'. Each of us has a 'twin' on Twin Earth, a molecule-for-molecule replica. The only difference between Twin Earth and Earth is that whereas the liquid that we call 'water' is H_2O, the liquid that Twin Earthlings call water is XYZ. The two liquids look, smell, and feel the same. Now assume that I and my Twin Earth replica both believe that water is wet. Do our two beliefs have the same content? If you answer 'yes' then you have a 'narrow' account of content identity (narrow because the world outside is not part of the content); if you answer 'no' then you have a 'wide' account. According to Stich, folk psychology is committed to a wide account, whereas science had better be committed to a narrow account. The folk psychologist will reason that my belief is true if and only if H_2O is wet, whereas my twin's belief is true if and only if XYZ is wet. Since the truth conditions differ, surely so must the content of belief. However, if we subscribe to this folk-psychological reasoning then we also have to accept that the contents of our beliefs do not supervene on our non-relational physical properties. That is to say, folk psychology allows that two humans can be in identical physical states while having different beliefs. And this – so the eliminativist thinks – violates the principles of good materialism (Stich 1996: 22–3).

8 A different semantic argument draws on folk-psychological 'meaning holism'. This is the doctrine according to which the content of any given belief depends on surrounding beliefs. Thus the content of my belief that today is Saturday depends on my belief that Saturday follows Friday and precedes Sunday; that my watch shows the day correctly; that yesterday was in fact Friday; and so on. Taken to its logical conclusion, according to folk-psychological meaning holism, two beliefs will only have identical content if they have identical 'doxastic surroundings'. But then no two beliefs are ever identical, and nomological generalisations, of the sort that science strives for, are impossible. If science is right about the possibility of finding generalisations that can be stated in context-free terms, then folk psychology must be wrong (Stich 1996: 25–6).

9 A further semantic argument starts from the observation that 'the belief tokens that folk psychology classifies as having the same content can be extremely heterogeneous' (Stich 1996: 26). Folk psychology allows that a genius, a retarded person, and a

dog can all have belief tokens with the same content, say, that a bowl with water has just been brought into the room. But, eliminativists insist, the three mind–brain states in question have very different causal powers, and thus are much too heterogeneous to be grouped together. Surely a future cognitive science must find nomological generalisations that cut deeper than folk-psychological content (Stich 1996: 27).

10 Another eliminativist line of attack starts from the idea that the folk-psychological intentional properties of beliefs (and other attitudes) cannot be reduced to, or identified with, physical properties. For instance, it is not possible to identify the thought that 2 + 2 = 4 with any one physical realisation. Humans can think this thought with their wetware, Martians with their greenware, and infinite kinds of computers with their hardware. Once this is granted, the eliminativist proceeds to demonstrating an epiphenomenalism with respect to intentional properties. If intentional properties are neither identical with, nor reducible to, physical properties, then they are causally irrelevant. This conclusion is compelling provided only that we add two plausible assumptions: that every event token having intentional properties is identical to some event token having physical properties; and that the causal powers of a physical event token are completely determined by its physical properties (Stich 1996: 27).

Against eliminativism

Eliminativism has provoked numerous criticisms. Some critics address specific eliminativist arguments. Others concentrate on showing, more generally, that eliminativism is incoherent, irrational, and self-refuting.

The charge of incoherence: Lynn Rudder Baker, Jerry Fodor, Peter Hacker, J. D. Trout

Peter Hacker charges the eliminativist with a number of incoherences. To begin with, the eliminativist claims to *believe* in the superiority of his or her philosophical stance over others. However, if as the eliminativist alleges, there are no beliefs, then the eliminativist cannot possibly *believe* in the virtues of his or her eliminativism. Moreover, the eliminativist must give reasons for his or her position. Such reasons must establish that folk psychology gives insufficient explanations of consciousness, memory, perception, sensation, learning, and understanding. But consciousness, memory, perception, sensation, learning, and understanding all are folk-psychological categories. Thus, in criticising folk psychology, the eliminativist is again forced to rely on folk psychology itself (1996: 540). Furthermore, the eliminativist seeks to provide superior *explanations*. But the very concept of *explanation* is inseparable from the folk-psychological concept of *understanding* (1996: 541). In addition, the eliminativist position is incoherent also because adopting it would lead to an elimination of scientific psychology and all 'human form of life'. Both scientific psychology and the human form of life are constituted by our folk psychology: 'There can be no such thing as a *psychological* science of creatures which lack beliefs, desires, and emotions'. And: 'We can no more abandon our vocabulary of sensation, perception, emotion and attitude…than we can cease at will to be human beings. These expressions and their use are constitutive of a human form of life' (1996: 541–2).[4]

Lynn Rudder Baker focuses on the social and intellectual consequences that would result from an elimination of folk psychology. As she sees it, if we were to give up folk psychology, then our social, moral, legal, and linguistic practices would all become senseless or mysterious. Social practices would be rendered unintelligible because our social interaction with others is built upon, and around, contentful, intentional states. Moreover, 'in the absence of contentful states, one could never do anything unintentionally, by accident, or by mistake' (1987: 130). Moral and legal practices would suddenly become senseless because we could no longer say, for instance, that someone had lied or been dishonest. To lie to someone implies trying to bring about a false *belief* in the other. And linguistic practices would become mysterious: 'it would be a total mystery why we say the things we do'; our saying the things we do could no longer be due to our beliefs and desires (1987: 131).

Baker even alleges that accepting the eliminativist position would amount to 'cognitive suicide' (1987: 134–48). She gives three reasons. First, without our folk psychology rational acceptability would be at risk. The eliminativist cannot distinguish between being justified and not being justified in accepting claims: both justification and acceptance are concepts of folk psychology (1987: 136–7). Second, eliminativism undermines the 'near platitude' that language is meaningful only if people can mean something in using language. But once this idea is given up, it no longer makes sense to speak of language being used for making assertions (1987: 140). And finally, the concepts of truth and falsity are linked to true and false attributions of intentional content to humans. If the latter goes, then so must the former. And thus the elimination of folk psychology leads to an elimination of truth and falsity (1987: 143).

Not all critics of eliminativism endorse Baker's and Hacker's charge of eliminativist inconsistency.[5] J. D. Trout calls Baker's and Hacker's charge 'naïve', and cites, with applause, Patricia Churchland's way of rejecting these inconsistency charges:

> If the eliminativist is correct in his criticisms, and if the old framework is revised and replaced, then by using the new vocabulary, the eliminativist's criticisms can be restated with greater sophistication and with no danger of pragmatic contradiction. (For example, the new eliminativist might declare: 'I gronkify beliefs,' where gronkification is a neuropsychological state defined within the mature new theory.) It would be foolish to suppose folk psychology must be true because at this stage of science to criticise it implies using it. All this shows is that folk psychology is the only theory available *now*.
>
> (Churchland 1986: 97; *cf.* Trout 1991: 383)

Trout himself favours a 'more sophisticated' charge of incoherence. It would be inconsistent to reject folk-psychological theory because it is constantly and successfully involved and tested (albeit implicitly tested) in scientific practice. Scientists presuppose this theory all the time in their co-operations, and no scientist has ever felt inclined to question folk psychology's usefulness as a medium of interaction (Trout 1991).[6]

Finally, close to the various charges of inconsistency or incoherence is Jerry Fodor's 'transcendental argument' in favour of folk psychology:

> What's relevant to whether common-sense psychology is worth defending is its dispensability *in fact*. And here the situation is absolutely clear. We have no idea of how to explain ourselves to ourselves except in a vocabulary which is *saturated* with belief-desire psychology. One is tempted to transcendental argument: What Kant said to Hume about physical objects holds, *mutatis mutandis*, for the propositional attitudes; we can't give them up *because we don't know how to*.
>
> (1987: 9–10)

The argument from evolution: Andy Clark[7]

Andy Clark shares one of the central premises of the eliminativists: to wit, that folk psychology is a theory. Clark gives this premise a specific twist, however. He proposes that folk-psychological theory is a so-called 'bedrock theory', a theory selected for by evolution: that is, a theory that is innate, universal, and compelling for our thinking.

Clark's insistence on folk psychology as a bedrock theory is meant to counter Stich's idea according to which folk psychology goes back to the speculations of 'ancient shepherds and camel drivers' (Stich 1983: 229). For Clark this is an unlikely scenario. Folk psychology is crucial to the survival of a species like ours, a species whose members live in groups. Thus it makes sense to assimilate folk psychology to naïve physics, and hypothesise that the former, like the latter, is by now 'in our genes' (Clark 1987: 145–7).

Seeing folk psychology as a bedrock theory can explain why folk psychology is universal; why it has not developed much over the past 2,500 years or so – the clocks of evolution run slowly; or why it does not adequately address the psychological intricacies of mental illness or of other species. Folk psychology simply was not selected for solving such problems (1987: 147).

Clark does not assume that folk psychology must be true simply because it is, in all likelihood, the product of evolution. At the same time, however, he urges that it is unlikely to be altogether false:

> A bedrock theory is simply one which is learned or arrived at by the employment of exceptionally well-tested (probably dedicated) cognitive competencies. And evolutionary demands of speed and cost-efficiency favour the development of competencies which yield theories which are *usually* (but not always) an *approximate* guide to the most vital facts about the domain. Bedrock theories, in short, are as unlikely to be absolutely correct as they are to be radically misguided.…Innateness and usefulness are at best *clues* to truth and cannot be expected to substitute for it.
>
> (1987: 146, 149)

Finally, Clark suggests that we might be trapped in our folk psychology. Because folk psychology is innate, we are *compelled* to use it, even as scientists. There are definite limits to any attempts to move beyond our evolution-based capacities (1987: 149).

The non-sequitur of the eliminativist argument: Stephen Stich (1996)

Conversions from Saulus to Paulus are relatively rare in science and philosophy, but the debate over folk psychology has witnessed one such case. Stephen Stich, who used to be one of the three perhaps most famous eliminativists, alongside Patricia and Paul Churchland, has recently become one of the most outspoken critics of eliminativism.

Stich's main criticism of his own – and the Churchlands' – earlier eliminativist argument concerns the move from showing that folk-psychological theory is (in good part) false to assuming that the theoretical posits of that folk-psychological theory do not exist. This move was earlier licensed by Lewis's account of theoretical entities. By the mid-1990s, Stich had lost his faith in Lewis's proposal. He now thinks that our verdict on the existence of beliefs or desires is pretty much predetermined by what kind of theory of reference we adopt: 'On the description theory, eliminativism is trivially true; on the causal-historical theory, eliminativism is trivially false' (Stich 1994: 360). Lewis's view of theories is of course a version of the description theory. Which of the two theories of reference is appropriate when evaluating the existence of folk-psychological posits? As Stich sees it, this question has no unique and compelling answer. In any case, none of the eliminativists has given an argument as to why we should favour the description theory over the causal–historical theory, and thus the whole eliminativist argument is question begging in one central respect (1996: 34–6).

Stich also contends that the study of the history of science has not as yet identified compelling 'principles of rational ontological inference that are strong enough to dictate what we should conclude about the entities invoked by a theory when we come to believe that the theory is seriously mistaken'. In the absence of such principles, Stich maintains, we have to accept that political and social factors will determine whether folk-psychological entities do, or do not, exist (1996: 71, 73).

The anti-stagnation arguments: Terence Horgan, John Searle, Kathleen Wilkes, James Woodward

Several authors take issue with Paul Churchland's claims according to which folk psychology has a bad track record as a research programme.

According to John Searle (1992: 48), it is simply a mistake to treat folk psychology as a 'research project'. But if it is not a research project then it also does not make sense to criticise it for its slow rate of change.

Terence Horgan and James Woodward (1991 [1985]) too argue that Churchland's stagnation claim is misplaced. For instance, the two co-authors remind us that whereas people of the eighteenth and nineteenth centuries were likely to explain human behaviour in terms of personality traits, people of the twentieth century are more likely to invoke situational factors (1991 [1985]: 152–3). Moreover, Churchland fails to acknowledge that much of scientific psychology is based upon folk psychology and that scientific psychology indeed *has* made major advances over the past fifty years.[8] Horgan and Woodward are unimpressed with Churchland's tacit assumption that any *scientific-psychological* theory based upon folk psychology must be flawed:

> This is rather like arguing that any sophisticated physical theory employing central forces must be false on the grounds that the ordinary person's notions of pushing and pulling have been empirically unprogressive.
>
> (1991 [1985]: 153)

A similar line of reasoning speaks against using the two-belief-systems hypothesis of social psychology as an argument against the folk-psychological concept of belief. While it is perhaps true that the idea of unconscious beliefs and inferences goes beyond folk psychology, it constitutes at most 'an extension and partial modification of traditional folk psychology', but not its outright rejection or refutation (Horgan and Woodward 1991 [1985]: 158–9).

Of course, the most direct line against the stagnation charge is to argue that there is no reason to tamper with a true theory! Or, as Simon Blackburn puts it, 'One man's stagnation is another man's certainty' (Blackburn 1991: 199; cf. Greenwood 1991: 88). Kathleen Wilkes makes a similar point by borrowing an idea of Peter Strawson. Strawson writes that 'there is a massive central core of human thinking which has no history…there are categories and concepts which, in their most fundamental character, change not at all' (quoted in Wilkes 1984: 360). Now if human mental life does not develop dramatically, and if folk psychology is tuned to that mental life, then obviously we should not expect folk psychology to show much change over time and place (Wilkes 1984: 360).

The anti-propositional modularity arguments: Richard Double, Frances Egan, John Heil, Terence Horgan, James Woodward

The arguments to be mentioned here all relate to the claim of Stich (1983), as well as Ramsey *et al.* (1996 [1991]), that folk-psychological theory takes propositional attitudes to be 'functionally discrete, semantically interpretable states that play a causal role in the production of other propositional attitudes and ultimately in the production of behaviour' (Ramsey *et al.* 1996 [1991]: 95).

The critics either reject the idea that folk psychology is committed to propositional modularity, or else deny that folk psychology's commitment to propositional modularity brings it in opposition to work on neural nets. Richard Double (1985) represents the first group. He bases his denial on empirical, questionnaire-type research. The subjects – students of Double's philosophy classes – did not see propositional modularity as essential to their concept of belief.[9] Double suggests therefore that failure of propositional modularity is no reason for eliminating the concept of belief. As Double sees it, the following conditions are constitutive of our common-sense concept of belief:

> (1) Beliefs are real mental conditions, that is, not instrumental fictions. (2) Beliefs represent facts, and thus, can be characterised sententially as possessing a whole range of semantic properties such as being true, false, justified, etc. (3) Beliefs often are produced by an interaction between things outside our skins and our nervous systems. (4) Beliefs connect up with other mental conditions, often to produce other beliefs and behaviour. (5) Beliefs are often knowable to us directly by simple reflection.
>
> (Double 1985: 424)

Double claims not to be able to find a single result of cognitive science or neuroscience that would undermine any of these five conditions.

Frances Egan (1995) argues that folk psychology's commitment to propositional modularity is innocuous. It is true: folk psychology does posit mental states that fulfil the criteria of functional discreteness, semantical interpretability and causal involvement. But folk psychology is ontologically minimalist with respect to these states: It does not make any claims or commitments with respect to the issue of how these states are physically or computationally realised: 'While the folk posit enduring, stable states that play causal roles in producing behaviour, there is no evidence that they share any particular views on underlying psychological or neural processes or mechanisms' (1995: 189).

A different line of defence of propositional modularity by John Heil (1991) argues that connectionist models do not undermine the claim that beliefs are functionally discrete states with specific causal roles. A belief can be widely distributed over the weightings of the whole network *and yet* have a discrete and restricted causal role (1991: 128).

Arguments relying on the rejection of the theory theory: Daniel Dennett, John Greenwood, Steven Horst

Of course anyone who rejects the idea that our folk-psychological capacities are due to the possession of an (empirical) theory will also have good reason to reject eliminativism. After all, if folk psychology is not an empirical theory – or no theory of any kind – then surely it cannot be false either (see Stich and Ravenscroft 1994, above). But if it is not false, then why replace it?

In similar vein, some critics have stressed that the existence of at least some of our mental states is proved by direct observation. Steven Horst makes such a claim for 'consciously accessible occurrent mental states such as conscious judgements and perceptual *Gestalten*'. Mental states of this kind are 'directly accessible and hence non-theoretical in character'. And this makes them unlikely candidates for elimination (1995: 141).

The same point is made more generally by John Greenwood:

> Even if we recognised that all our folk-psychological explanations of human action are inaccurate, we would still have good reason to maintain the existence of intentional psychological states. One reason is self-knowledge of such states.
>
> (1991: 79)

Dennett's account of folk psychology as a 'craft' or 'skill' (of making predictions based on principles of rationality) also can be used to answer the eliminativist challenge. First, a craft is not a theory. Second, a system of norms is not true or false. And third, folk psychology achieves its successes by treating beliefs and desires *instrumentally*, as '*abstracta*' (Dennett 1993 [1981]: 129). Of course, if beliefs, desires, and the other folk-psychological posits are abstracta, then none of the results of neuroscience need pose a threat to their (limited form of) existence.

Arguments against the reduction-or-elimination alternative: Simon Blackburn, John Dupré, Jeffrey Foss, Robert McCauley, John Searle

A number of critics have rejected the grim alternative that Paul Churchland first posed for folk psychology: to wit, that either folk psychology 'smoothly reduces' to neuroscience, or else its ontology should be eliminated (Blackburn 1991; Dupré 1993; Foss 1995; McCauley 1993 [1986]; Searle 1992). Suffice it here to summarise Foss's criticism.

Foss points out that theory reduction hardly ever occurs anywhere in science. Theory reduction was an important idea for positivist philosophy of science; here theory reduction was meant to explain the two most cherished features of science: its unity and progressiveness. Post Kuhn and Feyerabend, however, theory reduction has little role to play (1995: 411).

One of the mistaken ideas underlying the idea of theoretical reduction is that successful new scientific theories make older theories redundant. In fact such cases are extremely rare. Even in the most narrow of scientific subfields, scientists use a considerable number of irreducibly different theories of the same phenomenon. Each of these theories might be handy for a different purpose: 'One might give quick but rough answers, while another may be more accurate but harder to use, and so on' (1995: 418).

Another mistake of theory reductionism is that it overestimated the importance of laws in science. Most efforts of scientists go into discovering and determining particulars: 'the charge of mass ratio of the electron, the age of the universe, the nature of quasars, the evolutionary history of humankind, and so on' (1995: 137).

The most implausible aspect of eliminativism, however, is that it tries to extend the reduction-or-elimination opposition to the relationship between whole disciplines: scientific (folk-psychology-based) psychology and neuroscience. This is incredible not least because it 'reduces' whole disciplines to mere sums of theories: 'Method is completely ignored' (1995: 423).

In light of the above, it will no longer come as a surprise that Foss sees the reduction-or-elimination alternative as a 'red herring, a distraction'. The question cannot be whether folk psychology will reduce smoothly to neuroscientific theory, or not; the only meaningful question is whether folk psychology will continue to play some role or other in the 'information economy of science'. And that question seems difficult to answer negatively (1995: 425).

Individualism in the debate over folk psychology

Having provided above a summary of the main positions in the current debate over folk psychology, I can now turn to the task of identifying individualism, or individualistic tendencies, in the main theories and arguments of this debate.

Individualistic figures of thought shared by all positions

Almost everyone in the debate conceives of folk-psychological activity as taking place in the following type of social setting: an individual folk psychologist attributes mental states to another individual human being. The folk psychologist is able to engage in this attribution owing to possession of either a folk-psychological theory, or else a simulation device (or both). The interpreted human being is not thought of as explaining him- or herself to the folk-psychological interpreter; the interpreted being is conceived of as passive, and non-intervening. The individualism of this setting is clear enough. Knowledge is possessed by the individual; the individual knower is not internally related to other knowers; there is no interaction with the subject matter; and there is no feedback from the interpreted actor back to the interpreter.

Sometimes participants in the debate over folk psychology refer also to a second type of setting. This is the situation in which an individual folk psychologist identifies, or expresses, his or her own mental states. It hardly needs to be pointed out that this scenario is individualistic, too.

This is not to deny that many participants in the debate emphasise folk psychology's crucial role for human social life. But this emphasis as such does not yet weaken or undermine individualistic commitments. For as long as one sees folk psychology merely as the *precondition* of social life, one has not left individualism behind. After all, the individualist does not deny the existence of the social. But the individualist sees social life as resulting from the interaction between independent individuals. The collectivist, on the other hand, regards social life as constitutive of knowledge, and as constitutive even of the individual.

Individualism is often linked to, and backed up by, what might be called 'universalism'. The link between the two is this: if universal structures of the human mind fix its content, then the role of social factors and variables can only be negligible and trivial. Assuming that universalism is true, we can study individuals in isolation because they are from the start equipped with all that is needed to acquire knowledge.

One can find resonances of this individualistic universalism almost everywhere in the philosophy of folk psychology. Most participants in the debate over folk psychology take it for granted that folk psychology has changed little over the centuries, and that its basic structures can be found in all human cultures. The point is emphasised by all sides: by the theory theorists and by their opponents. The former insist on this point to emphasise either the truth of folk psychology, or else its inertia and stagnation as a research programme; the latter to highlight its ontological neutrality.

It might well be that all folk psychologies show some common, 'universal' traits, or that they all share a common core. Perhaps this common core is the triangle of belief, desire, and action. The collectivist does not need to doubt such commonalties to remain committed to the cause. All the collectivist will reject is the move from universalism to individualism. The fact that some features of folk psychology are universal is no licence for construing folk psychology as the possession of the individual mind. Despite the existence of universal features it could still be the case that folk psychology is a social institution, and that it has a collective subject.

I also regard several of the binary parameters of the debate over folk psychology as individualistic. That is to say, the very idea of having to choose between folk-psychological theory being either normative or descriptive–explanatory, either analytic or explanatory, and either true or false of an independent reality – all this is itself individualistic at least in so far as it is based on insufficient appreciation of collectivistic alternatives. But to make this claim stick, I need first to introduce a sociophilosophical account of folk psychology. I shall show in Chapter 9 that the listed alternatives cease to be exclusive of one another as soon as we interpret folk psychology as a social institution.

The individualism of theory theories

As seen earlier in this chapter, theory theorists model folk-psychological knowledge on scientific theories. That move is not, as such, individualistic. What *is* individualistic, however, is to construe theories as the possessions of individual minds or brains. In Part I of this book, I have tried to make plausible the idea that scientific theories are social entities; that is, that they are social institutions. To advocate such a view of scientific theories is tantamount to maintaining that scientific theories are held by collectives rather than individuals, and that individuals can be said to possess a theory only in so far as they are members of a theory-holding collective.

To bring this collectivistic perspective into play here, one would have to focus on the *relations between folk psychologists* and study how they form a collective of interpreters; how they maintain their shared theory; how they deal with unusual cases; how they sanction deviations from the theory; how they try to reach agreement on controversial aspects of the theory; how they train newcomers to the theory; how they collectively create routines and drills; and so on. Alas, none of these questions have played any role in the debate over folk psychology thus far.

Of course, theory theorists acknowledge that folk-psychological theory is 'shared': that is, that it is held by more than just one individual. But for theory theorists the 'shared' nature of folk psychology means no more than that copies or instances of the folk-psychological theory happen to be in many different brains or minds. And they happen to be there, either because these copies are innate ('in the genes'), or else because they result from observing, and theorising about, the same kind of social and psychological environment. Neither of these proposals is satisfactory. Neither innateness nor individual theorising can constitute the normativity of folk psychology, that is the strong intuitions that there are right and wrong applications of folk psychology to given cases, or that one ought to be able to account for one's actions in folk-psychological terms.

Turning to individualistic tendencies of specific versions of the theory theory, note that Churchland's move from theories *qua* accepted platitudes to theories *qua* configurations of synaptic connections of individual brains makes his individualism more pronounced. It is not difficult to build a bridge from shared platitudes to social institutions, but once theories have been turned into distributions of synaptic weightings the bridge seems to have been burnt. It is also strongly individualistic to say, as Churchland does, that the theory theory explains how the sceptical problem of the existence of other minds is overcome. To assume that the child infers the existence of other minds on the basis of observations and theories strikes me as absurdly individualistic, or even solipsistic.

The individualism that informs all versions of the theory theory is particularly visible and pronounced in the case of the work of developmental psychologists. That should not come as a surprise; after all, psychology is firmly rooted in individualism. (In Part I, we covered some of the historical causes for this phenomenon.) Be this as it may, the model of the child-scientist trying to develop theories of the folks around him or her is clearly individualistic. Many efforts in the history and sociology of science have been directed at demonstrating the social nature of scientific discovery and invention; some aspects of this work were illustrated by my historical case study above. In light of this work one must question whether the child's allegedly non-social theorising is analogous to the work of the scientist. Alternatively, one might suspect that there is indeed a deep parallel between scientists' and children's theory construction, but that identifying and developing this parallel demands lifting the individualistic framework within which children are studied by developmental psychologists. Abandoning this framework means redirecting psychological curiosity: to study interactions between the child and other people that might direct, and guide, the child's theorising; to model the child on a *social scientist* rather than a natural scientist, that is to allow for subject matters that interact with the child; or to consider typical social situations of children of different age groups.

Individualism in criticisms of, and alternatives to, the theory theory

First of all, I find individualistic the Wittgensteinian line of the primary expressive function of psychological predicates. Only if one works with a radically reduced social ontology can one come to believe that, say, all central uses of the term 'pain' are expressive. But this model breaks down as soon as we imagine two or more people discussing the pains of one of them.

A different form of individualism is dominant in some versions of the simulation theory. These theories seek to defend the idea of a privileged first-person access to one's mental states. On this view, only I can be right about my own mind. Obviously, this runs counter to the collectivistic claim according to which 'right' or 'wrong' are applicable only against the background of a collectively shared consensus concerning standards and paradigmatic cases.

Kathleen Wilkes is to be applauded for emphasising the many social functions to which folk psychology is put. Folk psychology is more than a device of explanation and prediction. But here too this important insight is not developed to the point where it would result in a collectivistic interpretation of folk psychology as a whole. For Wilkes, folk psychology is merely the precondition of these social processes; it is not itself constituted by them.

Both Dennett and Wilkes stress the normative nature of folk-psychological explanations. Above, I myself insisted that attending to the normativity of folk psychology leads one naturally into seeing it as a social institution. Have Dennett and Wilkes moved towards collectivism? Unfortunately, the answer has to be negative. These two authors locate only one normative aspect or dimension of folk psychology. Normative rationality provides the interpreter with a predictive strategy concerning the actions of the interpreted human being. But neither author investigates how individuals hit upon the concept of such normativity and rationality, and neither author focuses on the normativity that links together different folk psychologists (as a community of interpreters). That is to say, for Dennett and Wilkes the normative dimension is confined to the relationship between the individual and his or her environment – as seen and evaluated by the folk-psychological interpreter. And thus the further, and crucial, normative dimension – that is, the ways in which different folk psychologists hold one another to task for their interpretations and the ways in which they distinguish between right and wrong applications of the theory or the craft – does not come into view.

Individualism in the debate over eliminativism

This last point carries over – *mutatis mutandis* – into the debate over eliminativism. Much of the debate over eliminativism concerns the issue of whether folk-psychological theory is true or false, right or wrong. But here these normative notions are notions used *by the analyst* – that is, the philosopher or psychologist – with respect to the overall structure of categories used *by the actors* – the individualistically construed folk psychologists. Thus the essentially social–normative role of folk psychology itself, a role that folk psychology has in virtue of being collectively sustained, still remains invisible.

There is something individualistic also about the notion that we could decide to throw folk-psychological old-speak overboard.

Clearly the abandonment of folk psychology would be a complex social process, so much so that one might wonder whether it is at all *socially* possible. That this important issue remains hidden from the sight of eliminativists attests as much to their individualism as does their view of the social as no more than a hurdle for scientific progress (Churchland 1989 [1981]: 9).

In light of my earlier remarks of the individualism of the theory theory, it is perhaps understandable why eliminativists such as Churchland do not see much of a problem here. First, they assimilate folk psychology to a theory. Second, they construe 'possession of a theory' as a relation between an individual and a structure of platitudes (or a configuration of weights of synaptic connections). Third, it seems a straightforward matter for an individual to reject one theory, and adopt another. Therefore, and fourth, eliminativists have no difficulty conceiving of the possibility that a whole society might do the same.

Finally, I had better comment on Stephen Stich's idea according to which – in the absence of strong normative principles of rational ontological inference – we are entitled to rely on social and political considerations in deciding whether beliefs and desires exist. Stich's suggestion seems to me to be Janus faced: one face looks to the received individualistic views; the other looks ahead towards a collectivistic interpretation of theories in general, and folk psychology in particular. The backward-facing Stich commits himself to the old, received opposition between the rational and the social. Either we have rational principles, or else we have to make do with the social. That 'strong normative principles' could themselves be social entities is a possibility not considered by Stich. The forward-looking Stich rightly rediscovers the finitist aspect of our concepts: from the alleged failings of our folk-psychological theory we can either conclude that its posits do not exist, or else infer that folk psychology told us the wrong kind of story about beliefs and desires. The choice between these two alternatives depends ultimately on human interests and goals. From this insight into finitism it is not a big step to appreciate the idea that theories are social institutions. The step is small because social institutions too have a finitist character. This is an issue that I shall explain in the next chapter.

9 Folk psychology as a social institution

Introduction

I now turn to the third, final, and decisive step of my overall argument. In the Interlude, I introduced and explained one theory about social institutions. In Chapter 8, I gave a summary of the debate over folk psychology and suggested that all major positions in this debate are, tacitly or openly, committed to different brands and versions of individualism. In the present chapter, I shall connect the main themes of these two earlier chapters. I shall argue that folk psychology is a social institution in the sense explained in the Interlude. This means, first and foremost, that folk psychology has a self-referential and self-validating structure, and that folk psychology is a fundamental collective good. I shall also invoke evidence to the effect that our folk psychology is not universal in all its features: Even core elements of our Western folk psychology – such as our concept of 'belief' – are absent from the folk psychologies of other, non-Western cultures. Moreover, I shall make a specific proposal concerning several of the most basic intuitions that underlie conflicting theoretical positions in the folk psychology debate: to wit, that these diverging intuitions all become plausible once we understand folk psychology as a social institution. I shall conclude by suggesting that some central folk-psychological kinds are 'artificial kinds' – entities shaped and formed by folk-psychological practices – and that some of our most cherished intuitions concerning privileged access and self-creation are due to mistaken renderings of the 'grammar' of folk psychology.

In making my case, I shall draw on, and occasionally extensively quote from, the studies of a considerable number of philosophers, psychologists, and anthropologists. I am doing so partly to familiarise the philosophical reader with types of voices usually ignored in the philosophical debate (for instance, the voices of the anthropologists), and partly to give credit where credit is due. As concerns at least some of the main claims of this chapter, much of the key evidence has been around for some time already; all that remains for me to do is to connect it properly to my main thesis.

Folk psychology and self-referentiality

Social institutions are self-referring systems of beliefs and actions. I claim that folk psychology is a social institution. And thus I must be able to show that folk psychology essentially involves self-referential structures. My argument has four steps.

Step 1: folk psychology, language, laws

To start us off, we need to remember that the mastery of folk psychology is – or includes centrally – the mastery of a language; that is, on the one hand, the mastery of concepts such as 'belief', 'desire', 'action', 'love', or 'fatigue', and on the other hand, the ability to form and understand sentences such as 'Desires often lead to action', 'Oscar doesn't believe in the new government', 'Oscar is tired', 'Oscar is in love', or 'Oscar desires to read a good book'.

To master a (descriptive) language is to be able to distribute individuals into conventional similarity classes and to have knowledge about regular associations – that is, laws – between individuals falling into such classes. Without knowledge of such laws one would not be able to test and justify the application of learnt predicates in new observational contexts (Hesse 1970: 41; 1974: 14–15; Arbib and Hesse 1986: 8). For instance, to master the concepts 'cat' and 'mouse' is to be able to decide whether a newly encountered animal belongs to one, or neither, of these two classes; it is to be able to identify relevant similarities between the newly encountered animal and those animals that earlier were grouped into one of the two classes. Moreover, part of the mastery of the concepts of 'cat' and 'mouse' is knowledge of the law 'cats chase mice'. 'Being a cat' is regularly associated with mice-hunting and mice-eating behaviour; and 'being a mouse' is regularly associated with avoiding cats. Knowledge of such laws enables us to point to two animals running around the garden and say things like: 'The animal running away must be a mouse; just see how angrily the cat is chasing it'. Or: 'No, it cannot be a mouse for every now and again the two animals seem to alternate the roles of chaser and chased'.

What holds for concepts such as 'cat' and 'mouse' obviously applies also to folk-psychological concepts such as 'belief', 'pride', or 'love'. Here too the mastery of the concepts consists of the ability to identify newly encountered behaviours and mental states as relevantly similar to earlier encountered behaviours and mental states. For instance, I am a competent user of the word 'love' to the extent to which I am able to recognise similarities between Romeo's, Werther's, and Aïda's feelings. And here too, regular associations, or laws, serve an important function in prediction and correction. Thus we might say of a person: 'Look, he can't be in love: he is sleeping well at night; his hands don't sweat in the presence of his beloved; he avoids seeing her; he doesn't think much about her'; and so on.

The above remarks on language and folk psychology amount to what I regard as a *minimalist* theory theory. This version of the theory theory is not as ambitious as are Churchland's, Lewis's, Sellars' or Stich's proposals: the present minimalist theory theory is committed neither to theoretical entities as unobservables, nor to defining 'belief' or 'desire' on the basis of their causal roles. Nevertheless, my minimalist theory theory constitutes an alternative to those – such as Hacker – who draw a sharp dividing line between language and theory. Recall for instance Hacker's insistence that no theory is involved in saying that there are chairs and tables in my office. But this is true only if we use a fairly rich notion of theory. It is not true if one allows that the earlier-mentioned low-level regular associations may count as laws of sorts. After all, Hacker's observational statement 'There are two chairs and one table in my room' can be challenged with a critical remark like 'But there is no surface to sit on, and no surface to eat or write on anywhere in this room'. This critical remark does rely on regular associations between chairs and places to sit, or tables and places to eat at, or write on.

By the same token, there is no reason to be overly impressed with the claim that if we allow folk-psychological platitudes to count as 'laws' then the concept of law will become too thin or even empty. There may well be important differences between folk-psychological platitudes and some laws in physics. And there may well be contexts in which we might want to mark these differences terminologically. But all this does not rule out interesting similarities between scientific and folk laws and theories.

Step 2: platitudes, rationality, normativity

Folk-psychological laws or platitudes such as, say, 'Those who are thirsty seek to drink' or 'The feeling of success brings pride' have a number of important normative uses and dimensions. Such platitudes often function as standards of rationality.

Consider, first of all, how we as folk-psychological interpreters would think of individuals or groups that systematically violated such platitudes. Surely in the absence of good overriding reasons we would count as 'irrational' people who violated these platitudes. The thirsty person who avoids drinking for no good reason, or the successful speaker who denies feeling even the slightest degree of pride in his or her achievement, strike us as odd and irrational. We feel that – *ceteris paribus* – the thirsty and the successful are *obliged* to act in certain ways; we feel that they *ought* to act in certain manners. In emphasising this normative aspect of folk psychology, I side with Dennett in particular. Dennett insists that we predict others' actions by means of a calculation of what it would be rational for them to do. When these others then fail to act according to our predictive calculation, we will count them as irrational – unless, of course, we identify a mistake in our calculations.

Second, folk-psychological platitudes have a normative function not only with respect to how we as interpreters evaluate the actions of others. These platitudes also prescribe how we ourselves – we as interpreters – ought to, or are permitted to, act; and how we ourselves ought to, or are permitted to, justify our actions. I know what I *ought to* feel when claiming that I am in love with Aïda (I mean the Egyptian girl, not the opera); I know what I am *permitted to* feel when a paper of mine went down well with a difficult audience; and I know what kinds of justifications are appropriate if you challenge either the sincerity of my love for Aïda or my right to feel proud about my success.

Third, folk psychology has a normative dimension also in so far as it specifies how a folk-psychological hypothesis about someone's mental states *ought* to be supported by various types of evidence. Put differently, folk-psychological interpretations can be right or wrong, and standards of rationality constrain the interpreter's acts of making sense of others. We often air our interpretations of others' actions in the presence of other (potential) interpreters, and the latter frequently challenge our interpretations. We might be accused of having interpreted uncharitably, unfairly, or overgenerously and one-sidedly. When we are criticised in this way, some folk-psychological platitudes are usually cited and used as evidence against us. Someone might tell us that we have no right to claim that Oscar is in love with Aïda, that we *ought* to have inferred the very opposite, given that Oscar never thinks much of her, doesn't lose a second of sleep because of her, and so on.

Whereas the first normative dimension of folk-psychological platitudes has – as noted above – attracted attention before, the same cannot be said for the second and third dimensions. This is of course not surprising given the individualism that informs the philosophy of folk psychology. After all, my second and third point relate the folk-psychological *interpreter* to other interpreters and thus make that interpreter part of a community with institutions, standards, and norms.

Step 3: normativity, rationality and social institutions

'Obligation', 'ought', 'right', and 'wrong' have their 'logical home' in social institutions. 'Obligation' and 'ought' presuppose the existence of social institutions, and it is impossible to reduce their meaning to non-social, say biological or psychological, categories. Hume argued this point two and a half centuries ago, and much of the work of Durkheim, Wittgenstein, and their disciples can be said to have re-enforced and strengthened Hume's argument (Hume 1990; Durkheim 1971 [1915]; Wittgenstein 1953; Bloor 1997a).

This is not the place to give a detailed rational reconstruction of Hume's, Durkheim's, or Wittgenstein's arguments. The essential idea can, in any case, be grasped without a fine-grained analysis of their texts. Indeed, the crucial point has already been touched upon in the Interlude. There we saw that the possession of a concept implies the ability to distinguish between right and wrong applications of this concept, and that 'right' and 'wrong' presuppose an independent standard. As long as we remain on the level of the individual, there is no distinction between '*seems* right' and '*is* right', between '*seems* obligatory' and '*is* obligatory', between '*seems* rational' and '*is* rational'. Only a collective consensus can establish the '*is* right', '*is* obligatory', and '*is* rational'; and only when measured against this standard can individual actions and beliefs *be* right or wrong, rational or irrational, obligatory, permitted, or forbidden.

To emphasise that obligation, right and wrong, presuppose a collective consensus is of course just a different way of insisting that obligations and self-referentiality come hand in hand:

obligations are what we collectively take to be obligations; obligations are social institutions.

Step 4: drawing the conclusion

We now have all of the needed premises for drawing the conclusion that folk psychology is a social institution in so far as it essentially involves a self-referential and, by implication, self-validating component; because folk psychology is permeated by normative concepts (such as 'ought' or 'rational'), it too is characterised by self-referentiality:

Premise 1	Mastery of folk psychology involves mastery of a language.
Premise 2	Mastery of a language involves knowledge of platitudes.
Premise 3	Folk-psychological platitudes have several normative dimensions: they define how one *ought* to act.
Premise 4	'Oughts' [*sic*] are what we collectively take to be oughts; that is, 'oughts' are self-referential.
Conclusion	Folk psychology has a self-referential component.

Folk psychology, rationality, morality

The way in which I have used the concept of obligation to achieve this result raises the question of how folk psychology relates to folk morality. It seems natural to suggest that what I have done above is to liken, in some key respects, folk psychology to morality. I agree with this suggestion but feel that it needs to be qualified and developed further. I am not claiming that folk psychology *equals* morality. Moral codes define what it is to act in a morally good or morally bad way, but they do not concern the relation between, say, thirst and drinking (except contingently, i.e. when, say, we might be obliged to leave some liquid for others who are just as thirsty as we are ourselves). The relationship between thirst and drinking is a question of rationality, not of morality. We re-establish a deeper parallel between the two cases, however, by noting that the moral quality of an action relates to the moral code in the same way that the rational quality of an action (or belief, or desire) relates to the platitudes of folk psychology. The moral code defines what it is to be moral, and act morally, and folk psychology defines what it is to be rational, and act rationally.

This is of course tantamount to saying that rationality is itself a folk-psychological category. To recognise this is to free oneself from the mistaken perspective of those philosophers who write as if folk psychologists somehow tried to 'capture' an independent feature of reality called 'rationality'. Folk psychologists are not trying to 'capture' rationality in the way in which physicists are trying to 'capture' gravity. Dennett comes closest to the standpoint I am urging. He does so at least in those passages in which he treats rationality as an intentional-stance-dependent feature. Unfortunately, Dennett does not manage to hold on to this position consistently; more than once he falls back into conceiving of rationality as a stance-independent real factor (Baker 1994).

Folk psychology as fundamental social institution and collective good

I suggested in the Interlude that social institutions are collective goods in so far as they provide solutions to co-ordination problems. Does folk psychology fit this bill? The answer should be obvious: because folk psychology defines core principles of rationality, it is not only *a* social institution, it is *the most fundamental* – indeed 'the *bedrock*' – social institution. And thus it is also the ultimate collective good. All co-ordination of human behaviour presupposes at least this much: that there is a mechanism that allows reliable predictions as to when, where, and how the other human bodies are going to move. Folk psychology, or at least its more fundamental parts, does just this. Folk psychology creates (moderately) predictable humans; indeed it creates – and continuously recreates – the very individuals whose beliefs and actions sustain forms of life.

The importance of folk psychology in this respect can perhaps best be appreciated by reminding ourselves of our collective efforts in protecting it from empirical refutations. For instance, if I wish to be understood by others, and if I wish to enlist others for my purposes, then I *must* present myself in a way that conforms to folk-psychological platitudes, or laws. To explain my actions – even my most outrageous ones – to others is to find an appropriate set of folk-psychological platitudes in the light of which I am understandable and predictable. The same applies when I am interpreting the actions of someone, say Oscar, to a group of others, say my readers. If I wish to make Oscar's behaviour intelligible to you, then I must identify a set of folk-psychological platitudes that explain his actions. And when discussing his conduct with Oscar himself, he too will hold me to glossing it in terms of the folk psychology we both share.

Of course we do occasionally come across data that do not fit the folk-psychological laws. For instance, we encounter people who entertain both premises of a practical syllogism ('I desire *p*'; 'I know that to get *p*, I have to bring about *q*') without, however, drawing the conclusion (by setting themselves to bring about *q*). In such situations we do not, however, start doubting the truth of the folk-psychological principles; instead, we either look for further data that save the practical syllogism *ceteris paribus* (maybe I had a stronger desire *r* that overruled my desire for *p*), or else we will dismiss the actor as irrational. Groping for an analogy, we might say that folk psychology functions in our daily life like a paradigmatic theory functions in Kuhnian normal science; what is being tested is not the theory but the scientists. They have to prove their mettle by showing how true the theory is, and how successfully it can be applied.

Given the fundamental role of folk psychology, it should not come as a surprise that it is also the institution most strongly protected by sanctions. Inability to come up with acceptable folk-psychological accounts of ourselves and others is sanctioned by disapproval, lack of acceptance, or even referral to psychiatric services. No wonder, therefore, that social psychologists find their subjects 'telling more than they can know': rather than admit that they have no introspective access to many of their higher cognitive processes, subjects will tell folk-psychological stories of how their mind allegedly works (Nisbett and Wilson 1977). And because they draw on roughly the same folk psychology, their introspective reports will even largely coincide! As W. Lyons puts it:

> our 'introspections' will employ 'processes' and reveal 'processes' borrowed from this general 'folk psychological' viewpoint and not differ very much from everyone else's 'introspections', for everyone else will borrow in much the same way from much the same source and with much the same expectations in regard to the 'data of introspection'. In a culture with a highly developed and centralised education system, common language, readily accessible store of knowledge and speculation, and constant and widespread intercourse, this 'folk psychological' viewpoint will be continually buttressed by the fact that we learn much the same accounts, methods, proce-

> dures, and explanations in regard to anything of a cognitive sort.
>
> (Lyons 1986: 126)

In normal life neither the usefulness nor the truth of folk psychology is ever an issue, and free riding on this collective action is never a live option. All of our everyday talk continuously invokes folk psychology, and this talk re-establishes its usefulness in everyone's understanding. Note also that we cannot even formulate our views regarding folk psychology without already using it. Even those who wish to replace folk psychology by neuralese have to say that they *believe* it to be replaceable, that they find such replacement *desirable,* and that they will take *action* to encourage such replacement. In other words, even to think about folk psychology's pros and cons is to reaffirm one's adherence and commitment to it.

All this is not to say, of course, that folk psychology knows nothing of finitism and revisability. Note first of all that it will be a judgement of similarity to decide whether Oscar's behaviour at a given time should lead us to call him full of 'joy', or 'happiness', or 'love'. Nothing about the past uniquely determines the outcome of this decision.

Second, the results of scientific and philosophical enquiry constantly force us to make decisions concerning the application of concepts such as 'belief' or 'desire' to new, or newly discovered, circumstances. Let me mention two examples. Take first Stich's example of Mrs T, the old lady who gradually loses beliefs as the result of some degenerative disease (Stich 1983: 55–6; 1996: 24–5). Eventually, Mrs T no longer knows what 'assassination' means and who McKinley was, but she is stil able to answer the question 'What happened to McKinley?' with 'McKinley was assassinated'. Did she still *believe* that McKinley was assassinated given that she lacked knowledge of who he was, and that assassinated persons are dead? Different psychologists and philosophers of mind will have somewhat different intuitions on this issue; some might conclude that Mrs T *does believe* that McKinley was assassinated; some others might insist that she *does not believe* it; and still others might suggest introducing a new concept, say '*delieve*' (shorthand for 'degenerate form of belief') for her peculiar propositional attitude.

Another example is that of the 3-year-old child whose concept of 'belief' is not yet fully representational. According to the 3-year-old, for someone, say *X*, to believe that *p*, means either that *p* is the case and *X* knows it, or that *p* is not the case but *X* pretends it to be the case. The youngster of this age cannot grasp the idea of a mistaken belief, however; that is, the idea that *X* might believe that *p* even though, in fact, *p* does not obtain. Does the 3-year-old have the concept of belief? Or does the child have a mistaken conception of belief; or lack beliefs; or perhaps have 'preliefs' (Perner *et al.* 1994) where 'prelief' is a propositional attitude that covers the partial representational capacities of the 3-year-old? Again, nothing in the prior history of uses of 'belief' uniquely determines one particular solution.

And third, It is also worth mentioning that even the very fabric of our folk psychology is occasionally challenged in the realm of art. What makes many a novelist or poet great is that they stretch our folk-psychological concepts and platitudes to their breaking point. (Think for instance of the poetry of Paul Celan.) Lest the damage be too great, however, we also employ cohorts of hermeneuticists who will reduce even the most outlandish of deviation to a mere combination of folksy platitudes. (Think for instance of Hans-Georg Gadamer's interpretations of Celan's poetry (Gadamer 1997).)

Thus folk psychology *does* know of finitism and revisability. If it nevertheless changes little over time, it is partly because we all do our best to protect it, partly because most of the time we apply it in a routine and blind fashion. And we do both because our attention is usually elsewhere: we use our folk-psychological abilities for various purposes, and most of these purposes are best served by folk-psychological stability. And finally, because we are routinely successful in all this, we are completely oblivious to this extraordinary collective achievement.

Again a comparison with an aspect of science can perhaps be used to illuminate this point. Maybe we can appreciate the nature of folk psychology as being both a fundamental collective good and at the same time somehow 'invisible' by likening it to the work done to guarantee the ubiquity of the principal physical constants. An enormous effort goes into maintaining this constancy, according to some estimates three times the amount directly involved in science and technology (Hunter 1980). The scientist or engineer, or any lay person for that matter, who measures the length of a table, or makes an appointment for 5:00 p.m., is indebted to, and dependent on, this gigantic exercise. At the same time, these users also re-enforce, unconsciously as it were, this practice and extend the standardised units to ever more localities and situations.

Folk psychologies

To show that folk psychology is a social institution it suffices to demonstrate that it involves self-referential structures of actions and beliefs, and that it is a collective good. This I have done above. Although not equally essential to my main thesis, I now turn to arguing that at least some parts or aspects of folk psychology are not universal; that is, that we have reason to speak of folk psycholo-*gies* rather than of *one single* folk psychology.

Indigenous psychologies

A first important source of evidence here is research into so-called 'indigenous psychologies'. Both anthropologists and social psychologists have contributed to this literature (e.g. Heelas and Lock 1981a; Kim and Berry 1993). By 'indigenous psychologies' these authors mean 'the cultural views, theories, conjectures, classifications, assumptions and metaphors – together with the notions embedded in social institutions – which bear on psychological topics…"indigenous"… can be contrasted with "specialist" psychologies, those developed by academic psychologists' (Heelas 1981a: 3).

Interest in indigenous psychologies is motivated by sheer anthropological curiosity on the one hand, and a certain amount of scepticism towards mainstream academic psychology on the other hand. This scepticism is based on the suspicion that general psychology might be both 'culture blind' and 'culture bound': it is culture blind in so far as it disregards the possibility that the development and exercise of human mental faculties might be decisively shaped by culture; and it is culture bound in so far as many of its theories are specific to, and true only of, members of Western cultures (Berry and Kim 1993: 277). Contributors to research into indigenous psychologies seek to overcome culture blindness and boundedness, and wish to be open for the possibility that 'given the cross-cultural variety of the indigenous, it is unlikely that we alone have got it right' (Heelas and Lock 1981b: xiv).

Much of the 'indigenous psychologies' research programme has concentrated on conceptions of the self, the mind, and the person. One early study by Gottfried Lienhardt investigated the Dinka (of the southern Sudan) (Lienhardt 1961). Lienhardt suggested that

> the Dinka have no conception which at all closely corresponds to our popular modern conception of the 'mind' as mediating

> and, as it were, storing up the experiences of the self. There is for them no such interior entity to appear, on reflection, to stand between the experiencing self at any given moment and what is or has been an exterior influence upon the self.
>
> (Lienhardt 1961, quoted from Heelas 1981a: 9–10)

Moreover, Lienhardt claimed that for the Dinka the sentence 'I recall that...' is ill-formed because recollection is no activity of any I: a Dinka man who calls his child 'Khartoum' (because the man has been imprisoned there) does conceive of Khartoum as the *agent* of memory, and of himself as the object acted upon (Lienhardt 1961, quoted from Heelas 1981a: 10).

Other authors report that Confucius did not know of a self with its private states (Needham 1981: 74); that the Chewong (of Malaysia) make no conceptual distinction between 'thoughts' and 'feelings' (Howell 1981: 139), or that the Maori (of New Zealand) see the self as the passenger rather than the driver of the body:

> If the self in the Western view can be seen as the driver of the car, then in the Maori view it must be seen as the passenger in its body – the body consisting of numerous parts which, if they are kept oiled and serviced through correct ritual observance, should run smoothly under their own power and in their own predestined way.
>
> (Smith 1981: 158)

For members of these alien cultures their folk psychology is no less universal and neutral than ours is for us, and their difficulties in understanding our folk psychology mirror our difficulties in grasping theirs. Said one Eskimo in bewilderment: 'If thoughts were in the mind, how could they do anything?' (Harré 1981: 85).

In light of the psychological debates covered in Part I of this book, it is perhaps also worth mentioning that introspection seems to presuppose a Western conception of a private self. As William Lyons insists:

> The Balinese, for example, or at least those still untouched by Western culture, seem to have no terminology for a private self with an inner private mental life and so no terminology for introspecting the events of that inner life.
>
> (1986: 98)

> Introspection...is a myth of our culture, an invention of our 'folk psychology'.
>
> (1986: 104)

Cultural psychology

Fairly close to the 'indigenous psychologies' research programme is anthropological work done under the title of 'cultural psychology' (Stigler *et al*. 1990; Shweder 1991). Here too the focus is on the influence of culture upon human psychology, and on folk models of the mind.

One intriguing study of this genre is Richard A. Shweder's and Edmund J. Bourne's paper 'Does the Concept of the Person Vary Cross-Culturally?' (Shweder and Bourne 1991). The test case is the difference between the Western individualistic, *ego*centric notion of the person, and the non-Western *socio*centric conception. According to the Western conception, the person is a

> bounded, unique, more or less integrated motivational and cognitive universe, a dynamic centre of awareness, emotion, judgement, and action organised into a distinctive whole and set contrastively both against other such wholes and against a social and natural background.
>
> (Clifford Geertz, quoted from Shweder and Bourne 1991: 122)

The non-Western personality excludes the idiosyncratic and individualistic from his or her expressions. What counts is the never-changing circle of life, not the individual:

> It is dramatis personae, not actors, that endure; indeed it is dramatis personae, not actors, that in the proper sense really exist. Physically men come and go – mere incidents in a happenstance history of no genuine importance, even to themselves. But the masks they wear, the stage they occupy, the parts they play, and most important, the spectacle they mount remain and constitute not the facade but the substance of things, not least the self.
>
> (Clifford Geertz, quoted from Shweder and Bourne 1991: 123)

To support the reality of this difference, Shweder and Bourne studied person descriptions made by the Oriyas (of Bhubaneswar, India), and Americans. Members of both communities were asked to provide descriptions of other members of their respective groups. The statements of the Oriyas and the Americans differed significantly. Oriyas focused on concrete contexts, relations, temporal fixpoints, and concrete cases ('She brings cakes to my family on festival days', 'He curses at his neighbours'), the Americans on abstract traits ('She is friendly', 'He is aggressive') (Shweder and Bourne 1991: 135–7). The Oriyas see the individual defined in and through his or her relations to others, the Americans see the individual as an autonomous and atomistic unit.

Shweder and Bourne test and reject a number of received hypotheses meant to explain the differences between the Western and the non-Western ways of thinking. Such hypotheses locate the causes of the difference – between the concrete and abstract conceptions of persons – in variance in formal schooling, in literacy level, in socio-economic status, in linguistic structure, or in extent of contact with other cultures. In each case the authors show that the data do not support these hypotheses. For instance, just to mention their criticism of the last proposal, it is not true that the Oriyas are parochial; the complicated caste system within which they live exposes them to many different life-styles (Shweder and Bourne 1991: 144).

Shweder and Bourne's own explanation emphasises the holism of the Indian world-view in general, and social life in particular. The character of the individual can only be grasped by understanding his or her role and function within the community at large; the individual is essentially defined in and through his or her relations to others. The Indian 'person-in-society' is therefore not an autonomous individual: 'He or she is regulated by strict rules of interdependence that are context specific and particularistic, rules governing exchanges of services, rules governing behaviour to kinsmen, rules governing marriage, and so on' (Shweder and Bourne 1991: 149). The individual therefore cannot be separated from the status he or she occupies, and does not have intrinsic moral value in abstraction from that status. This contrasts of course sharply with the Western notion according to which each person is conceived of as 'a particular incarnation of abstract humanity' (Louis Dumont, quoted in Shweder and Bourne 1991: 151).

Cross-cultural psychology

Cross-cultural psychology is a third research programme that must be mentioned when talking about folk psycholog*ies*. Cross-cultural psychologists too are interested in psychological diversity but their interest reaches further than the identification and explanation of indigenous psychological classifications. Cross-cultural psychologists take standard psychological testing methods to the field and compare the performance of different peoples. For instance, cross-cultural psychologists have done extensive research into different peoples' susceptibility to visual illusions (see e.g. Berry *et al.* 1992: 147–9). Cross-cultural psychologists are also keen to establish the precise distance between different folk-psychological categorisations. Folk-psychological categorisations of emotions have figured particularly prominently in this field. I shall here summarise one cautious review of the field (Russell 1991).

A first observation arising from the ethnographic record is that the languages of the world differ widely in the number of words they provide for categorising emotions. English has more than 2,000, Dutch 1,501, Taiwanese 750, Malay 230, Ifalukian 58, and Chewong just 7 (Russell 1991: 428).

Second, the very word *emotion* itself does not seem to be universal. The Tahitians, the Bimin-Kuskusmin of Papua New Guinea, the Gidjingali aborigines of Australia, the Ifalukians of Micronesia, the Chewong of Malaysia, and the Samoans – they all do not have a word for *emotion* (1991: 429).

Third, some English emotion words have no equivalent in some other language. For instance, Gidjingali, an Australian aboriginal language, has just one word, *gurakadj,* where English has *terror, horror, dread, apprehension,* and *timidity*. In some African languages the same word covers what English distinguishes as *anger* and *sadness* (1991: 430). The Ilongot use *betang* to cover *shame, timidity, embarrassment, awe, obedience,* and *respect*. Many non-Western cultures have no word for *depression,* the Eskimos have none for *anxiety,* the Quichua of Ecuador lack one for *remorse,* and the Ifaluk do not have one for *surprise* (431).

Fourth, some languages have emotion words missing in English. The German *Schadenfreude* is too well known to need an explanation. The Japanese *amae* means a pleasant feeling of dependence upon someone (432). Some words from other languages are even impossible to paraphrase in English: the Ifaluk feel *fago*

> when someone dies, is needy, is ill, or goes on a voyage, but *fago* is also felt when in the presence of someone admirable or when given a gift. *Fago* is used in some situations in which

> English speakers would use *love, empathy, pity, sadness,* and *compassion* – but not in all such situations.
>
> (1991: 433)

Fifth, cross-cultural psychologists have not only studied the ethnographic data, they have also conducted various emotion-recognition experiments. In such studies, people from different cultures are asked to identify the different emotions expressed in a series of pictures of faces that, say, smile, frown, scowl, or sneer. Typically, such studies constrain the subjects by making them pick the respective emotion from a fixed and short list, or else let the subjects choose their own words. There is considerable similarity amongst different cultures, but there are also important differences. For instance, in one study of seven facial expressions, American students categorised four as *contempt* and three as *disgust*. Japanese students reversed the classification of three of the seven (1991: 436). Differences were more marked in comparison between English speakers and members of preliterate societies. For example,

> the number of facial expressions for which 70% or more of the observers agreed with one another was 23 out of 24 for English speakers, 6 out of 24 for the Fore [of New Guinea], and 6 out of 23 for the Sadong [of Borneo].
>
> (1991: 436)

Sixth, and finally, Russell considers a number of hypotheses concerning the nature of emotion categories. The one that he regards as most plausible conceives of emotion categories as 'scripts'. A script is an ordered sequence of features, that is 'the causes, beliefs, feelings, physiological changes, desires, overt actions, and vocal and facial expressions' (1991: 442). The script hypothesis can explain both cross-cultural similarities and differences:

> Those languages with fewer emotion categories would have more general scripts: Each script would have fewer features and cover a broader range of phenomena. Languages with many emotion categories have more specific scripts: Each script would have more features and cover a narrower range of phenomena. Moreover, some features are culture specific, and others are pancultural. Or, better, *culture specific* and *pancultural*

define two ends of a continuum. Some features may be limited to few cultures; others found in all or almost all.

(1991: 443)

Similarities and differences

Above, I have sampled the research into folk psycholog*ies* to show that there are indeed grounds for using the plural rather than the singular. This is not, of course, to deny the obvious fact that psychological and anthropological studies have unearthed similarities as well as differences. Such similarities are to be expected because we all share – more or less – the same biology.

It is not deeply controversial that there are *at least some* significant differences in the ways in which different cultures think of personhood or classify emotions. Most writers on folk psychology are happy to grant such differences. Their complacency with respect to the anthropological and psychological record will only be threatened if it can be shown that cultural variation can be found 'all the way down', that is even with respect to concepts such as desire or belief. I shall therefore introduce some of the anthropological data on belief in the next section.

Belief

Up to this point, I have written about folk psychology in general as a social institution rather than about belief, desire, or action as social institutions. But clearly the latter talk can be justified as well. In this section I shall propose a number of considerations relating to the category of belief. The view that I wish to make plausible is that 'belief' and 'believer' – that is, someone or something capable of entertaining, and being able to attribute, beliefs – are social statuses. And this means of course that they are social institutions: someone is a believer if they are collectively taken to be a believer; and something is a belief if it is collectively taken to be a belief.

Theories about belief as social institutions

Although this is not – strictly speaking – a compelling reason for taking belief to be a social institution, it is perhaps not altogether superfluous to point out in passing that the various theories about belief – theories proposed by philosophers and psychologists – all are, or strive to become, social institutions. In other words, the analyses of belief proposed by, say, advocates and defenders of eliminativism can undoubtedly be subjected to the same type of interpretation as I have carried out for theories of thought in Part I of this book.

Take for instance philosophers and scientists with considerable past investments in physiology and the neurosciences on the one hand, and little training in conventional mainstream psychology on the other hand. Surely such philosophers and scientists will not feel particularly motivated to seek out ways in which the results of the neurosciences can perhaps be made out to cohere with (folk-) psychological old-speak. The neuro-folks' research programme will capture much more attention – and resources – on the academic marketplace if they are able to present their work as a revolution *vis-à-vis* our ways of thinking and defining ourselves, or *vis-à-vis* rationality, truth, and morality. In this corner of the academic world, researchers will find the parallels between folk psychology and phlogiston theory utterly convincing, and they will feel little inclination to advocate the causal–historical theory of reference.

Alternatively, think of mainstream social psychologists, for instance people who have spent twenty or thirty years studying emotion or prejudice. Undoubtedly, given their work and their training, such scientists will have the gravest difficulties making sense of the idea that beliefs and desires might not exist! Indeed, they are either easily drawn towards a Fodorian view according to which beliefs and desires are right there in the mind–brain, or attracted to a Clarkian conception according to which we cannot help but use the belief–desire framework. Confronted with the eliminativist arguments, our social psychologists will react with incredulity: They will fail to see the parallel between belief and phlogiston, and they will find the eliminativists' employment of the description theory of reference unconvincing.

The general argument

A more direct way of showing that 'belief' and 'believer' are social statuses is to rely on an argument used above (in the first section of this chapter) for showing that folk psychology involves self-referentiality. Making the necessary adjustments for the category of belief, the argument takes the following form:

Premise 1	Full mastery of belief-ascription involves mastery of belief-talk.
Premise 2	Mastery of belief-talk involves knowledge of platitudes.
Premise 3	Platitudes concerning beliefs have several normative dimensions: they define how one *ought* to act, or what one ought to believe.
Premise 4	'Oughts' [*sic*] are what we collectively take to be oughts; that is, 'oughts' are self-referential.
Conclusion	Belief-talk is self-referential.

I take it that the premises of this argument are no longer controversial at this point. Lest my proposal is misunderstood, be it noted that I am not claiming that all belief-attribution is inevitably linguistic. But it seems to me a plausible conjecture that our paradigm cases of belief-attribution are cases in which such attribution happens *viva voce*. Platitudes concerning beliefs are, for instance: 'You cannot believe two contradictory claims at the same time', or 'If you really believe that *p*, then you must not act as if you believed that not *p*'.

An impoverished ontology

Another observation worth making is this. Note that different philosophers offer widely differing accounts of what constitutes a belief. For some authors a belief is a type of physical state of the brain; for others it is a functional state of the brain or mind; still others conceive of beliefs as psychological or abstract, or more or less fictitious entities. I suspect that this variety is grist to the mill of the sociophilosopher. To see this we need only remember that the tacit ontology underlying most modern philosophy is radically impoverished: It consists of only three levels: the physical, the psychological and the abstract Platonistic. Social entities do not easily fit into this three-part division. No wonder philosophers cannot agree on whether beliefs are primarily material, psychological, or abstract entities. Beliefs are none of the above. They are irreducibly social entities.

Social metaphors

Moreover, it is worth noting that some philosophers come fairly close to identifying the social character of beliefs. For instance,

Jonathan Bennett argues succinctly for the conclusion that 'What *is* a belief does not have an answer' within the traditional ontological distinctions. However, at the same time, he insists that beliefs do figure in causal explanations. Trying to cut through this Gordian knot, Bennett helps himself to an intriguing analogy that deserves quotation in full:

> Folk psychology does insist that attributions of beliefs and desires must help to *explain* behaviour. If you like, say that they must aid in *causally explaining* behaviour.... You still haven't implied that there is any such item as a belief, or as a desire....For a simple analogy, think of the causal explanatoriness of statements about shortages ('There is a shortage of food in Ethiopia; there is no shortage of oil in Mexico'); such statements can have explanatory power without our reifying shortages, treating them as though they were particular items in the world – negative storage bin, perhaps.
>
> (1991: 19–20)

The striking feature of this comparison is of course that 'shortage' is a socio-economic category. Notice how in reaching out for an apt comparison Bennett intuitively chooses to utilise an example that is clearly both social and self-referential.

True believers

Another author who comes close to anticipating the idea that 'belief' and 'believer' are social statuses is Dennett. Here I am of course referring to his idea that the property of being a 'true believer' is dependent on the 'intentional stance'. That is to say, my property of being a true believer becomes visible only once my behaviour is being predicted and interpreted through the folk-psychological stance. Add to this Dennett's one-time insistence that humans are '*communal*' rather than '*individual* folk psychologists' (1991b: 142), and the following conclusion clearly suggests itself: it is a collective judgement that makes someone a true believer; it is a collective judgement that determines whether one is worthy of being interpreted on the basis of rationality and humanity.

We can begin to understand some of the processes and criteria involved in this collective judgement by drawing on Stich's analysis of folk-psychological belief-attribution (1983: Ch. 5). Stich points out that for a person *A* to attribute a belief *p* to another person *B* is for *A* to make a similarity judgement. That is to say, when *A* says that '*B* believes that *p*', *A* means roughly the following:

> *p*.
> *B* is in a belief state similar to the one which would play the typical causal role if my utterance of that had had a typical causal history.
>
> (1983: 88)

We folk psychologists assess the similarity of beliefs on a variety of different grounds. One central feature figuring in this assessment, however, is 'ideological similarity', that is the extent to which the two beliefs are embedded in similar networks of beliefs. One consequence of this feature is that our willingness to count someone as a true believer decreases as the overlap between their and our networks of beliefs gets smaller. Thus hearing of the Nuer people's claim that a sacrificial cucumber used in certain rituals is actually an ox makes us wonder whether they can believe this at all, whether they qualify at all as 'true believers' in the sense of our folk psychology (cf. Stich 1983: 98–100). In other words, we grant the social status of 'believer' in good part on the basis of how much agreement we find between the views of humans (and perhaps other creatures) that we encounter on the one hand, and members of our own respective community of folk psychologists, on the other.

When does the child possess the concept of belief?

As will be recalled, according to the child psychologists, acquiring the full concept of belief is a process of maturation and/or individual theorising. How can this view be squared with the above insistence on 'belief' and 'believer' as social statuses? The answer lies in the realisation that the psychologists have not in fact identified the border between lacking and possessing the concept of belief. The border they have studied and detected is a different border: to wit, the border between different stages in the development of the child's ability to *conform* to central aspects of our – the grown-ups' – practices of belief-attribution. *Conforming to a practice* here contrasts with *following a practice*. To follow a practice (or rules) means to be able to do more than just conform to that practice. To follow a practice is to be able to distinguish between right and wrong ways of going on; it is to be able to characterise actions as belonging to a given practice; it is to be able to argue with others over whether or not a given way of 'going on' is acceptable; it is to be able to sanction others if they fail to go on in the right sort of way; and it is to be able to self-adjust one's own doing in the light of others' feedback (Bloor 1996b, 1997a).

Clearly, the false-belief experiments check for conformity alone; they never seek to determine the presence of any of the characteristics of following a practice just mentioned. Now, do we want to say that a person possesses a concept if that person merely *conforms* to the practice of applying the concept, or do we want to say that concept possession presupposes the ability to *follow* the practice? Surely, our normal practice of attributing concept possession or concept mastery points towards the latter alternative. We would not normally say that a person possesses a concept unless the person understands the concept; and we would not say that the person understands the concept unless that person can engage in the sorts of activities that I have mentioned above as being involved in rule following.

We can put this point into the terminology of the Interlude by saying that the child psychologists treat the child's successive belief-attribution mechanisms as a sequence of simple and isolated *N*-devices. Under various empirical circumstances the child either 'attaches' or 'signals', or fails to 'attach' or 'signal', the belief-pattern. What is missing here is obvious: the aspects of normativity, and the notion that the primary 'possessor' of a concept is a normative system of classifiers.

In light of the above, one can perhaps reformulate the so-far-identified, three stages of the child's development in the following way:

Stage 1	The child under the age of 2 *is not conforming* to our (grown-up) practice (or rules) of attributing true or pretence beliefs, and *is not conforming* to our practice (or rules) of attributing false beliefs.
Stage 2	The 3-year-old *is conforming* to our practice (or rules) of attributing true or pretence beliefs, but *is not conforming* to our practice (or rules) of attributing false beliefs.
Stage 3	The 5-year-old *is conforming* to our practice (or rules) of attributing true or pretence beliefs, and *is*

conforming to our practice (or rules) of attributing false beliefs.

Somewhat speculatively, we can perhaps distinguish the following phases as following from Stage 3 above:

Stage 4 The 5+-year-old *is conforming* to our practice (or rules) of attributing beliefs; the child's dispositions have been shaped by feedback from others (parents, siblings, teachers, etc.), but the child is not yet able to sanction (assess) others' belief-attributions, nor able to defend his or her own.

Stage 5 At a later stage the child has developed the dispositions (of his or her community) with respect to both belief-attribution and sanctioning, but is not yet able to articulate the rules fully.

Stage 6 The child has developed such verbal understanding. The child is now treated as a 'true believer' – this is because being a 'true believer' and being a 'true belief-attributer' are one and the same social status.

It would be fascinating if child psychologists eventually subject Stages 4 to 6 to the same kind of detailed scrutiny that they have come to give routinely to Stages 1 to 3.

'Belief' is not universal

As documented earlier, social anthropologists and cultural psychologists have for some time now investigated indigenous classifications of emotions and conceptions of the self. Unfortunately, core concepts of folk psychology – concepts such as 'belief', 'action', or 'desire' – have not attracted similar attention and careful scrutiny. The major exception to this rule is Rodney Needham's book *Belief, Language, and Experience* (1972). This study was mentioned in passing in Stich (1983), but otherwise it has not figured much in debates over folk psychology. In part this might be due to the fact that Needham's book is difficult to read and even more difficult to categorise. Although the work of a social anthropologist, the book defines its line of enquiry as belonging to 'empirical philosophy' (xiv); it has long chapters on etymology, on counterparts of 'belief' in other languages, on philosophical theories about belief, and on much else besides.

Needham's main contention is that it is wrong to assume that all languages (and thus all linguistic communities) have a word that can function as the natural translation for the English words 'belief' and 'to believe'. Here are his main arguments in support of this radical thesis.

First, he points out that a 'bewildering variety of senses' attach to words in foreign languages that have been used as translations for the English 'believe'. For example, words of the Nuer that have been used as translations include '*liaghè*' (to believe, have faith, praise, glorify, be proud), '*ngath*' (to trust), '*ngac momo ke loc*' (to know something with heart), '*ngath ke loc*' (to trust with heart), '*butè*' (to think, trust, believe), '*dhoonge*' (to think, trust, believe, hope), '*luèngè*' (to agree, consent, obey), '*ngaadhè*' (to hope, trust, wait), '*nhoghè* (*ro*)' (to be clear, acknowledge, (with *ro*) to trust), '*nyèthè*' (to follow the example of another), '*ruaatè*' (to trust, believe, think), '*thèghè*' (to esteem highly, honour, revere, adore), and '*waanè*' (to imagine, think, suspect, impute) (1972: 24–27). The important point here is that none of these words coincide precisely with the English 'believe'. This makes it questionable whether a straightforward translation of any of these words by the English word 'believe', or else the straightforward translation of any English 'believe'-sentence by means of these Nuer words, is ever fully adequate.

Second, Needham cites Evans-Pritchard's view according to which the Nuer do not in fact have 'our' concept of belief at all. The Nuer express their faith with expressions such as '*kwoth a thin*' (God is present) but not with such expressions as 'I believe in God' or 'There is a God':

> That would be for Nuer a pointless remark. God's existence is taken for granted by everybody. Consequently when we say, as we can do, that all Nuer have faith in God, the word 'faith' must be understood in the Old Testament sense of 'trust' (Nuer *ngath*) and not in the modern sense of 'belief' which the concept came to have under Greek and Latin influences. There is in any case, I believe, no word in the Nuer language which could stand for 'I believe'.
>
> (Evans-Pritchard, quoted in Needham 1972: 22)

Third, Needham reminds his readers of the etymology of the English word 'belief' and of the changes in the Western concept of belief. Needham's point here is to show that our concept of belief has a definite history; that there is nothing 'natural' about the concept of belief. As far as the etymology of 'believe' is concerned, the Indo-European root '*laub*' originally meant 'love' or 'desire'; from the Gothic (*galaubjan*) through to Old English (*gelefan*) and Middle English (*bileven*), it came to mean 'trust', and later the assessment of something as true. In other Indo-European languages the development went quite differently; for instance, the Spanish '*libidine*' (lust, lewdness) too goes back to the same root (1972: 43). The ideational history is no less complex. In the Jewish Bible – that is, the Old Testament of the Christian Bible – the word used for 'believe' is the Hebrew '*he'emin*'. In its profane uses, '*he'emin*' means 'to believe a report', or 'to accept as true', and 'to trust'. In its biblical uses, '*he'emin*' designates a special and unique relationship between *Yahweh* and His chosen people. '*He'emin*' is never used for the relationship with any other gods, only for the relationship with *Yahweh*. When the Bible was translated into Greek, '*he'emin*' was translated as *pisteúein*. *Pisteúein* usually meant 'to trust', 'to feel loyalty', or 'to have a firm conviction'. *Pisteúein* was not the word, however, that the earlier Greeks had used for expressing their beliefs in the gods. The earlier Greeks had used *nomízein* (to recognise, acknowledge) instead. Eventually, *pisteúein* replaced *nomízein* and became a technical term in religious language. It now meant to believe in God on the basis of the right kind of upbringing and education, and informed by the appropriate feelings of piety. The elements of trust and loyalty were no longer equally prominent. *Pisteúein* also often meant 'putting faith in the words of God', and therefore applied to the scriptures in particular. Owing to Paul's efforts, *pisteúein* also came to imply obedience; to believe was to obey (1972: 45–8). Needham concludes: 'Our conception of belief, so far as it has been religiously moulded, is demonstrably an historical amalgam, composed of elements traceable to Judaic mystical doctrine and Greek styles of discourse' (1972: 49).

Fourth, Needham claims that the word 'belief' has a plethora of different uses in modern English, uses that cannot be reduced to one specific core. In other words, there is not *one* concept of belief even in the modern English language. He finds here the explanation for why philosophers have failed to agree on one single account:

> Each philosopher adopts, more or less arbitrarily, a different acceptation of belief from among the kaleidoscopic representations in current usage; and because they write about the same word, or about similar statements, they then sound as though they were advancing different theories about the same (if complex) thing.
>
> (1972: 126–7)

Fifth, Needham denies the English language any privileged position with respect to the contents of the mind. The mere fact that the English language inclines us to speak of beliefs is no reason for assuming that beliefs exist. Indeed, because other linguistic communities seem to get by without our category of belief, and because beliefs have no qualia – a point on which Needham sides with Wittgenstein – Needham concludes that beliefs do not in fact exist as 'natural resemblances' amongst humans (1972: Ch. 8). It follows that 'when other peoples are said, without qualification, to "believe" anything, it must be entirely unclear what kind of idea or state of mind is being ascribed to them' (1972: 188). Such ascription can be justified only to the extent that these 'other people' have a social institution of belief-talk, a social institution that resembles our own:

> Belief-propositions implicate, in part at least, jural features of the social world – statuses, moral obligations, norms of co-operation, etc. – and this means that sociological analysis is required if the propositions are to be correctly construed. This analysis, in its turn, cannot be confined entirely to local significations but will need ultimately to be cast in terms of abstraction derived from the comparative study of institutions. In construing statements about belief, therefore, it is necessary to combine formal analysis, the linguistic translation of cultural singularities, and general sociology...in order to comprehend the setting to which such statements relate.
>
> (1972: 168–9)

Needham's claims are of course grist to the mill of the sociophilosophical theory of folk psychology. Not only does Needham emphasise the normative aspects of belief, but he also clearly sees that belief is a social category. One point is worth stressing once more, however. To side with Needham, and to accept that there are indigenous folk psychologies that do not have *our* concept of belief, does not amount to denying that in other folk psychologies there are concepts that carry out functions similar to our concept of belief. This much is to be expected on the basis of our common biological history. For the purposes of making plausible that 'belief' is a social institution, it suffices, in any case, if *some* degree of variation and convention can be displayed.

Saving conflicting intuitions

Above, in Chapter 8, I gave a rough summary of the debate over folk psychology. We saw there that different writers disagree over whether the platitudes of folk psychology are normative or descriptive, synthetic or analytic, theoretical or expressive. We also encountered the arguments for and against the theory theory, the controversy over innateness and modules, and the dispute over the eliminability of folk psychology as a whole. I now want to revisit some of the more prominent of these disagreements and comment on them from the standpoint of my sociophilosophy of folk psychology. I want to suggest that the interpretation of folk psychology as a social institution can throw new light on these contested issues in general and the conflicting intuitions that inform the different positions in particular. Put differently, in good Aristotelian fashion, I want to suggest that the collectivistic rendering of folk psychology can 'save the phenomena': it can show how the seemingly contradictory intuitions can all be partly justified. At the same time, however, to conceive of folk psychology as a social institution and a collective good allows us to overcome some of the binary parameters of the debate over folk psychology.

Platitudes

I begin with some observations on the conflicting accounts of folk-psychological platitudes. In the present chapter, I have myself put special emphasis on the normative aspects of these platitudes. Does this mean that I favour a normative *over* a descriptive, or an analytic over a synthetic, rendering of these platitudes? My answer is a clear 'no'. The suspicion that my own interpretation of folk psychology is neo-Rylean is only very partially justified. My position is rather that we need not choose between folk-psychological theory being either normative or descriptive, or either analytic or explanatory. The platitudes of folk psychology, it turns out, are descriptive–explanatory *and* analytic *and* normative. It all depends on their use in a given context. That is to say, depending on their use in a given situation, folk-psychological platitudes, such as 'someone who is thirsty will try to get something to drink', or 'the feeling of success brings pride', can be descriptions of how people typically behave, prescriptions of how people ought to behave, and sentences defining what we mean by 'thirst', by 'the feeling of success', and by 'pride'.

To bring this out, think of a social institution such as the institution of promising. Clearly, for those subscribing to this institution, the sentence 'promises are to be kept' has undoubtedly a normative meaning; sentences such as these are usually invoked to remind others of their promises. It is equally obvious that for someone who is committed to this institution, the sentence is analytic (or highly analytic, if we accept Quine's thesis that analyticity is a question of degree). But neither the sentence's normativity nor the sentence's analyticity rules out that the very same sentence can also function as a descriptive–explanatory law. And the reason why it can function in this latter way is related to its normative or analytic uses. In their daily communicative actions, members of the institution routinely remind one another of the institution's importance, and they sanction those that try to free-ride. Each one has a vested interest in all of their interlocutors taking their promises seriously. Or to put it another way, each one has a vested interest in the sentence 'promises are kept' being analytic for all their interlocutors. Yet precisely because people are by and large successful in this feat, the platitude 'promises are (usually) kept' is in accordance with the facts – that is, is true. The case is no different for folk-psychological laws such as 'the feeling of success brings pride'. The sentence does state how one ought, or is permitted, to feel, and the link between success and pride is analytic at least to some considerable degree. But again, because we collectively establish this link, and because we are by and large successful at it, therefore the law is also descriptively true.

It hardly needs a special argument to make plausible the view that the opposition between folk psychology as a means of expression and folk psychology as a means for explanation is also misleading. A sentence such as 'I am in pain' might well express my pain rather than denote it. But it would be insensitive to the diversity of uses to which language can be put to suggest that this is the only function of the word 'pain'. Moreover, 'I am in pain' is meaningful only against the background of large portions of folk psychology in its explanatory-cum-normative dimension. To utter 'I am in pain' and to smile joyously at the same time, or to say 'I have pain in my arm' while holding my foot – these are violations of our folk-psychological platitudes. The reason why we can use language to express mental states to others is that we share with them the knowledge of what these mental states are like, and how they are (or ought to be) linked to one another.

Theory theory versus craft or simulation theory

The collectivistic construal of folk psychology can also help us overcome the somewhat sterile dichotomy of theory theory versus simulation theory. No doubt, simulation – or *Einfühlung,* as it used to be called[1] – does play an important part in our folk-psychological abilities. To see others in agony makes us feel awful; to see others joyful makes us feel good. No doubt some of these reactions are instinctual and shared with many animals. But there is also a sociological perspective on this intuition. It is central to many institutions that they create in their members a sense of community, of common concern, of 'we'. Such 'we' expresses itself in common emotional reactions, in feeling disgust, shame, or guilt in like circumstances, and in saving one another's face.[2] Sharing an institution with others makes us – in varying degrees – both like them and concerned for them. It even creates the possibility of states of mind in which collective intentionality precedes individual intentionality (Searle 1990). For instance, much collective deliberation concerns what 'we' must do to achieve what 'we want'. Again, this might have a biological basis – after all, lions or hyenas are able to hunt as a group[3] – but no doubt this biological basis gets moulded and restructured through specifically human forms of social interaction. The point is, in any case, that in many forms of 'we', including collective intentionality, we experience some form of immediate unity with others. In these states of 'we' we do not form hypotheses about the others' mind; we know immediately what they feel because we know what we feel ourselves. Of course, simulation in this sense does not take us beyond folk psychology as a social institution and network of platitudes. Turning ourselves successfully into measuring instruments for what others feel and think presupposes that we have been primed or calibrated in like ways, *and* that we have at least similar interpretation functions, that is functions that give meaning to the instrument reading.[4] And what could these functions be if not the categories and platitudes of our folk psychology?

Folk psychology as a social institution 'licenses' the intuition that leads to the simulation theory; it 'supports' the intuitions behind the theory theory; and it 'provides for' the intuition that folk psychology is first and foremost a craft. Something like the theory theory arises naturally from our daily experience as communal folk psychologists: we constantly explain the behaviour of some people to some other people; we try to make explicit to our children how to go about making sense of their friends, teachers, or parents; and we attempt to formulate, for our own future purposes, inductive generalisations concerning the behaviour of our fellow humans. At the same time, however, we also note time and time again that practice inevitably outruns theory, that explicit theorising is possible only against the backdrop of our blind and unreflective acting on the basis of drills and skills. In this respect, the social institution of folk psychology is no different from other social institutions. It exists only as an institution-in-use.

Bedrock versus speculation

Another interesting clash of intuitions concerns the question whether folk psychology is an innate 'bedrock' theory, or else the result of speculation. One feels reminded here of political battles over social institutions such as marriage or the family; in such battles one often finds two camps: one camp insists on the fundamental, perhaps genetically fixed, need for the institution; the other camp insists on the historically contingent nature of the institution. One camp concentrates its attention on universal and fundamental features of the institution and then extrapolates from them. The other camp focuses on variable characteristics and then goes on to model other aspects of the institution on these protean features.

Now it might easily appear as if my collectivistic interpretation of folk psychology must oppose the idea of folk-psychological, innate *modules*. Indeed, authors defending such modules (Tooby and Cosmides 1992, 1995) often write as if innate modules and social factors stand in some sort of trade-off, or zero-sum relationship: If folk-psychological modules exist, then social institutions have little left to do. If evolutionary biology and evolutionary psychology are justified in positing such modules, then folk-psychological capacities can be accounted for on the level of the isolated individual, and social interactions can be ignored as irrelevant or epiphenomenal.

My response to this suggestion relies on what I have said before about normativity and pattern matching. It does not matter how much of the folk-psychological work is done by innate modules, for however much these modules end up doing, they can never deliver folk-psychological *concepts*. However complex the isolated modules are theorised to be, they will always and inevitably remain devices on the level of isolated *N*-devices. And thus they cannot arrive at a conception of right or wrong ways of interpreting human behaviours. This is because the notions of 'right' and 'wrong' presuppose an independent standard, a standard that can be constituted only by a collective. Of course, the capacities that enable a human being to become, or be, part of a normative system or collective in turn are – at least in part – innate. But this does not make the interaction epiphenomenal or reducible. After all, the interaction might have results and consequences that cannot be predicted or explained on the basis of what the individuals bring to it.

Elimination

What I have said above about the clash of intuitions over the issue 'bedrock theory versus speculation' obviously applies also to the dispute over the question whether folk psychology can ultimately be eliminated or not. Indeed, the parallel to disputes over the eliminability of certain political institutions – such as the family or the state – is often striking. I take it that this in itself provides some indirect evidence for my main thesis: that is, that folk psychology is a social institution. If defenders and eliminators [*sic*!] of folk psychology employ the same sorts of argumentative tools as do defenders and eliminators of paradigmatic social institutions, then it is plausible to assume that the similarities between folk psychology and these other social institutions run pretty deep.

Note for instance the rhetoric employed by Paul Churchland and Lynn Rudder Baker in their respective attack and defence of folk psychology. According to Churchland, folk psychology has failed 'on an epic scale', and its history is one of 'retreat, infertility, and decadence' (see above p. 303). It has not made any progress in more than 2,000 years and cannot really help us overcome the barriers between us. Neurophilosophy, on the other hand, will lead to a better future in which humans will have new and revolutionary ways of understanding one another and themselves. Science, art, law, and medicine will all fundamentally change in and through the neuroscientific 'conceptual revolution' (Churchland 1995). Whereas Churchland's voice is that of the revolutionary, Baker's voice is that of the conservative who seeks to remind us of just how much we stand to lose if we were to overthrow the received folk-psychological order. To abandon folk psychology would be to abandon truth, reason, morality, and understanding; in short, it would be 'cognitive suicide'. In a style that is often reminiscent of Edmund Burke, Baker insists that we have no reason to mistrust 'the wisdom of our ancestors'.

Kinds of kinds

The sociophilosophy of folk psychology can also help us avoid the uncomfortable choice between treating folk-psychological kinds either as mere 'mythical posits' or else as cutting the mind–brain at its joints. I submit that many folk-psychological kinds are *artificial kinds,* such as 'typewriter', rather than *natural kinds,* such as 'tiger', or *social kinds,* such as 'authority'.[5]

Social, natural, artificial, and human kinds

I have introduced this tripartite distinction earlier in this book (Part I, Conclusion; cf. Interlude), but the idea is perhaps important – and difficult – enough to justify a further explanation. Natural, social, and artificial kinds are all social institutions because they all have a self-referential component. But in each case the self-referential component functions somewhat differently. In the case of a social kind, such as 'authority', talk referring to authority, and actions structured by, or informed by, beliefs about authority, ultimately refer to further such talk and further such actions. The referent ('the authority') is constituted by such authority-talk and such authority-action, and thus, in the final analysis, the referent is that talk and action itself. The reference here is, as it were, 'exhausted' by the self-reference. One way to make this intuitively clear is to consider what would happen if the talk (and the action) were to cease. Obviously, in this case the referent would disappear; the authority *qua* authority would no longer exist – although the physical entity upon which the status 'authority' had been imposed might well persist.

In the case of natural kind terms such as 'mountain', reference is *not* exhausted by self-reference. 'Mountain' has *both* a self-referential and an '*alter*-referential' component. The self-referential component consists of the criteria that language users employ for classifying physical entities as mountains, that is mountain models, paradigms, and prototypes of mountains. And models and paradigms have a self-referential aspect because nothing is a model for anything in and of itself; a model or paradigm is what is taken to be a model or paradigm by a linguistic community. The *alter*-referential component of a natural kind term such as 'mountain' consists of the fact that speakers use this term to refer – away from the term – towards individuals in the physical world, individuals that exist independently of the reference, individuals that are not changed by the reference.

The idea that natural kind terms do not change the things classified is central to all philosophers who have emphasised the concept of natural kinds. Most of them take this idea for granted. In the recent literature, its explicit formulation is provided by Ian Hacking:

> In the case of natural kinds, classifying in itself does not make any difference to the things classified. Thinking of things as a kind, and giving a name to that class of things, does not of itself make any difference to the things so distinguished.
>
> 'Of itself?' The idea is not precise. The fact that we notice a class of entities as interesting may lead us to use or change things in that class. So naming has consequences. But the mere formation of the class, as separable in the mind, and in language, our continuing use of the classification, our talk about it, our speculation using the classification, does not 'of itself' have the consequences.
>
> Mary Douglas objected to an earlier statement of mine that microbes adapt to our classifications by becoming immune to our anti-bacterial agents, and so we have to reclassify them....Who says there are no feedback effects! Well, if there is one, it works not at the level of individuals but through a great many generations.
>
> (1992: 189–90)

That the individuals are independent of the classification of course implies that they will continue to exist once our classificatory activities cease.

Finally, let us turn to artificial kinds. This category is something of a hybrid of the first two. But it is important enough to be treated separately and labelled distinctively. First of all, artificial kinds differ from social kinds in that they have *alter*-reference: talk about typewriters is not just talk about talk; typewriter-talk is about physical things 'out there' in the world. Second, artificial kinds do have – like social and natural kinds – a self-referential component: there definitely are models, paradigms, and prototypes of what a typewriter ought to look like and how it ought to function. Third, artificial kinds do not, however, pass the 'Hacking criteria' for natural kinds: feedback effects between naming and changing are intended and wanted! Typewriters do not exist independently of the activity of classifying typewriters. Instead, typewriter-talk and typewriter-action (i.e. typing) are intertwined with social processes of production (i.e. typewriter production and typewriter repair) that bring the entities classified as typewriters into existence. In the case of artificial kinds, human action makes it so that the individuals referred to fit the prototypes, rather than vice versa. And finally, consider what happens to an artificial kind once the classificatory activity comes to an end. Unlike the individuals falling into social kinds, individuals falling into artificial kinds will persist. But unlike the individuals falling into natural kinds, they might well deteriorate quickly and not reproduce (like species) or reoccur (like mountains or rivers).

My category of artificial kinds is a generalisation of Hacking's 'human kinds'. Hacking distinguishes between natural kinds – as defined in the quote above – and human kinds. The decisive difference between the two cases is that feedback loops only exist in the case of human kinds:

> That is where some human kinds begin to differ from those routinely thought of as natural kinds. The classification of people and their acts can influence people and what they do directly. And I believe this is true only of people. Trivial enough: only people can understand when they are called and how they are described, so only people can react to being named and sorted. But it becomes an important difference in kinds when we realise that entities – people and their acts – of a kind can change in response to being so grouped.
>
> (1992: 190)

Folk-psychological kinds as artificial kinds

With the tripartite distinction in place, we can now turn to address the question of what kinds of kinds are folk-psychological kinds such as belief, desire, joy, and so on. Surely, for the folk-psychological realist, such as Fodor, 'belief' is a natural kind term, like 'mountain' or 'electron'. But what about such authors as Stich or Churchland, writers who contemplate the possibility that folk-psychological kinds are 'mythical posits'? It seems that their view is best reconstructed on the model of Emile Durkheim's sociological theory of religion (1971 [1915]). Durkheim famously suggests that in their gods, societies celebrate themselves. In speaking about their gods, in attributing superhuman powers to them, humans in fact speak about their own community and attribute 'super-individual' powers to their community. But they do all this without realising it. Likewise, if folk-psychological kinds are mere 'mythical posits', then whatever the respective kind terms refer to, they do not refer to real psychological and physiological processes. Most likely they *actually* refer again to social processes, but under a psychological disguise.

The category of artificial kinds, as introduced here, allows for a third view. To apply this category to folk-psychological kinds is to give a reconstruction of the sometimes-heard claim that our psychological classifications are constitutive of our mental states and events. Our psychological vocabulary does not classify mental states and events that exist wholly independently of the vocabulary. Instead, having *this* vocabulary rather than *that* vocabulary causes us to have – or be more likely to have – one kind of mental experience rather than another. Now if folk-psychological kinds are artificial kinds, if being trained in a given folk psychology causes one to have mental experiences corresponding to this folk psychology, then obviously folk-psychological kinds are more than mere 'mythical posits'. They are *real – real as artefacts*.

Moreover, note an interesting difference between the creation of typewriters and the creation of, say, states of mind such as *Schadenfreude* (malicious joy) – that famously untranslatable German emotion (word). In the case of the typewriter, the (physical) activity needed to create the referent is, as it were, 'ontologically' separable from the activity of classification. A robot could carry out the one without being able to carry out the other. Contrast this with the case where a mother teaches her child the meaning of *Schadenfreude*. To train the child in the use of the concept is, by the same token, partially to restructure or enrich the emotional life of the child. The action of creating a new emotion in the child's mental experience, and the action of enabling the child to classify the newly learnt emotion, are intertwined to the point of being indistinguishable.

The 'artificial kinds' hypothesis in anthropology

As already indicated, the thesis that folk-psychological kinds are artificial kinds is not altogether new. Although earlier authors did not have a sufficiently clear conception of the distinction between social, natural, and artificial kinds, they did anticipate the thesis in various forms. Indeed, I take it to be strong support for the thesis that writers from a variety of disciplines and schools of thought have arrived at closely related formulations of it (e.g. Dennett 1991a; Hacking 1995; Harré 1981; Johnson 1987; Kobes 1993; Heelas 1981a; Mischel 1977; Sharpe 1990; Taylor 1977). I shall here confine myself to mentioning just one or two striking examples.

In the literature on indigenous psychology, the distinction between mythical, natural, and artificial kinds is stated clearly by Paul Heelas and Andrew Lock:

> Whether indigenous notions describe actual psychological states and processes (in which case classifications of emotions can be judged against something), whether they organise, indeed constitute what is psychologically existent (in which case different classifications are equally true), or whether they operate as 'mythological' items within broader institutions (in which case they are not really to do with generic psychological nature), are alternatives which have not yet been fully investigated
>
> (1981b: xv)

Later, in the same book, Heelas criticises some anthropologists – for example, Mary Douglas – for presenting 'an oversocialised view of man'. In other words, according to the critics, Douglas and some other anthropologists deny that folk-psychological talk is ever about the mind; that is, they construe all talk about the mind as being *actually* about society (Heelas 1981b: 60). Judged from the point of view of my tripartite distinction, it is easy to understand how anthropologists slide into this exaggerated position. They rightly note that folk-psychological kinds are social institutions but they draw the wrong conclusion that folk-psychological terms must ultimately, or exhaustively, be referring to (aspects of) social institutions.

Heelas himself sides with the view that indigenous psychologies as bodies of knowledge are partly constitutive of lived experience: 'Indigenous psychologies have much to tell us about psychological nature: Constitutive of it, they mark or indicate actual psychological processes and states of affairs' (1981a: 16). At the same time, Heelas cautions his readers against interpreting the 'constitutive argument' too widely. On the one hand, there certainly are psychological phenomena that are not influenced by how they are conceptualised. For instance, the way we retrieve words from memory might well be such a process. On the other hand, 'there is the possibility that some indigenous formulations are adventitious with respect to man's psychological nature'. Here Heelas refers to Needham's aforementioned book on the concept of belief. Both Heelas and Needham take the latter's study to show that since 'belief' is not a universal category in all folk psychologies, therefore beliefs do not really exist as psychological phenomena (Heelas 1981a: 17a; Needham 1972: 146).

Interestingly enough, Rom Harré has challenged Needham's (and thus also Heelas's) conclusion by drawing attention to artificial kinds as the forgotten third alternative:

> This conclusion does not follow....Needham's [conclusion] is still universal. No one has beliefs. If we consider Needham's reasons for drawing his conclusion, they are all reasons for taking belief to be a non-universal psychological property, a much more interesting hypothesis....It seems to me that the correct conclusion to draw...is that belief is a mental state, a grounded disposition, but is confined to people who have certain social institutions and practices.
>
> (1981: 82)

David Martel Johnson's account of belief as cultural achievement

Harré does not give much of an account of how belief could be an artificial kind. Johnson (1987)[6] does just that. His thesis is

> that it is unnecessary to insist that the notion of belief (a) either applies to every human at every time, or (b) has never applied to any of them, because (c) belief is not a category universal to humans, like having a chin or being capable of speech, but a cultural achievement like the discovery of agriculture or the use of fire.
>
> (1987: 319)

Johnson suggests that it was the Greeks who invented our notion of belief. The Greeks introduced a new ideal of thought, an ideal of 'complete mental consistency'. This ideal was – and still is today – inseparable from the ability to believe something:

> In effect...to ask 'Do you *really* believe so and so?' is to ask whether the purported believer is able to maintain this so and so consistently with all the other points relevant to it which he also proposes to accept. In more explicit terms, I maintain that no one believed anything, strictly speaking, until Greek thinkers of the sixth century B.C. showed people how to do this....My view is that the clear separation – made possible by this ideal – of literal truth on one side, from myth, religious faith, poetry, political obligation, etc. on the other, is what first allowed people to have beliefs in a way that had not been possible before.
>
> (1987: 323–4)

Brain, society, sensations

Up to this point I have pointed to ways in which mental states might be regarded as social artefacts. At least briefly it is worth noting that something similar can be said also for states of the brain.

To begin with, the thesis that at least some brain states are social artefacts – in the above sense – is of course trivially true for the materialist. If at least some mental states are social artefacts, and if mental states are physical states of the brain, then at least some brain states are social artefacts.

Moreover, and leaving aside the materialist's in-principle argument, culture does of course shape up physically in a myriad of ways. Indeed, we must beware not to fall into the idealistic trap of reducing institutions to being mere beliefs about beliefs: institutions essentially involve action; they involve physical intervention as much as they involve representation.[7] To make up people's minds is not just to give them categories, it is also to exercise their bodies, to restrain, dress, nourish, starve, stimulate, hurt, train, distribute, as well as mutilate their bodies. In other words, cultures and their folk psychologies make up minds by remaking bodies.[8]

I insist on this Foucauldian point regarding social institutions in general, and folk psychology in particular, not least because it permits us to appreciate the possibility that there might not be a universal language of sensations.[9] Culture shapes our nervous system: it increases sensitivity in one area, and desensitises us in others. For instance – and to draw on my own limited cross-cultural experiences – Finns hardly feel the very mosquito bites that drive the British or German tourists mad; some of their favourite foods taste revolting to us; and their ability to stand the cold seems miraculous. Or, to move from the anecdotal to the scientific, some psychological evidence suggests that the thresholds for pain caused by heat differ significantly between people from different cultures (Chapman and Jones 1944; Nelkin 1986; Grahek 1995).

It is because culture shapes our bodies that it shapes our brains; and it is because culture shapes our brains – literally causing some areas to grow more, others less (Kandel and Hawkins 1993; Harré and Gillett 1994) – that at least some states of the brain might well be called *social states*. They are social because they are real artefacts of our culture, and social also because they predispose us to differ in the intensity, quality, and duration of some of our sensations.

No doubt these last remarks take us beyond the issue of the self-validation of folk-psychological kinds. The remaking of our bodies can increase and reinforce the degree to which our mental experience conforms to folk-psychological predicates. But of course our innate physiology ultimately restricts the scope of those mental experiences that culture is able to shape. And what is more, although folk psychology can make itself true with respect to central aspects of our mental experience, it cannot possibly make itself true with respect to our brain hardware. If the brain happens to have a parallel architecture, then no cultural classification and body-bashing will give the brain a von Neumannesque structure (barring the futuristic possibility of introducing chips into the brain).

I-talk

My suggestion of conceptualising folk psychology as a social institution will not seem convincing to many philosophers unless I can account for our intuitions regarding privileged first-person access, and the logic of 'I'.

'I'-talk as a social institution

The natural starting point here is a question that prima facie takes us beyond the realm of the philosophy of folk psychology and into the realm of the philosophy of language. I mean the question of whether the first person, the 'I', is a referring expression. Elizabeth Anscombe famously argues for a negative answer (Anscombe 1994 [1975]; cf. Malcolm 1995 [1979]). In her article 'The First Person', she considers various proposals on how, and to which entities, the 'I' might be taken to refer. Thus she discusses the possibilities that the 'I' might be a proper name, a pronoun, or a demonstrative, and that it might refer to a Cartesian ego, the body, the person, or the self. In each case Anscombe shows compellingly that the proposed solution is deeply unsatisfactory. Her conclusion is therefore: ' "I" is neither a name nor another kind of expression whose logical role is to make a reference, *at all*' (1994 [1975]: 154).

I agree with Anscombe's critical assessment of the proposed accounts, but I disagree with her conclusion. Her conclusion would follow only if her list of possible solutions were indeed exhaustive. But it is not. We can identify the missing alternative by noting that all of the possibilities considered by Anscombe are cases of *alter*-reference. Anscombe does not reckon with the possibility that 'I' might be self-referential, that is that 'I' might refer to the 'I'-talk. To put it another way, Anscombe does not contemplate the ideas that the use of 'I' might be a social institution; that ' "I"-talker' might be a social status; and that 'I'-talk takes different forms in different societies. This is the notion that I wish to make plausible here.[10]

'I'-talk is self-referential in two ways. First, it is self-referential in the way in which every social institution is self-referential. We have seen that both natural kind terms and social kind terms have a collectively sustained, self-referential, component. We find that component in the case at hand, too. The meaning of 'I', its correct and incorrect uses, rests with the collective, and it is continuously negotiated by its members. The meaning of 'I' is what we collectively take to be this meaning. Call this the *collective* self-referentiality of the 'I'.

'I'-talk is self-referential, however, in a second way. 'I'-talk is self-referential also on the level of the individual. It is part and parcel of the collectively sustained meaning of 'I' that 'I' can (and must) be used by a member of the collectivity to ascribe states and events to him- or herself. This self-ascription happens in 'I'-statements. The 'I' in such statements does not refer to some mythical entity like a self, or a noumenal ego; it refers to the individual who has the social status of 'I'-talker, a status constituted by the individual's I-talk – as well as the confirming 'you–he–she' talk of other 'I'-talkers.

It is hard to imagine a society of human beings that does not possess some social institution of 'I'-talk. Undoubtedly, members of different cultures ascribe to themselves rather different attributes, states, and events. But that they *do* self-ascribe some attributes, states, and events or other is probably universal. This social institution is no doubt underwritten by some biological universals, but it does not equate with them. The distinction between right and wrong usage presupposes the social institution. Biological facts cannot deliver this good.

That some form of 'I'-talk is ubiquitous does not, of course, mean that there cannot be substantial differences between the 'I'-talking of different cultures (Mühlhäusler and Harré 1990: 105–14). Such differences correlate with differences in folk-psychological conceptions of the self – strongly suggesting that 'I'-talk is itself part and parcel of folk psychology.

The social role of 'I'-talker comes with rights, responsibilities, and obligations. The most important right is membership in the institution in the first place. Recall that there were times when servants were not allowed to address their masters in the first person; that is, instead of saying 'At what time shall I bring your shaving water in the morning, milord?', they were supposed to say 'At what time will your Lordship be requiring his shaving water?', or some such. What was it that the servants were refused here? Undoubtedly they were denied being an independent agent with his or her individual perspective on the world, and being a centre of wishes, desires, and actions. In brief, they were of course refused the right to 'speak their mind'.

As far as responsibilities and obligations are concerned, note that 'I'-talk is often involved in indicating the speaker's or writer's degree of commitment to statements or actions. This comes out most clearly in negative cases; that is, in cases where the speaker is concerned with excusing him- or herself from such commitments. 'I wasn't thinking properly when I was doing that'; 'I wasn't really me when acting in this way', 'I was beside myself' – all these expressions indicate that our folk model of responsibility assumes that usually the 'I', or the 'I think', accompanies those actions for which one can be held responsible. The same tendency is visible also in the stylistic phenomena of 'hedging'; these are linguistic ways of avoiding appearing to be too strongly committed to the claims of one's (scientific) papers or books. Many hedging moves amount to replacing first-person pronouns with third-person constructions (Kusch and Schröder 1989).

Privileged access

The doubly self-referential structure of 'I'-talk is a strange and unique phenomenon. It is difficult to keep in focus and prone to give rise to conflicting intuitions. It is one of those phenomena that easily sets us up for a linguistic holiday. Here I wish to focus on two individualistic intuitions that have become central in our Western folk psychology and its philosophical interpretation.

Take first the intuition of privileged first-person access: that is, that only *I* can know who *I* am and what goes on in *my* mind. This intuition might perhaps arise in the following way. In using 'I'-statements I am using 'I's that are self-referring ('I' refers to 'I'-talk) and self-validating ('I' has a referent because of its repeated use). 'I'-statements thus cannot fail to refer. And because 'I'-statements cannot fail to refer when uttered by the self-ascriber, therefore – it easily seems to us – they must be true when uttered or thought by the self-ascriber.

Unfortunately, this line of reasoning constitutes a case of language going on holiday. The mistake consists in extending the self-reference and self-validation from the logical subject, the 'I', to the whole 'I'-statement, that is to the logical subject *and* to the logical predicate. In a sentence such as 'I am thinking', it is the 'I' that is self-referential and self-validating, but not the 'am thinking'. *What* I ascribe to myself is not as such self-validating. *What* I ascribe to myself can be self-validating only if 'I'-talk has first constituted, and continues to maintain, a referent of which the thinking can be predicated in a self-referring and self-validating fashion.

We can both secure this analysis and push it further by briefly considering the Cartesian *cogito* and some of its interpretations and criticisms.

First, we can see why *cogito, ergo sum* seems so compelling to us. It is the grammar of 'I'-talk that makes it appear in this way. The 'I'-statement '*cogito*' self-refers to the 'I'-talk of whoever utters or thinks this statement. But this very self-reference is possible only if there are repeated uses of the 'I', both by the same speaker and by other members of the linguistic community. The 'I'-statement '*cogito*' thus already refers to the 'I'-talk of someone who exists as the referent for further self-ascriptions. Perhaps we can put this point by saying that '*cogito, sed non sum*' would be a *performative* contradiction: It denies performing an act of self-reference *by* performing an act of self-reference.[11]

Second, my comments (a) that we have to distinguish between

subject and predicate of self-ascription, and (b) that self-validation applies only to the subject, that is to the 'I', were anticipated by Gassendi in one of his objections to Descartes. Gassendi objected that any self-ascription could have been used by Descartes to prove his existence: that is, that *ambulo, ergo sum* could have done the job of *cogito, ergo sum*. Whatever the route by which Gassendi came to this conclusion – and I can claim no competence regarding this issue – the upshot of his criticism is essentially on target.[12]

Third, Gassendi's objection is the inverse of Lichtenberg's famous objection: 'One should not say "I think", but rather "it thinks" [*es denkt*], just as one says "it thunders"' (1958: 458). Lichtenberg's point seems to have been that phenomenologically we do not experience thinking as an activity carried out by us. This interpretation is supported by the following further aphorism: 'Just as we believe that things happen outside us without our intervention, so also the ideas of them can occur in us without our aid. Indeed, we have also become what we are without assistance from ourselves' (1958: 482). Lichtenberg was right about the fact that our folk psychology does not fit all that well with the phenomenology of thinking. Indeed, we might think of Lichtenberg as an early challenger of our Western folk psychology! At the same time, however, Lichtenberg failed to see the deeper social significance of the 'I think' formula: to wit, that it expresses a responsibility condition, and that it is essentially involved in constituting the speaker as a referent for further self-ascriptions.

Fourth, in light of my emphasis on the twofold self-referentiality of the institution of 'I'-talk, it is intriguing to note in which form the *collective* self-referentiality of 'I'-talk appears in Descartes' distorted perspective. If the overall argument of this book is correct, then it is the collectivity (and collective self-reference) from which the relative stability of meaning derives. This is of course just a different way of expressing the idea that meanings are social institutions. Descartes obviously did not have this concept of social institution, and therefore had to anchor stability of meaning elsewhere. As is well known, in Descartes' system it is God who makes meanings stable, truths persistent, and reality objective (Ethemendy 1981).

At first sight, this seems like a very different solution from the one suggested here. And yet there *is* a way to bridge the gap between Cartesius and us. This bridge is Durkheim's suggestion that in their gods societies celebrate themselves and their achievements. In celebrating his God, Cartesius was celebrating collectively sustained use (Durkheim 1971 [1915]).

Self-creation

Having at least briefly dealt with the intuition of privileged first-person access, let us now turn to the intuition of first-person creation. I mean the intuition that I am whatever I take myself to be, and that I can thus radically change who I am by changing my self-ascriptions. This intuition is of course central in Western culture, and like the intuition of first-person access, it informs much of the resistance to sociological theories of knowledge, of the mind, or of the self. After all, such resistance often turns on the idea that we are free to reject society and recreate ourselves in opposition to society. That the idea of radical self-creation is central in our Western culture hardly needs much arguing either. Most of us were brought up to admire famous converts to whatever religion we were taught, and many of us will have fallen, at some stage or other, under the spell of existentialist philosophies of self-creation – be it in Kierkegaard's, Nietzsche's, Heidegger's, Sartre's, or later Foucault's versions. The vision of contemporary eliminative materialists, to wit, that we can radically recreate ourselves by performing an intellectual–existentialist leap from folk-psychological old-speak to future neuralese, is but the most recent expression of this intuition.

I submit that the intuition of radical self-creation might be due to a conflation of individual self-referential 'I'-talk with the collectively sustained institution of 'I'-talk. Collective self-reference is, as it were, *absolute* rather than relative. It is not checked by anything outside of the collectivity: whoever *is* the authority depends solely on whoever is collectively taken to be the authority, and whatever *is* the paradigm, prototype, or model for 'electron', 'tree', or 'typewriter' also depends on the self-referring belief of the collectivity. Contrast this with the case of individual 'I'-talk: Although this 'I'-talk is self-referring and self-validating, it lacks independence and absoluteness. Not only does individual 'I'-talk depend on the social institution of 'I'-talk, but what individuals can rightfully ascribe to themselves is also restricted by the communities in which they live. How could we possibly come to overlook this relativity and dependence in favour of the notion of radical self-creation? I suspect that the answer lies again in the direction of the Durkheimian analysis of God as the collectivity in disguise. We have already seen that Western metaphysical thought (e.g. Descartes) takes the collectivity to be some sort of superego, as omniscient, omnipotent, and radically free to choose. Although the perspective is distorted, it is easy to see how these predicates derive from the peculiar 'absoluteness' of the collective self-reference. Because collective self-reference is thus already misconstrued as the individual writ large it is perhaps not surprising that we feel inclined to take the further step of assimilating our individual, dependent, and relative 'I'-talk to collective, independent, and absolute self-reference.

Conclusion

My remarks on 'I'-talk and individualistic intuitions might easily appear to have moved us well beyond the main themes of this Part II, that is the social rendering of folk psychology, and the criticism of individualism in the philosophy of folk psychology. But this appearance is deceptive. What the argument of the last section shows is this. Individualism in the philosophy of folk psychology stands in a tradition of philosophising about the self. This tradition consists of philosophical renderings of certain intuitions regarding the self. These intuitions can only be understood in light of the above theory of social institutions. These intuitions are easily misleading, and they have *in fact* led astray both the philosophical tradition in general and the philosophy of folk psychology in particular. And finally, it is only by understanding the self-referential, institutional foundation of these intuitions that we can see not only from where they derive their persuasive force, but also to what extent they may be warranted.

The individualistic rendering of these intuitions has made philosophers blind to the ideas that folk psychology might be a social institution and a collective good, that self-ascription is a social process, and that individual self-reference presupposes collective self-reference. Spurious models or notions such as the individual child-scientist with his or her static, non-social subject matter; the reduction of folk-psychological language to its expressiveness; the claim that we need to 'solve' the problem of other minds; and the belief that we can radically redefine ourselves in terms of an altogether new scientific language – all these are but so many symptoms of the same underlying disease. The same can obviously be said about the standard set of binary choices regarding folk psychology: to wit, that folk psychology must be either normative or explanatory, either analytic or synthetic, either expressive or descriptive. As we have seen, the appearance of contradictory opposition is due to an insufficient appreciation of the nature of social institutions.

I conclude with four general remarks.

1 *What difference does it make?*

Assume someone reacted to my argument in this Part II by saying 'OK, granted that folk psychology is a social institution. Why should we care? What difference would it make if you were right?'

I would like to think that my answer to this challenge has already been given along the way. But it is perhaps worthwhile making it explicit once more and for a final time. I take it that the debate over folk psychology is meant to establish the nature and structure of this body of common-sense knowledge. What kind of knowledge are we dealing with here? How is this knowledge maintained and sustained? How does it relate to its users, on the one hand, and other scientific or common-sense bodies of knowledge, on the other hand? How, and to what extent, is folk psychology learnt? Is it a ubiquitous or universal body of knowledge, or is it – wholly or in part – tied to specific cultures? Is it likely to be replaced by 'neuralese'? And so on.

I think it should be obvious that my sociophilosophy of folk psychology either suggests, or else supports, such answers to these questions as are not mainstream and already fully accepted in the folk-psychology debate. Just to mention perhaps the most important point, it surely does make a big difference whether we think of the exercise of folk-psychological knowledge on the model of a simple pattern-matching device – say one of Barnes' *N*-devices – or whether we think of it on the model of a social institution – say one of Barnes' arrays of interacting *S*-devices. If we adopt the latter perspective, our scientific and philosophical curiosity will be redirected: we will want to investigate phenomena of self-presentation, negotiation, parental training, and sanctioning, and we will want to do all this with a historical and cross-cultural perspective. We will also want to study what the newly born child must bring to interactions for that child eventually to learn proto-normative and normative behaviour. In short, it amounts to replacing an individualistic with a collectivistic perspective.

2 *Folk psychology and scientific psychology*

I have now completed the overall argument of this book. In Part I, I argued by means of a historical case study that *scientific* psychological bodies of knowledge are social institution, and in the present Part II, I sought to establish the same thesis with respect to *folk*-psychological knowledge. If I was successful in all this, then

Barnes', Bloor's, and Searle's theory of social institution has passed two crucial tests. If it can successfully illuminate phenomena of both the *durée longue* and the short term, then surely it must be on the right track.

There are good reasons for thinking about folk psychology and scientific psychology in tandem. First, philosophers and psychologists alike often model the folk psychologist on the scientific psychologist. Second, as I have tried to show, however, if we wish to engage in this modelling, we had better have an adequate conception of the scientific psychologist! If our conception of the latter is individualistic, then our theory of the folk psychologist is unlikely to come out any differently. I showed in Part I how *scientific* theories and practices are social; this is the model that needs to be carried over into our analysis of the folk psychologist. Third, it is not just the case that a collectivistic understanding of scientific psychological knowledge supports a collectivistic understanding of folk-psychological knowledge. The reverse is true as well! If folk psychology is a social institution, then of course all scientific psychological knowledge that sticks to the concepts of folk psychology essentially remains within that same institution.

3 *Change and stability*

Throughout most of the argument of this Part II,[13] I have made one enormous concession to the received philosophy of folk psychology. I have gone along with the assumption that folk psychology is by and large static and unchanging. My reason for this is as follows. It is often said – not so much in written work as in the 'oral tradition'[14] within the academia – that sociological perspectives can be pertinent only where there is change, and that social studies come into their own only where competing, historically contingent, social–political interests can be identified. The sociophilosophical theory of folk psychology presented here refutes this picture. Even if we hold that our Western folk psychology has been, as it were, 'diachronically universal', even if we assume that it has not changed much through the centuries, even then we can still give a social account of this folk psychology. Having made my point concerning the *durée longue,* I am now ready to grant that folk psychology does change over time. Anyone doubting this point need do no more than read the work of Lucien Febvre, Marc Bloch, or other *Annales* historians, for example.[15] Or think of the ways in which psychoanalysis and its terminology has become part and parcel of our self-understanding.[16]

Changes that have occurred, however, fall well short of outright elimination of one complete folk psychology by another. The only cases of outright elimination of folk psychologies I can think of are cases in which the cultures themselves – and usually their people – were destroyed as well. And yet, nothing I have said rules out the possibility that sometime in the future our great-great-grandchildren might speak neuralese. Sociophilosophy of folk psychology does not legislate a priori against this possibility. What the above does show, nevertheless, is that it must be a mistake to model the elimination of folk psychology on the elimination of particular scientific theories, such as phlogiston theory. To eliminate folk psychology is to eliminate the social institution that is the basis for all others, and it is to eliminate the social institution that most of us have least interest in destabilising. Moreover, because folk psychology is a social institution as well as constitutive of our mental experience, to replace our folk psychology by neuralese is to change radically both social life and consciousness.

4 *The consolations of folk psychology*

Finally, amidst all these dazzling future possibilities, only one thing is certain: whatever the mind–brain–behaviour talk of the future is going to be like, it will always have to be a social institution of sorts. And thus it will always need both the sociology of knowledge and sociophilosophy to explain its structure and character. That is no small comfort in this time and age when the sociologist of knowledge is everyone's favourite whipping boy.

Notes

Introduction to Part I

1 'Misleading' because the existence of imageless thought was only one of the contested issues.
2 On Charcot's and Galton's schools see Danziger (1990a: Ch. 4).
3 See e.g. Harwood (1993).
4 Useful models here are Ash (1995), Galison (1987), Whitley (1984), and Smith and Wise (1989).
5 Of writers on the sociology of scientific knowledge, I have been most influenced by Bloor (1976, 1983), Collins (1985), Edge and Mulkay (1976), Harwood (1993), MacKenzie (1981), Shapin and Schaffer (1985), and Shapin (1990, 1994). I am also indebted to historians of the 'research school' tradition; see especially Fruton (1990), Geison (1981), Geison and Holmes (1993), Morrell (1972), and Turner (1993, 1994). I hope it will be clear throughout how much I owe to the work of three historians of psychology in particular: Ash (1980a, 1980b, 1995), Danziger (1979, 1980a, 1980b, 1983, 1990a, 1990b, 1997), Richards (1989, 1992, 1996).

1 The Würzburgers

1 This list is based on Cordes (1898), Elsenhans (1897), Lipps (1883, 1903), Meumann (1907c), Oesterreich (1910: Ch. 9), Schrader (1903, 1905), and Wundt (1888a).
2 For descriptions and pictures of these instruments, see Myers (1911), Caudle (1983), and Sokal *et al.* (1976).
3 These numbers are calculated on the basis of Mayer and Orth (1901), Marbe (1901), Orth (1903), Külpe (1903, 1904), Nanu (1904), Ach (1905), Taylor (1905/6), Watt (1905), Dürr (1906), Messer (1906), Schultze (1906), Wertheimer (1906), Bühler (1907b, 1908c, 1908d), Grünbaum (1908), Segal (1908), Ach (1910), Beer (1910), Wartensleben (1910), Feuchtwanger (1911), Schanoff (1911), Westphal (1911), Koffka (1912), Haering (1913), Selz (1913), Lindworsky (1916), Rangette (1916), K. Bühler (1918).
4 Interestingly enough, in so arguing, Külpe was also rejecting his own earlier *Outline of Psychology* (*Grundriß der Psychologie*, 1893) and

adopting, almost verbatim, the viewpoint of G. Martius, one of the reviewers of his *Grundriß* (Martius 1896: 40).

5 Here Külpe acknowledged that Stumpf (1906) had urged a similar line, but claimed to have arrived at his position independently (1907b: 603).

6 Külpe was also the examiner of Bühler's *Habilitationsschrift* (Bühler 1907b, 1908c, 1908d). His report was again very positive. See O. Külpe, *Gutachten über das Habilitationsgesuch des Herrn Dr. med. et phil. Karl Bühler*, 30 March, 1907. Archiv des Rektorats und Senats der Universität Würzburg, ARS 397 (*Personalakte* Bühler).

2 Friends and foes

1 Ach is criticised only in passing. Wundt had a high opinion of Ach. In a letter of 1 November 1907, Wundt wrote to Meumann: 'Of the younger people in Würzburg, I deem Ach to be the most important; I hope that he will abandon this wrong track soon.' (Wundt Archiv der Universität Leipzig)

2 It seems that Ach was the only Würzburger that Müller thought worthy of a detailed criticism – Ach later succeeded Müller in Göttingen. In 1927 Müller proposed to his fellow academicians of the *Göttinger Akademie der Wissenschaften* that Ach be accepted as a new member. In his letter of recommendation (21 October 1927, Müller characterised Ach as someone 'who deserves praise for a certain originality in thinking' (Akte 'Pers 25 149' of the Archiv der Göttinger Akademie der Wissenschaften). I find it hard to decide whether this endorsement was half-hearted. It certainly contrasted negatively with David Hilbert's letter of recommendation for Müller in 1911. Hilbert claimed that Müller occupied 'the leading position' in psychophysics (ibid., Akte 'Pers 16 200'). Müller's membership in the academy was finally secured in 1911; earlier bids in 1893 (supported by Felix Klein, Akte 'Pers 16 204') and 1901 had been unsuccessful. Ach was not accepted.

3 Meumann's criticism of the Würzburgers' work was solicited by Wundt. On 1 November 1907, Wundt wrote to Meumann:

> I am glad to hear that you agree with my comments on the interrogation experiments. But I would be very disappointed if this agreement were to stop you from publishing your own critical comments. On the one hand, my paper lacks the direct experimental proof that you would be able to provide; and, on the other hand, this method of experimentation has unfortunately spread so widely that it cannot hurt if it is investigated from more than one angle.
>
> (Wundt Archiv der Universität Leipzig)

Wundt was highly critical of Meumann's *Intelligenz und Wille*, and published a long critical review of the book (Wundt 1910c). Wundt informed Meumann of his forthcoming criticism in a letter of 9 April 909 (Wundt Archiv).

3 Recluse or drillmaster versus interlocutor and interrogator

1 The best proof of the all-pervasiveness of this theme in much of German university philosophy was the attacks upon it at a slightly later stage. See e.g. Spengler (1918) and Jaspers (1919).
2 Müller (in Göttingen) had a similar habit (Misch 1935: 45).
3 For Wundt's own account of the running of his laboratory, see Wundt (1910a).
4 That is, it functioned according to the 'principle of creative resultants' (Wundt 1903a: 778).
5 I shall deal with Wundt's *Völkerpsychologie* in Chapter 5 at greater length. For Wundt's programme, see Wundt (1888b); for good accounts and discussions, see Sganzini (1913), Danziger (1983), Schneider (1990), and Oelze (1991).
6 I am grateful to Ed Haupt for numerous email exchanges on Müller. See Misch (1935) and Sprung and Sprung (1997) for general summaries of Müller's life and work.
7 Perhaps one can say that Wundt's strategy of reserving *Völkerpsychologie* for himself proved self-defeating: Because he did not involve his students in this research programme, he ended up without an audience and without a following for it.
8 Interestingly enough, at least in one context, Külpe later argued differently. When Külpe petitioned the university in Bonn for more rooms for his institute in 1911 (29 June) he insisted that experimental subject and experimenter ought not to be in the same room. It was for this reason that the institute needed extra rooms. (Külpeana V: 12).
9 Wundt wrote to the editor that 'Marbe has compromised experimental psychology with his latest work to such a degree that I do not want to have the Archive compromised by him as well'; letter to E. Meumann, 5 June 1903, quoted from Bringmann and Ungerer (1980: 54).

4 Purist versus promiscuist

1 Wundt's bitterness comes out vividly in a private letter to a certain A. Sichler (27 June 1908):

> Külpe's misconceptions are [compared with those of other of my critics] more severe and less understandable. One has every right to expect that an author – especially one who writes *ex officio* about a topic – has first thoroughly acquainted himself with that topic. But Külpe seems to judge on the basis of a vague memory.... And what is most regrettable is that other philosophers draw their knowledge of the most recent philosophy from Külpe's little book. Of course it has to be admitted that this book has the advantage of being short, and of not demanding much intellectual effort from its readers.
>
> (Wundt-Archiv)

5 Collectivist versus individualist

1 This criticism had also come from within Wundt's own school. See Krüger (1915: 209).
2 The analysis of this section was inspired by Brunner (1995).

6 Protestant versus Catholic

1 For one authoritative account of the whole issue, see Nipperdey (1990: Ch. XII).
2 Interestingly enough, Wundt's main neo-Thomist critic, Clemens Gutberlet, also linked *Meister* Eckehart to Luther. Both Eckehart and Luther suffered from excessive 'German inwardness'. In Eckehart's case this led him to believe that by turning inwards, the soul could fuse with God. To Gutberlet this constituted pantheism (Gutberlet 1916: 132).
3 All these documents are in the '*Personalakte* Külpe' (ARS Nr 607), Universitätsarchiv Würzburg.

7 Conclusions

1 A different way of summarising my findings pictorially could have taken its model from N. Wise's paper 'Mediations: Enlightenment Balancing Acts, or the Technologies of Rationalism' (1993).
2 Here I draw heavily on Barnes (1983, 1988), Bloor (1997a), and Searle (1995). Cf. Kusch (1997), and Chapter 9 below.
3 See the literature cited in note 1 above.
4 This link between Collins' Experimenters' Regress and the theory of institutions was first suggested to me by D. Bloor (personal communication). It cannot be stressed enough that these circularities are highly *mediated* by calculations, methods, and theoretical assumptions. This often disguises the circle.
5 The analysis of social and natural kind terms given here owes almost everything to Barnes (1983, 1988) and Bloor (1997a). The idea that natural kind terms do not change the things that are being classified comes from Hacking (1992, 1995).
6 I borrow the concept of artificial kind from Guttenplan (1994).
7 For a brief summary of laboratory studies within the sociology of scientific knowledge see Knorr Cetina (1994). For the most important statements on the interrelations between social order and natural order see Bloor (1976, 1983), Collins (1985), Shapin and Schaffer (1985), and Shapin (1994).
8 These topics are central also in P. Galison's work. See e.g. Galison (1985).
9 This position is of course modelled on the Churchlands' eliminative materialism. See Churchland (1988: 43–9).

8 The folk psychology debate

1 R. A. Sharpe (1990: 60–73) criticises the theory theory along the same lines.
2 See especially the anthologies edited by Hirschfeld and Gelman (1994), Davies and Stone (1995a, 1995b), and Carruthers and Smith (1996).
3 From here on I follow the excellent summary of eliminativist arguments in Stich (1996: 18–31).
4 Sharpe (1990: 69–70) argues the related thesis that our folk-psychological concepts are constitutive of our mental life.
5 The charge is endorsed by John Searle (1992: 48):

> The commonsense objection to eliminative materialism is just that it seems to be crazy. It seems crazy to say that I never felt thirst or desire, that I never had a pain, or that I never actually had a belief, or that my beliefs and desires don't play any role in my behaviour.

6 Garfield (1988: 112–15) anticipates Trout's argument.
7 For another evolutionary argument, see Graham (1993 [1987]).
8 This point is made at great length also in Greenwood (1991) and in Hewson (1996).
9 Hewson (1996) has used questionnaire methods to challenge another eliminativist's claim about the folk-psychological concept of belief, that is the claim that the folk favour a 'wide' account of content.

9 Folk psychology as a social institution

1 By German hermeneutic philosophers of the nineteenth century, such as Schleiermacher or Dilthey. See e.g. Gadamer (1975 [1960]).
2 These ideas are central in ethnomethodology. See e.g. Garfinkel (1984 [1967]).
3 A point much emphasised in Searle (1995: 24).
4 For work on the calibration of instruments, see Jardine (1986). For the concept of 'interpretation function', see e.g. Churchland (1979: 108–10).
5 For a related attempt to distinguish natural kinds from psychological kinds, see Danziger (1997: 181–93).
6 Liisa Kanerva first drew my attention to Johnson's paper, years before I ever got interested in these issues.
7 I take the 'representing–intervening' opposition from Hacking (1983).
8 The distinction between 'making' and 'remaking' too comes from I. Hacking: 'We remake the world, but we make up people'. I insist on the obvious point that people are part of the world. See Hacking (1984: 124).
9 No book has impressed this point on me as strongly as Desjarlais (1992).
10 Other critics have argued that Anscombe's list of possible solutions is incomplete in other respects. C. McGinn and J. Glover miss a consideration of the possibility that 'I' is like 'now' or 'here'. See e.g. McGinn (1983: 54), and Glover (1988: 206). There are some affinities between my account and that of Harré (1987).

11 I take the idea of performative contradiction from Hintikka (1962).
12 For a discussion of Gassendi's objections, see Merrill (1979).
13 With the exception of my use of Needham (1972) and Johnson (1987).
14 This happy phrase is Searle's. See Searle (1995: *passim*).
15 See e.g. Febvre (1982 [1942]) and Bloch (1962 [1936]). See also Altschule (1965).
16 Paul Churchland invokes the case of psychoanalysis against those who doubt 'that the vocabulary of a sophisticated science could ever gain general use'. See Churchland (1995: 323). I am not convinced that the case of psychoanalysis can support Churchland's hopes concerning neuralese. After all, psychoanalysis is still fairly close to our received belief-desire folk psychology.

Bibliography

Unpublished sources: consulted archives

Archiv der Göttinger Akademie der Wissenschaften
Archiv der Universität Göttingen
Archiv der Universität Leipzig
Archiv der Universität Würzburg
Bayrische Staatsbibliothek München, Handschriftenabteilung
Universitätsbibliothek Göttingen, Handschriftenabteilung

Published sources

Ach, N. (1899) *Über die Beeinflussung der Auffassungsfähigkeit durch einige Arzneimittel,* Diss. Würzburg.

Ach, N. (1905) *Über die Willenstätigkeit und das Denken,* Göttingen: Vandenhoeck & Ruprecht.

Ach, N. (1910) *Über den Willensakt und das Temperament: Eine experimentelle Untersuchung,* Leipzig: Engelmann.

Ach, N. (1911) 'Willensakt und Temperament. Eine Widerlegung', *Zeitschrift für Psychologie* 58: 241–70.

Altschule, M. D. (1965) 'Acedia: Its Evolution from Deadly Sin to Psychiatric Syndrome', *British Journal of Psychiatry* 111: 117–19.

Anschütz, G. (1911) 'Über die Methoden der Psychologie', *Archiv für die Gesamte Psychologie* 20: 414–98.

Anschütz, G. (1912a) 'Spekulative, exakte und angewandte Psychologie. Eine Untersuchung über die Prinzipien der psychologischen Erkenntnis II', *Archiv für die gesamte Psychologie* 24: 1–30.

Anschütz, G. (1912b) 'Tendenzen im psychologischen Empirismus der Gegenwart: Eine Erwiderung auf O. Külpes Ausführungen "Psychologie und Medizin" und "Über die Bedeutung der modernen Denkpsychologie" ', *Archiv für die Gesamte Psychologie* 25: 189–207.

Anschütz, G. (1913a) *Die Intelligenz: Eine Einführung in die Haupttatsachen, die Probleme und die Methoden zu einer Analyse der Denktätigkeit,* Osterwieck: Zickfeldt.

Anschütz, G. (1913b) 'Einige Bemerkungen zu meiner Kritik von O. Külpes Ausführungen "Psychologie und Medizin" und "Über die Bedeutung der modernen Denkpsychologie" ', *Zeitschrift für Psychologie* 66: 155–60.
Anschütz, G. (1913c) 'Zusatz zu meinen Bemerkungen', *Zeitschrift für Psychologie* 67: 506.
Anscombe, G. E. M. (1976) 'The Question of Linguistic Idealism', *Acta Philosophica Fennica* 28: 188–215.
Anscombe, G. E. M. (1994 [1975]) 'The First Person', in Q. Cassam (ed.) *Self-Knowledge*, Oxford: Oxford University Press, 140–59.
Arbib, M. A. and M. B. Hesse (1986) *The Construction of Reality*, Cambridge: Cambridge University Press.
Arps, G. F. (1921) 'In Memory of Wilhelm Wundt', in G. S. Hall *et al.* (1921) 'In Memory of Wilhelm Wundt', *Psychological Review* 3: 153–88, here 185–6.
Ash, M. G. (1980a) 'Academic Politics in the History of Science: Experimental Psychology in Germany, 1879–1941', *Central European History* 8: 255–86.
Ash, M. G. (1980b) 'Wilhelm Wundt and Oswald Külpe on the Institutional Status of Psychology: An Academic Controversy in Historical Context', in W. G. Bringmann and R. D. Tweney (eds) *Wundt Studies: A Centennial Collection*, Toronto: Hogrefe, 396–421.
Ash, M. G. (1995) *Gestalt Psychology in German Culture 1890–1967: Holism and the Quest for Objectivity*, Cambridge: Cambridge University Press.
Aster, E. v. (1908) 'Die psychologische Beobachtung und experimentelle Untersuchung von Denkvorgängen', *Zeitschrift für Psychologie* 49: 56–107.
Avenarius, R. (1888) *Kritik der reinen Erfahrung*, vol. 1, Leipzig: Fues.
Avenarius, R. (1890) *Kritik der reinen Erfahrung*, vol. 2, Leipzig: Reisland.
Baade, W. (1913a) 'Über die Registrierung von Selbstbeobachtungen durch Diktierphonographen', *Zeitschrift für Psychologie* 66: 81–93.
Baade, W. (1913b) 'Über Unterbrechungsversuche als Mittel zur Unterstützung der Selbstbeobachtung', *Zeitschrift für Psychologie* 64: 258–76.
Baade, W. (1918) 'Selbstbeoachtung und Introvokation', *Zeitschrift für Psychologie* 79: 68–96.
Baerwald, R. (1908) 'Die Methode der vereinigten Selbstwahrnehmung', *Zeitschrift für Psychologie* 46: 174–98.
Baeumker, C. (1916) 'Nachruf auf Oswald Külpe', *Jahrbuch der Königlich Bayerischen Akademie der Wissenschaften*, München: Franz, 73–107.
Baeumker, C. (1921 [1913]) *Anschauung und Denken*, Paderborn: Schöningh.
Baker, L. R. (1987) *Saving Belief*, Princeton, NJ: Princeton University Press.
Baker, L. R. (1994) 'Instrumental Intentionality', in S. Stich and T. A. Warfield (eds) *Mental Representation: A Reader*, Oxford: Blackwell, 332–44.
Balzer, W. (1993) *Soziale Institutionen*, Berlin: Walter de Gruyter.

Barnes, B. (1983) 'Social Life as Bootstrapped Induction', *Sociology* 17: 524–45.

Barnes, B. (1988) *The Nature of Power*, Cambridge: Polity Press.

Barnes, B. (1995) *The Elements of Social Theory*, London: UCL Press.

Barnes, B., D. Bloor, and J. Henry (1996) *Scientific Knowledge: A Sociological Analysis*, London: Athlone Press.

Baron-Cohen, S. (1995) *Mindblindness*, Cambridge, MA: MIT Press.

Beer, M. (1910) 'Die Abhängigkeit der Lesezeit von psychologischen und sprachlichen Faktoren', *Zeitschrift für Psychologie* 56: 264–98.

Ben-David, J. and R. Collins (1966) 'Social Factors in the Origins of a New Science: The Case of Psychology', *American Sociological Review* 31: 451–65.

Bennett, J. (1991) 'Analysis Without Noise', in J. D. Greenwood (ed.) *The Future of Folk Psychology: Intentionality and Cognitive Science*, Cambridge: Cambridge University Press, 93–119.

Berry, J. W. and U. Kim (1993) 'The Way Ahead: From Indigenous Psychologies to a Universal Psychology', in U. Kim and J. W. Berry (eds) *Indigenous Psychologies*, London: Sage, 277–80.

Berry, J. W., Y. H. Poortinga, M. H. Segall, and P. R. Dasen (1992) *Cross-Cultural Psychology: Research and Applications*, Cambridge: Cambridge University Press.

Betz, W. (1910) 'Vorstellung und Einstellung, I. Über Wiedererkennen', *Archiv für die gesamte Psychologie* 17: 266–96.

Binet, A. (1903) *L'étude expérimentale de l'intelligence*, Paris: Schleicher.

Blackburn, S. (1991) 'Losing Your Mind: Physics, Identity, and Folk Burglar Prevention', in J. D. Greenwood (ed.) *The Future of Folk Psychology: Intentionality and Cognitive Science*, Cambridge: Cambridge University Press, 196–225.

Bloch, M. (1962 [1936]) *Feudal Society*, London: Routledge & Kegan Paul.

Bloor, D. (1976) *Knowledge and Social Imagery*, London: Routledge & Kegan Paul.

Bloor, D. (1983) *Wittgenstein: A Social Theory of Knowledge*, New York: Columbia University Press.

Bloor, D. (1995a) 'Idealism and the Social Character of Meaning', Mimeo, Science Studies Unit, University of Edinburgh.

Bloor, D. (1995b) 'Idealism and the Sociology of Knowledge', Mimeo, Science Studies Unit, University of Edinburgh.

Bloor, D. (1996a) 'Collective Representations', Mimeo, Science Studies Unit, University of Edinburgh.

Bloor, D. (1996b) 'Wittgenstein and the Priority of Practice', Mimeo, Science Studies Unit, University of Edinburgh.

Bloor, D. (1997a) *Wittgenstein, Rules and Institutions*, London: Routledge.

Bloor, D. (1997b) 'Wittgenstein's Behaviourism', Mimeo, Science Studies Unit, University of Edinburgh.

Blumenthal, A. (1970) *Language and Psychology*, New York: Wiley.

Blumenthal, A. (1985a) 'Shaping a Tradition: Experimentalism Begins', in C. Buxton (ed.) *Points of View in the Modern History of Psychology*, New York: Academic Press, 51–83.
Blumenthal, A. (1985b) 'Wilhelm Wundt: Psychology as the Propaedeutic Science', in C. Buxton (ed.) *Points of View in the Modern History of Psychology*, New York: Academic Press, 19–49.
Bolgar, H. (1964) 'Karl Bühler: 1879–1963', *American Journal of Psychology* 77: 674–8.
Boring, E. G. (1935) 'Georg Elias Müller: 1850–1934', *American Journal of Psychology* 47: 344–8.
Boring, E. G. (1950) *A History of Experimental Psychology*, New York: Appleton-Century-Crofts.
Boring, E. G. (1953) 'A History of Introspection', *Psychological Bulletin* 50: 169–89.
Brentano, F. (1924 [1874]) *Psychologie vom empirischen Standpunkt*, 1st vol., ed. O. Kraus, Hamburg: Meiner.
Bringmann, W. G. and G. Ungerer (1980) 'Experimental vs. Educational Psychology: Wilhelm Wundt's Letters to Ernst Meumann', *Psychological Research* 42: 57–73.
Brock, A. (1991) 'Imageless Thought or Stimulus Error? The Social Construction of Private Experience', in W. R. Woodward and R. S. Cohen (eds) *World Views and Scientific Discipline Formation*, Dordrecht: Kluwer, 97–106.
Brönner, W. (1911) 'Zur Theorie der kollektiv-psychischen Erscheinungen', *Zeitschrift für Philosophie und philosophische Kritik* 141: 1–40.
Brown, J. R. (1989) *The Rational and the Social*, London: Routledge.
Brunner, J. (1995) *Freud and the Politics of Psychoanalysis*, Oxford: Blackwell.
Brunswig, A. (1910) *Das Vergleichen und die Relation*, Leipzig: Teubner.
Bugenthal, J. F. T. *et al.* (1966) 'Karl Bühler's Contribution to Psychology', *Journal of General Psychology* 75: 181–219.
Bühler, C. (1918) 'Über Gedankenentstehung. Experimentelle Untersuchungen zur Denkpsychologie', *Zeitschrift für Psychologie* 80: 129–200.
Bühler, C. (1919) 'Über die Prozesse der Satzbildung', *Zeitschrift für Psychologie* 81: 181–206.
Bühler, K. (1905) *Studien über Henry Homes*, Diss. Strassburg, Bonn: Bach.
Bühler, K. (1907a) 'Remarques sur les problèmes de la psychologie de la pensèe', *Archives de Psychologie* 6: 376–86.
Bühler, K. (1907b) 'Tatsachen und Probleme zu einer Psychologie der Denkvorgänge, I: Über Gedanken', *Archiv für die Gesamte Psychologie* 9: 297–365.
Bühler, K. (1907c) *Theses quas ad veniam legendi impetrandam defendet Carolus Bühler*, VII Junii MCMVII, Wirceburgi: Stürtz.
Bühler, K. (1908a) 'Nachtrag: Antwort auf die von W. Wundt erhobenen Einwände gegen die Methode der Selbstbeobachtung an experimentell erzeugten Erlebnissen', *Archiv für die Gesamte Psychologie* 12: 93–123.

Bühler, K. (1908b) Review of Stumpf (1906), *Archiv für die gesamte Psychologie* 11: 1–5.

Bühler, K. (1908c) 'Tatsachen und Probleme zu einer Psychologie der Denkvorgänge. II. Über Gedankenzusammenhänge', *Archiv für die Gesamte Psychologie* 12: 1–23.

Bühler, K. (1908d) 'Tatsachen und Probleme zu einer Psychologie der Denkvorgänge. III. Über Gedankenerinnerungen', *Archiv für die Gesamte Psychologie* 12: 24–92.

Bühler, K. (1909a) 'Über das Sprachverständnis vom Standpunkt der Normalpsychologie aus', in F. Schumann (ed.) *Bericht über den III. Kongress für experimentelle Psychologie in Frankfurt a. Main vom 22. bis 25. April 1908*, Leipzig: Barth, 94–130.

Bühler, K. (1909b) 'Zur Kritik der Denkexperimente', *Zeitschrift für Psychologie* 51: 108–18.

Bühler, K. (1911) 'Kinderpsychologie', in H. Vogt and W. Weygandt (eds) *Handbuch der Erforschung und Fürsorge des jugendlichen Schwachsinns unter Berücksichtigung der psychischen Sonderzustände im Jugendalter*, Jena: Fischer, 120–94.

Bühler, K. (1913) *Die Gestaltwahrnehmungen: Experimentelle Untersuchungen zur psychologischen und ästhetischen Analyse der Raum- und Zeitanschauung*, vol. 1, Stuttgart: Spemann.

Bühler, K. (1918) *Die geistige Entwicklung des Kindes*, Jena: Fischer.

Bühler, K. (1919) 'Einige Bemerkungen zu der Diskussion über die Psychologie des Denkens', *Zeitschrift für Psychologie* 97–101.

Bühler, K. (1922) 'Oswald Külpe', *Lebensläufe aus Franken* 2: 244–55.

Bühler, K. (1929 [1926]) *Die Krise der Psychologie*, 2nd edn, Jena: Fischer.

Bühler, K. (1933) *Ausdruckstheorie: Das System an der Geschichte aufgezeigt*, Jena: Fischer.

Burt, C. (1962) 'The Concept of Consciousness', *British Journal of Psychology* 53: 229–42.

Carruthers, P. (1996) 'Simulation and Self-Knowledge: A Defense of Theory-Theory', in P. Carruthers and P. K. Smith (eds) *Theories of Theories of Mind*, Cambridge: Cambridge University Press, 22–38.

Carruthers, P. and P. K. Smith (eds) (1996) *Theories of Theories of Mind*, Cambridge: Cambridge University Press.

Cassirer, E. (1910) *Substanzbegriff und Funktionsbegriff*, Berlin: Cassirer.

Caudle, F. M. (1983) 'The Developing Technology of Apparatus in Psychology's Early Laboratories', in *Annals of the New York Academy of Sciences* 412: 19–56.

Chapman, W. P. and C. M. Jones (1944) 'Variations in Cutaneous Pain Sensitivity in Normal Subjects', *The Journal of Clinical Investigation* 23: 81–91.

Churchland, P. M. (1979) *Scientific Realism and the Plasticity of the Mind*, Cambridge: Cambridge University Press.

Churchland, P. M. (1988) *Matter and Consciousness: A Contemporary Introduction to the Philosophy of Mind*, Cambridge, MA: MIT Press.

Churchland, P. M. (1989 [1981]) 'Eliminative Materialism and the Propositional Attitudes', in P. M. Churchland, *A Neurocomputational Perspective*, Cambridge, MA: MIT Press, 1–22.

Churchland, P. M. (1989 [1988]) 'Folk Psychology and the Explanation of Human Behaviour', in P. M. Churchland, *A Neurocomputational Perspective*, Cambridge, MA: MIT Press, 111–27.

Churchland, P. M. (1989) *A Neurocomputational Perspective*, Cambridge, MA: MIT Press.

Churchland, P. M. (1994) 'Folk Psychology (2)', in S. Guttenplan (ed.) *A Companion to the Philosophy of Mind*, Oxford: Blackwell, 308–16.

Churchland, P. M. (1995) *The Engine of Reason, the Seat of the Soul: A Philosophical Journey into the Brain*, Cambridge, MA: MIT Press.

Churchland, P. S. (1986) *Neurophilosophy*, Cambridge, MA: MIT Press.

Clark, A. (1987) 'From Folk Psychology to Naive Psychology', *Cognitive Science* 11: 139–54.

Cohen, H. (1902) *Logik der reinen Erkenntnis*, Berlin: Cassirer.

Cohn, J. (1906) Review of Ach (1905), *Deutsche Literaturzeitung* 27: 1684–6.

Cohn, J. (1913) 'Grundfragen der Psychologie', in M. Frischeisen-Köhler (ed.) *Jahrbücher der Philosophie: Eine kritische Übersicht der Philosophie der Gegenwart*, vol. 1, Berlin, Mittler, 200–35, 374–5.

Collins, H. M. (1985) *Changing Order: Replication and Induction in Scientific Practice*, London, Sage.

Coon, D. J. (1993) 'Standardizing the Subject: Experimental Psychologists, Introspection, and the Quest for a Technoscientific Ideal', *Technology and Culture* 34: 757–83.

Cordes, G. (1898) *Psychologische Analyse der Thatsache der Selbsterziehung*, Berlin: Reuther & Reichard.

Danziger, K. (1979) 'The Positivist Repudiation of Wundt', *Journal of the History of the Behavioral Sciences* 15: 205–30.

Danziger, K. (1980a) 'The History of Introspection Reconsidered', *Journal of the History of the Behavioral Sciences* 16: 241–62.

Danziger, K. (1980b) 'Wundt's Psychological Experiment in the Light of His Philosophy of Science', *Psychological Research* 42: 109–22.

Danziger, K. (1983) 'Origins and basic principles of Wundt's *Völkerpsychologie*', *British Journal of Social Psychology* 22: 303–13.

Danziger, K. (1990a) *Constructing the Subject: Historical Origins of Psychological Research*, Cambridge: Cambridge University Press.

Danziger, K. (1990b) 'Generative Metaphor and the History of Psychological Discourse', in D. E. Leary (ed.) *Metaphors in the History of Psychology*, Cambridge: Cambridge University Press, 331–56.

Danziger, K. (1997) *Naming the Mind: How Psychology Found its Language*, London: Sage.

Davidson, D. (1984) *Inquiries into Truth and Interpretation,* Oxford: Clarendon Press.

Davidson, D. (1993) 'Thinking Causes', in J. Heil and A. Mele (eds) *Mental Causation,* Oxford: Clarendon Press, 3–18.

Davidson, D. (1994) 'Donald Davidson', in S. Guttenplan (ed.) *A Companion to the Philosophy of Mind,* Oxford: Blackwell, 231–6.

Davies, M. and T. Stone (eds) (1995a) *Folk Psychology,* Oxford: Blackwell.

Davies, M. and T. Stone (eds) (1995b) *Mental Simulation,* Oxford: Blackwell.

Dennett, D. C. (1987) *The Intentional Stance,* Cambridge, MA: MIT Press.

Dennett, D. C. (1991a) *Consciousness Explained,* London: Penguin.

Dennett, D. C. (1991b) 'Two Contrasts: Folk Craft versus Folk Science, and Belief versus Opinion', in J. D. Greenwood (ed.) *The Future of Folk Psychology: Intentionality and Cognitive Science,* Cambridge: Cambridge University Press, 135–48.

Dennett, D. C. (1993 [1981]) 'Three Kinds of Intentional Psychology', in S. M. Christensen and D. R. Turner (eds) *Folk Psychology and the Philosophy of Mind,* Hillsdale, NJ: Lawrence Erlbaum, 121–43.

Desjarlais, R. R. (1992) *Body and Emotion: The Aesthetics of Illness and Healing in the Nepal Himalayas,* Philadelphia: University of Pennsylvania Press.

Deuchler, G. (1909) 'Beiträge zur Erforschung der Reaktionsformen', *Psychologische Studien* 4: 353–430.

Diamond, S. (1980) 'Wundt before Leipzig', in R. W. Rieber (ed.) *Wilhelm Wundt and the Making of a Scientific Psychology,* London: Plenum Press, 3–70.

Dorsch, F. (1963) *Geschichte und Probleme der angewandten Psychologie,* Stuttgart: Huber.

Double, R. (1985) 'The Case Against the Case Against Belief', *Mind* 94: 420–30.

Douglas, M. (1986) 'The Social Preconditions of Radical Scepticism', in J. Law (ed.) *Power, Action and Belief: A New Sociology of Knowledge?,* London: Routledge & Kegan Paul, 68–87.

Driesch, H. (1913) *Die Logik als Aufgabe: Eine Studie über die Beziehung zwischen Phänomenologie und Logik, zugleich eine Einleitung in die Ordnungslehre,* Tübingen: Mohr.

Dubs, A. (1911) *Das Wesen des Begriffs und des Begreifens: Ein Beitrag zur Orientierung in der wissenschaftlichen Weltanschauung,* Halle: Niemeyer.

Düker, H. (1972) 'Heinrich Düker', in L. J. Pongratz *et al.* (eds) *Psychologie in Selbstdarstellungen,* Bern: Huber, 43–86.

Dupré, J. (1993) *The Disorder of Things: Metaphysical Foundations of the Disunity of Science,* Cambridge, MA: Harvard University Press.

Durkheim, E. (1971 [1915]) *The Elementary Forms of the Religious Life,* London: Allen & Unwin.

Dürr, E. (1903) 'Über die Frage des Abhängigkeitsverhältnisses der Logik von der Psychologie', *Archiv für die gesamte Psychologie* 1: 527–44.

Dürr, E. (1906) 'Einige Grundfragen der Willenspsychologie', *Der Gerichtssaal: Zeitschrift fuer Strafrecht, Strafprozess und die ergaenzenden Disziplinen* 69: 168–93.

Dürr, E. (1907) *Die Lehre von der Aufmerksamkeit*, Leipzig: Quelle & Meyer.

Dürr, E. (1908a) *Einführung in die Pädagogik*, Leipzig: Quelle & Meyer.

Dürr, E. (1908b) 'Erkenntnispsychologisches in der erkenntnistheoretischen Literatur der letzten Jahre', *Archiv für die gesamte Psychologie* 13: 1–42.

Dürr, E. (1908c) 'Über die experimentelle Untersuchung der Denkvorgänge', *Zeitschrift für Psychologie* 49: 313–40.

Dürr, E. (1910) *Erkenntnistheorie*, Leipzig: Quelle & Meyer.

Ebbinghaus, H. (1885) *Über das Gedächtnis*, Leipzig: Duncker & Humblot.

Edge, D. and M. Mulkay (1976) *Astronomy Transformed: The Emergence of Radio Astronomy in Britain*, New York: Wiley.

Egan, F. (1995) 'Folk Psychology and Cognitive Architecture', *Philosophy of Science* 62: 179–96.

Ehrhard, A. (1902) *Der Katholizismus und das zwanzigste Jahrhundert im Lichte der kirchlichen Entwicklung der Neuzeit*, 3rd edn, Stuttgart: Roth.

Eisler, R. (1902) *Wundts Philosophie und Psychologie in ihren Grundlehren dargestellt*, Leipzig: Barth.

Elsenhans, T. (1897) *Selbstbeobachtung und Experiment in der Psychologie: Ihre Tragweite und ihre Grenzen*, Tübingen: Mohr.

Elsenhans, T. (1902) *Das Kant-Friesische Problem*, Heidelberg: Hörning.

Elsenhans, T. (1912) *Lehrbuch der Psychologie*, Tübingen: Mohr.

Erdmann, B. (1892) *Logische Elementarlehre*, Halle: Niemeyer.

Erdmann, B. (1900) 'Umrisse zur Psychologie des Denkens', in B. Erdmann *et al.*, *Philosophische Abhandlungen: Christoph Sigwart zu seinem siebzigsten Geburtstage*, Tübingen: Mohr.

Erdmann, B. (1907) *Logische Elementarlehre*, 2nd rev. edn, Halle: Niemeyer.

Ethemendy, J. (1981) 'The Cartesian Circle: *Circulus ex tempore*', *Studia Cartesiana* 2: 5–42.

Eucken, R. (1901) 'Thomas von Aquino und Kant: Ein Kampf zweier Welten', *Kantstudien* 6: 1–18.

Febvre, L. (1982 [1942]) *The Problem of Unbelief in the Sixteenth Century: The Religion of Rabelais*, Cambridge, MA: Harvard University Press.

Feuchtwanger, A. (1911) 'Versuche über Vorstellungstypen', *Zeitschrift für Psychologie* 58: 161–99.

Fodor, J. (1987) *Psychosemantics*, Cambridge, MA: MIT Press.

Foss, J. E. (1995) 'Materialism, Reduction, Replacement, and the Place of Consciousness in Science', *The Journal of Philosophy* 92: 401–29.

Foucault, M. (1984) 'Nietzsche, Genealogy, History', in P. Rabinow (ed.) *The Foucault Reader: An Introduction to Foucault's Thought*, London: Penguin, 76–100.

Frede, M. (1987) *Essays in Ancient Philosophy*, Duluth, MN: University of Minnesota Press.

Frede, M. (1988) 'The History of Philosophy as a Discipline', *The Journal of Philosophy* 85: 666–72.

Frege, G. (1893) *Grundgesetze der Arithmetik, Begriffsschriftlich abgeleitet*, vol. 1, Jena: Pohle.

Frege, G. (1894) Review of Husserl (1891), *Zeitschrift für Philosophie und philosophische Kritik* 103: 313–32.

Frege, G. (1934 [1884]) *Grundlagen der Arithmetik*, Breslau: Marcus.

Friedrich, J. (1897) 'Einflüsse der Arbeitsdauer und der Arbeitspausen auf die geistige Leistungsfähigkeit der Schulkinder', *Zeitschrift für Psychologie* 13: 1–53.

Friedwalt, A. (A. Messer) (1905) *Katholische Studenten*, Stuttgart: Greiner & Pfeiffer.

Frischeisen-Köhler, M. (1913) 'Philosophie und Psychologie', *Die Geisteswissenschaften* 1: 371–3, 400–3.

Frischeisen-Köhler, M. (1918) *Grenzen der experimentellen Methode*, Berlin: Reuther & Reichard.

Fröbes, J. (1961) 'Joseph Fröbes', in C. Murchison (ed.) *A History of Psychology in Autobiography*, vol. 3, New York: Russell & Russell, 121–52.

Fruton, J. S. (1990) *Contrasts in Scientific Style: Research Groups in the Chemical and Biochemical Sciences*, Philadelphia: American Philosophical Society.

Gadamer, H.-G. (1975 [1960]) *Truth and Method*, London: Sheed & Ward.

Gadamer, H.-G. (1997) *'Who Am I and Who Are You?' and Other Essays*, tr. R. Heinemann and B. Krajewski, Albany, NY: State University of New York Press.

Galison, P. (1985) 'Bubble Chambers and the Experimental Workplace', in P. Achenstein and O. Hannaway (eds) *Observation, Experiment, and Hypothesis in Modern Physical Science*, Cambridge, MA: MIT Press, 309–73.

Galison, P. (1987) *How Experiments End*, Chicago: University of Chicago Press.

Ganzer, K. (1982) 'Die Theologische Fakultaet der Universität Würzburg im theologischen und kirchenpolitischen Spannungsfeld der zweiten Hälfte des 19. Jahrhunderts', in P. Baumgart (ed.) *Vierhundert Jahre Universität Würzburg: Eine Festschrift*, Neustadt an der Aisch: Degener, 317–73.

Gardner, S. and G. Stevens (1992) *Red Vienna and the Golden Age of Psychology, 1918–1938*, New York: Praeger.

Garfield, J. L. (1988) *Belief in Psychology: A Study in the Ontology of Mind*, Cambridge, MA: MIT Press.

Garfinkel, H. (1984 [1967]) *Studies in Ethnomethodology*, Cambridge: Polity Press.

Geison, G. L. (1981) 'Scientific Change, Emerging Specialities, and Research Schools', *History of Science* 19: 20–40.

Geison, G. L. and F. L. Holmes (eds) (1993) *Research Schools: Historical Reappraisals*, *Osiris*, 2nd Ser., 8.

Gemelli, A. (1913) 'Die Realisierung', *Philosophisches Jahrbuch der Görres-Gesellschaft* 26, 360–79.

Geuter, U. (1986) *Daten zur Geschichte der deutschen Psychologie*, Göttingen: Hogrefe.

Geyser, J. (1902) *Grundlegung der empirischen Psychologie*, Bonn: Hanstein.

Geyser, J. (1908a) 'Die Vorzüge und Schwächen der neueren Untersuchung der Denkvorgänge durch das Ausfrageexperiment', *Philosophisches Jahrbuch der Görres-Gesellschaft* 21: 90–102.

Geyser, J. (1908b) *Lehrbuch der allgemeinen Psychologie*, Münster: Schöningh.

Geyser, J. (1909a) *Einführung in die Psychologie der Denkvorgänge: Fünf Vorträge, gehalten im April 1909 auf dem pädagogischen Kursus in Cöln*, Münster: Schöningh.

Geyser, J. (1909b) *Grundlagen der Logik und Erkenntnislehre: Eine Untersuchung der Formen und Prinzipien objektiv wahrer Erkenntnis*, Münster: Schöningh.

Geyser, J. (1910) 'Einige Bemerkungen zu dem Aufsatze von G. Moskiewicz', *Archiv für die gesamte Psychologie* 19: 545–53.

Geyser, J. (1913) 'Beiträge zur logischen und psychologischen Analyse des Urteils', *Archiv für die gesamte Psychologie* 26: 361–91.

Geyser, J. (1914a) *Die Seele: Ihr Verhältnis zum Bewusstsein und zum Leibe*, Münster: Schöningh.

Geyser, J. (1914b) 'Eine neue experimentelle Untersuchung der Vorstellungen und ihrer Beziehung zum Denken', *Philosophisches Jahrbuch der Görres-Gesellschaft* 27: 180–214.

Geyser, J. (1916) *Neue und alte Wege der Philosophie: Eine Erörterung der Grundlagen der Erkenntnis im Hinblick auf Edmund Husserls Versuch ihrer Neubegruendung*, Münster: Schöningh.

Gigerenzer, G. (1991) 'From Tools to Theories: A Heuristic of Discovery in Cognitive Psychology', *Psychological Review* 98: 254–67.

Gigerenzer, G. and D. G. Goldstein (1996) 'Mind as Computer: The Birth of a Metaphor', *Creativity Research Journal* 9: 131–44.

Glover, J. (1988) *I: The Philosophy and Psychology of Personal Identity*, London: Allen Lane.

Goldman, A. (1993) 'The Psychology of Folk Psychology', *Behavioral and Brain Sciences* 16: 15–28.

Goldman, A. (1995 [1989]) 'Interpretation Psychologised', in M. Davies and T. Stone (eds) *Folk Psychology*, Oxford: Blackwell, 74–99.

Gopnik, A. (1993) 'How We Know Our Minds: The Illusion of First-Person Knowledge of Intentionality', *Behavioral and Brain Sciences* 16: 1–14.

Gopnik, A. (1996) 'Theories and Modules: Creation Myths, Developmental Realities, and Neurath's Boat', in P. Carruthers and P. K. Smith (eds) *Theories of Theories of Mind*, Cambridge: Cambridge University Press, 169–83.

Gopnik, A. and H. M. Wellman (1994) 'The Theory Theory', in L. A. Hirschfeld and S. A. Gelman (eds) *Mapping the Mind: Domain Specificity in Cognition and Culture*, Cambridge: Cambridge University Press, 257–93.

Gordon, R. M. (1995 [1986]) 'Folk Psychology as Simulation', in M. Davies and T. Stone (eds) *Folk Psychology*, Oxford: Blackwell, 60–73.

Gordon, R. M. (1995 [1992]) 'The Simulation Theory: Objections and Misconceptions', in M. Davies and T. Stone (eds) *Folk Psychology*, Oxford: Blackwell, 100–22.

Gordon, R. M. (1996) ' "Radical" Simulationism', in P. Carruthers and P. K. Smith (eds) *Theories of Theories of Mind*, Cambridge: Cambridge University Press, 22–38.

Grabmann, M. (1913) *Der Gegenwartswert der geschichtlichen Erforschung der mittelalterlichen Philosophie*, Wien: Herder.

Grabmann, M. (1916) 'Der kritische Realismus Oswald Külpes und der Standpunkt der aristotelischscholastischen Philosophie', *Philosophisches Jahrbuch der Görres-Gesellschaft* 29: 333–69.

Gracia, J. (1992) *Philosophy and Its History: Issues in Philosophical Historiography*, Albany, NY: State University of New York Press.

Graf, F. W. (1989) 'Rettung der Persönlichkeit: Protestantische Theologie als Kulturwissenschaft des Christentums', in R. v. Bruch *et al.* (eds) *Kultur und Kulturwissenschaften um 1900: Krise der Moderne und Glaube an die Wissenschaft*, Stuttgart: Steiner, 103–31.

Graham, G. (1993 [1987]) 'The Origins of Folk Psychology', in S. M. Christensen and D. R. Turner (eds) *Folk Psychology and the Philosophy of Mind*, Hillsdale, NJ: Lawrence Erlbaum, 200–20.

Grahek, N. (1995) 'The Sensory Dimension of Pain', *Philosophical Studies* 79: 167–84.

Grane, L. (1987) *Die Kirche im 19. Jahrhundert*, Göttingen: Vandenhoeck & Ruprecht.

Greenwood, J. D. (1991) 'Reasons to Believe', in J. D. Greenwood (ed.) *The Future of Folk Psychology: Intentionality and Cognitive Science*, Cambridge: Cambridge University Press, 70–92.

Grünbaum, A. (1908) 'Über die Abstraktion der Gleichheit: Ein Beitrag zur Psychologie der Relation', *Archiv für die gesamte Psychologie* 12: 340–478.

Grünholz, E. (1913) 'Eine kritische Untersuchung über das Denken im Anschluss an die Philosophie Wilhelm Wundts', *Philosophisches Jahrbuch der Görres-Gesellschaft* 26, 305–27.

Grünholz, E. (1914) 'Das ontologische Prinzip in Wundts Erkenntnislehre', *Philosophisches Jahrbuch der Görres-Gesellschaft* 27: 1–20.

Gutberlet, C. (1891) 'W. Wundt's System der Philosophie', *Philosophisches Jahrbuch der Görres-Gesellschaft* 4: 281–96, 341–59.

Gutberlet, C. (1892) 'Die Willensfreiheit und die physiologische Psychologie', *Philosophisches Jahrbuch der Görres-Gesellschaft* 5: 172–87.

Gutberlet, C. (1896) 'Ist die Seele Thätigkeit oder Substanz?', *Philosophisches Jahrbuch der Görres-Gesellschaft* 9: 1–17, 133–70.

Gutberlet, C. (1898) 'Die "Krisis in der Psychologie"', *Philosophisches Jahrbuch der Görres-Gesellschaft* 11: 1–19, 121–46.

Gutberlet, C. (1899) *Der Kampf um die Seele: Vortraege ueber die brennenden Fragen der modernen Psychologie,* Mainz: Kirchheim.

Gutberlet, C. (1901) 'Eine neue actualistische Seelentheorie', *Philosophisches Jahrbuch der Görres-Gesellschaft* 14: 353–65.

Gutberlet, C. (1904) Review of Wundt (1903), *Philosophisches Jahrbuch der Görres-Gesellschaft* 17: 324–31.

Gutberlet, C. (1905) *Psychophysik: Historisch-kritische Studien über Experimentelle Psychologie,* Mainz: Kirchheim.

Gutberlet, C. (1907) *Die Willensfreiheit und ihre Gegner,* 2nd edn, Fulda: Actiendruckerei.

Gutberlet, C. (1908) 'Der gegenwärtige Stand der psychologischen Forschung', *Philosophisches Jahrbuch der Görresgesellschaft* 21: 1–32.

Gutberlet, C. (1911) 'Religionspsychologie', *Philosophisches Jahrbuch der Görresgesellschaft* 24: 147–76.

Gutberlet, C. (1914) 'Die philosophische Krisis der Gegenwart', *Philosophisches Jahrbuch der Görres-Gesellschaft* 27: 321–39.

Gutberlet, C. (1915) *Experimentelle Psychologie mit besonderer Berücksichtigung der Pädagogik,* Paderborn: Schöningh.

Gutberlet, C. (1916) 'Rudolph Eucken', *Philosophisches Jahrbuch der Görres-Gesellschaft* 29: 113–37.

Gutberlet, C. (1923) 'Constantin Gutberlet', in R. Schmidt (ed.) *Philosophie der Gegenwart in Selbstdarstellungen,* vol. 4, Leipzig: Barth (pages are not numbered consecutively).

Guttenplan, S. (1994) 'Natural kind', in S. Guttenplan (ed.) *A Companion to the Philosophy of Mind,* London: Blackwell, 449–50.

Güttler, K. (1896) *Psychologie und Philosophie,* München: Piloty & Löhle.

Habrich, J. (1914) *Über die Entwicklung der Abstraktionsfähigkeit von Schülerinnen,* Diss. München.

Hacker, F. (1911) 'Systematische Traumbeobachtungen mit besonderer Berücksichtigung der Gedanken', *Archiv für die Gesamte Psychologie* 21: 1–131.

Hacker, P. (1996) 'Methodology in Philosophy and Psychology', Mimeo, Department of Psychology, University of Edinburgh.

Hacking, I. (1983) *Representing and Intervening: Introductory Topics in the Philosophy of Natural Science,* Cambridge: Cambridge University Press.

Hacking, I. (1984) 'Five Parables', in R. Rorty *et al.* (eds) *Philosophy in History,* Cambridge: Cambridge University Press, 103–24.

Hacking, I. (1992) 'World-Making by Kind-Making: Child-Abuse for Example', in M. Douglas and D. Hull (eds) *How Classification Works: Nelson Goodman among the Social Sciences,* Edinburgh: Edinburgh University Press, 180–238.

Hacking, I. (1995) *Rewriting the Soul: Multiple Personality and the Sciences of Memory,* Princeton, NJ: Princeton University Press.

Haering, T. (1913) 'Untersuchungen zur Psychologie der Wertung (auf experimenteller Grundlage) mit besonderer Berücksichtigung der methodologischen Fragen. Erster Teil', *Archiv für Psychologie* 26: 269–360.

Hall, G. S. (1921) 'In Memory of Wilhelm Wundt', in G. S. Hall *et al.* (1921), 'In Memory of Wilhelm Wundt', *Psychological Review* 3: 153–88, here 154–5.

Hall, G. S. *et al.* (1921) 'In Memory of Wilhelm Wundt', *Psychological Review* 3: 153–88.

Hammer, S. (1994) *Denkpsychologie–Kritischer Realismus: Eine wissenschaftshistorische Studie zum Werk Oswald Külpes*, Frankfurt am Main: Lang.

Harré, R. (1981) 'Psychological Variety', in P. Heelas and A. Lock (eds) *Indigenous Psychologies: The Anthropology of the Self*, London: Academic Press, 79–103.

Harré, R. (1987) 'The Social Construction of Selves', in K. Yardley and T. Honess (eds) *Self & Indentity: Psychosocial Perspectives*, Chichester: Wiley, 41–52.

Harré, R. and G. Gillett (1994) *The Discursive Mind*, London: Sage.

Harris, P. L. (1991) 'The Work of Imagination', in A. Whiten (ed.) *Natural Theories of Mind*, Oxford: Blackwell, 283–304.

Harris, P .L. (1994) 'Thinking by Children and Scientists: False Analogies and Neglected Similarities', in L. A. Hirschfeld and S. A. Gelman (eds) *Mapping the Mind: Domain Specificity in Cognition and Culture*, Cambridge: Cambridge University Press, 294–315.

Harris, P. L. (1995 [1992]) 'From Simulation to Folk Psychology: The Case for Development', in M. Davies and T. Stone (eds) (1995) *Folk Psychology*, Oxford: Blackwell, 207–31.

Harwood, J. (1993) *Styles of Scientific Thought: The German Genetics Community, 1900–1933*, Chicago: University of Chicago Press.

Haugeland, J. (1990) 'The Intentionality All-Stars', in J. E. Tomberlin (ed.) *Philosophical Perspectives, IV: Philosophy of Mind and Action Theory*, Ataxadero, CA: Ridgeview, 383–427.

Heal, J. (1995 [1986]) 'Replication and Functionalism', in M. Davies and T. Stone (eds) (1995) *Folk Psychology*, Oxford: Blackwell, 45–59.

Heelas, P. (1981a) 'Introduction: Indigenous Psychologies', in P. Heelas and A. Lock (eds) *Indigenous Psychologies: The Anthropology of the Self*, London: Academic Press, 3–18.

Heelas, P. (1981b) 'The Model Applied: Anthropology and Indigenous Psychologies', in P. Heelas and A. Lock (eds) *Indigenous Psychologies: The Anthropology of the Self*, London: Academic Press, 39–63.

Heelas, P. and A. Lock (eds) (1981a) *Indigenous Psychologies: The Anthropology of the Self*, London: Academic Press.

Heelas, P. and A. Lock (1981b) 'Preface', in P. Heelas and A. Lock (eds) *Indigenous Psychologies: The Anthropology of the Self*, London: Academic Press, xiii–xvi.

Heidegger, M. (1978 [1913]) 'Die Lehre vom Urteil im Psychologismus: Ein kritisch-positiver Beitrag zur Logik', in M. Heidegger, *Frühe Schriften*, ed. F.-W. von Herrmann, *Gesamtausgabe*, vol. 1, Frankfurt am Main: Klostermann, 59–188.

Heil, J. (1991) 'Being Indiscrete', in J. D. Greenwood (ed.) *The Future of Folk Psychology: Intentionality and Cognitive Science*, Cambridge: Cambridge University Press, 120–34.

Hellpach, W. (1948) *Wirken in Wirren*, 2 vols, vol. 2, Hamburg: Wegner.

Henning, H. (1919) 'Assoziationslehre und neuere Denkpsychologie', *Zeitschrift für Psychologie* 82: 219–26.

Hensel, P. (1909) 'Die Aussichten der Privatdozenten für Philosophie', *Frankfurter Zeitung*, 29 July: 1–2.

Herbart, J. F. (1968 [1850]) *Psychologie als Wissenschaft, neu gegründet auf Erfahrung, Metaphysik und Mathematik, Zweiter, analytischer Theil*, Amsterdam: E. J. Bonset.

Hesse, M. (1970) 'Is There an Independent Observation Language?', in R. G. Colodny (ed.) *The Nature and Function of Scientific Theories: Essays in Contemporary Science and Philosophy*, Pittsburgh: University of Pittsburgh Press, 35–78.

Hesse, M. (1974) *The Structure of Scientific Inference*, London: Macmillan.

Hewson, C. M. (1996) *Putting Psychology into the Folk Psychology Debate*, Doctoral Dissertation, University of Edinburgh.

Heymans, G. (1890) *Die Gesetze und Elemente des wissenschaftlichen Denkens. Ein Lehrbuch der Erkenntnistheorie in Grundzügen*, vol. 1, Leipzig: Harrassowitz.

Heymans, G. (1894) *Die Gesetze und Elemente des wissenschaftlichen Denkens. Ein Lehrbuch der Erkenntnistheorie in Grundzügen*, vol. 2, Leipzig: Harrassowitz.

Hintikka, J. (1962) '*Cogito, Ergo Sum*: Inference or Performance?', *Philosophical Review* 71: 3–32.

Hirsche, W. (1935) 'Gedenkrede, gehalten am Sarge von Georg Elias Müller, in Göttingen, am 27. Dezember 1934', *Zeitschrift für Psychologie* 134: 145–9.

Hirschfeld, L. A. and S. A. Gelman (eds) (1994) *Mapping the Mind: Domain Specificity in Cognition and Culture*, Cambridge: Cambridge University Press.

Hönigswald, R. (1913) 'Prinzipien der Dankpsychologie', *Kantstudien* 18: 205–45.

Horgan, T. and J. Woodward (1991 [1985]) 'Folk Psychology is Here to Stay', in J. D. Greenwood (ed.) *The Future of Folk Psychology: Intentionality and Cognitive Science*, Cambridge: Cambridge University Press, 149–75.

Horst, S. (1995) 'Eliminativism and the Ambiguity of Belief', *Synthese* 104: 123–45.

Horwicz, A. (1882) Review of *Philosophische Studien*, vol. I, *Philosophische Monatshefte* 18: 497–502.

Howell, S. (1981) 'Rules Not Words', in P. Heelas and A. Lock (eds) *Indigenous Psychologies: The Anthropology of the Self*, London: Academic Press, 133–43.

Hume, D. (1990) *A Treatise of Human Nature*, Oxford: Clarendon Press.

Humphrey, G. (1963) *Thinking: An Introduction to its Experimental Psychology*, New York: Wiley.

Hunter, J. S. (1980) 'The National System of Scientific Measurement', *Science* 210: 867–75.

Husserl, E. (1970 [1891]) 'Philosophie der Arithmetik. Logische und psychologische Untersuchungen', in E. Husserl, *Philosophie der Arithmetik*, Husserliana XII, ed. L. Eley, The Hague: Nijhoff, 5–283.

Husserl, E. (1975 [1900]) *Logische Untersuchungen. Erster Band: Prolegomena zur reinen Logik*, Husserliana XVIII, ed. E. Holenstein, The Hague: Nijhoff.

Husserl, E. (1984 [1901]) *Logische Untersuchungen. Zweiter Band: Untersuchungen zur Phänomenologie und Theorie der Erkenntnis*, Husserliana XIX, ed. U. Panzer, The Hague: Nijhoff.

Husserl, E. (1987 [1911]) 'Philosophie als strenge Wissenschaft', in E. Husserl, *Aufsätze und Vorträge (1911–1921)*, Husserliana XXV, ed. T. Nenon and H. R. Sepp, Dordrecht: Nijhoff, 3–62.

Itkonen, E. (1978) *Grammatical Theory and Metascience: A Critical Investigation into the Methodological and Philosophical Foundations of 'Autonomous' Linguistics*, Amsterdam: Benjamins.

Itkonen, E. (1983) *Causality in Linguistic Theory: A Critical Investigation into the Philosophical and Methodological Foundations of 'Non-Autonomous' Linguistics*, London: Croom Helm.

Jaensch, E. R. (1935) 'Was wird aus dem Werk? Betrachtungen aus dem Gesichtpunkt der Kulturwende über G.E. Müllers Wesen und Werk und das Schicksal der Psychologie', *Zeitschrift für Psychologie* 134: 191–218.

James, W. (1983 [1890]) *The Principles of Psychology*, Cambridge, MA: Harvard University Press.

Jardine, N. (1986) *The Scenes of Inquiry*, Oxford: Clarendon Press.

Jasanoff, S., G. E. Markle, J. C. Petersen, and T. Pinch (eds) (1994) *Handbook of Science and Technology Studies*, London: Sage.

Jaspers, K. (1912) 'Die phänomenologische Forschungsrichtung in der Psychopathologie', *Zeitschrift für die Gesamte Neurologie und Psychia-trie*, 9: 391–408.

Jaspers, K. (1919) *Psychologie der Weltanschauungen*, Berlin: Springer.

Jaynes, J. (1974) 'Müller, Georg Elias', in C. C. Gillispie (ed.) *Dictionary of Scientific Biography*, vol. 9, New York: Scribner's, 561–3.

Jerusalem, W. (1905) *Der kritische Idealismus und die reine Logik: Ein Ruf im Streite*, Wien: Braumüller.

Johnson, D. M. (1987) 'The Greek Origins of Belief', *American Philosophical Quarterly* 24: 319–27.

Judd, C. H. (1921) 'In Memory of Wilhelm Wundt', in G. S. Hall *et al.* (1921), 'In Memory of Wilhelm Wundt', *Psychological Review* 3: 153–88, here 173–8.

Jung, M. (1973) *Der Neukantianische Realismus von Alois Riehl*, Diss. Bonn.

Kandel, E. R. and R. D. Hawkins (1993) 'The Biological Basis of Learning and Individuality', in *Mind and Brain: Readings from Scientific American Magazine*, New York: Freeman, 40–53.

Katz, D. (1936) 'Georg Elias Müller', *Acta Psychologica: European Journal of Psychology* 1: 234–40.

Katz, R. (1972) 'Rosa Katz', in L. J. Pongratz *et al.* (eds) *Psychologie in Selbstdarstellungen*, Bern: Huber, 103–25.

Kenny, A. (1993) *Aquinas on Mind*, London: Routledge.

Kiesow, F. (1961 [1930]) 'F. Kiesow', in C. Murchison (ed.) *A History of Psychology in Autobiography*, vol. 1, New York: Russell & Russell, 163–90.

Kim, U. and J. W. Berry (eds) (1993) *Indigenous Psychologies*, London: Sage.

Knorr Cetina, K. (1994) 'Laboratory Studies: The Cultural Approach to the Study of Science', in S. Jasanoff *et al.* (eds) *Handbook of Science and Technology Studies*, London: Sage, 140–66.

Kobes, B. W. (1993) 'Self-Attributions Help Constitute Mental Types', *Behavioral and Brain Sciences* 16(1): 54–6.

Koch, A. (1913) *Experimentelle Untersuchungen über die Abstraktionsfähigkeit von Volksschulkindern*, Diss. Bonn.

Koffka, K. (1912) *Zur Analyse der Vorstellungen und ihrer Gesetze: Eine experimentelle Untersuchung*, Leipzig: Quelle & Meyer.

Köhler, W. (1913a) Review of Anschütz (1912), *Zeitschrift für Psychologie* 64: 441.

Köhler, W. (1913b) 'Schlussbemerkung', *Zeitschrift für Psychologie* 67: 506.

Köhler, W. (1913c) 'Zu den Bemerkungen von G. Anschütz', *Zeitschrift für Psychologie* 66: 319–20.

Köhnke, K. C. (1986) *Entstehung und Aufstieg des Neukantianismus: Die deutsche Universitätsphilosophie zwischen Idealismus und Positivismus*, Frankfurt am Main: Suhrkamp.

König, E. (1901) *Wilhelm Wundt: Seine Philosophie und Psychologie*, Stuttgart: Fromanns.

Kroh, O. (1935) 'Georg Elias Müller: Ein Nachruf', *Zeitschrift für Psychologie*, 134: 150–90.

Krohn, W. O. (1893) 'The Laboratory of the Psychological Institute at the University of Göttingen', *American Journal of Psychology* 5: 282–4.

Krüger, F. (1915) *Über Entwicklungspsychologie: Ihre sachliche und geschichtliche Notwendigkeit*, Leipzig: Engelmann.

Kuhn, T. (1962) *The Structure of Scientific Revolutions*, Chicago: University of Chicago Press.

Külpe, O. (1893) *Grundriß der Psychologie, auf experimenteller Grundlage dargestellt*, Leipzig: Engelmann.

Külpe, O. (1898) 'Über die Beziehungen zwischen körperlichen und seelischen Vorgängen', *Zeitschrift für Hypnotismus, Psychotherapie sowie andere psychophysiologische und psychopathologische Forschungen* 7: 97–120.

Külpe, O. (1902) *Die Philosophie der Gegenwart in Deutschland: Eine Charakteristik ihrer Hauptrichtungen nach Vorträgen, gehalten im Ferienkurs für Lehrer 1901 zu Würzburg*, Leipzig: Teubner.

Külpe, O. (1903) 'Ein Beitrag zur experimentellen Aesthetik', *American Journal of Psychology* 14: 213–31.

Külpe, O. (1904) 'Versuche über die Abstraktion', in F. Schumann (ed.) *Bericht über den I. Kongress für experimentelle Psychologie in Giessen vom 18. bis. 21. April 1904*, Leipzig: Barth, 56–68.

Külpe, O. (1907a) *Immanuel Kant: Darstellung und Würdigung*, Leipzig: Teubner.

Külpe, O. (1907b) Review of Ach 1905, *Göttingische gelehrte Anzeigen* 169: 595–608.

Külpe, O. (1910a) *Einleitung in die Philosophie*, 5th edn, Leipzig: Hirzel.

Külpe, O. (1910b) 'Erkenntnistheorie und Naturwissenschaft', *Physikalische Zeitschrift* 23: 1025–35.

Külpe, O. (1912a) *Die Realisierung: Ein Beitrag zur Grundlegung der Realwissenschaften*, Leipzig: Hirzel.

Külpe, O. (1912b) 'Psychologie und Medizin', *Zeitschrift für Psychopathologie* 1: 187–267.

Külpe, O. (1912c) 'Über die moderne Psychologie des Denkens', *Internationale Monatsschrift für Wissenschaft, Kultur und Technik* 6: 1069–110.

Külpe, O. (1912d) 'Wilhelm Wundt zum 80. Geburtstag', *Archiv für die Gesamte Psychologie* 24: 105–10.

Külpe, O. (1914) 'Über die Methoden der psychologischen Forschung', *Internationale Monatsschrift für Wissenschaft, Kultur und Technik* 8: 1053–70, 1219–32.

Külpe, O. (1915a) *Die Ethik und der Krieg*, Leipzig: Hirzel.

Külpe, O. (1915b) *Zur Kategorienlehre* (Sitzungsberichte der Koeniglich Bayerischen Akademie der Wissenschaften, Philosophisch-philologische Klasse, 5. Abhandlung, Vorgetragen am 6. Februar 1915), München: Verlag der Koeniglich Bayerischen Akademie der Wissenschaften.

Külpe, O. (1920a) *Die Realisierung: Ein Beitrag zur Grundlegung der Realwissenschaften*, vol. 2, ed. A. Messer, Leipzig: Hirzel.

Külpe, O. (1920b) *Vorlesungen über Psychologie*, ed. K. Bühler, Leipzig: Hirzel.

Külpe, O. (1923a) *Die Realisierung: Ein Beitrag zur Grundlegung der Realwissenschaften*, vol. 3, ed. A. Messer, Leipzig: Hirzel.

Külpe, O. (1923b) *Vorlesungen über Logik*, ed. O. Selz, Leipzig: Hirzel.

Kusch, M. (1989) *Language as Calculus versus Language as the Universal Medium: A Study in Husserl, Heidegger, and Gadamer*, Dordrecht: Kluwer.

Kusch, M. (1991) *Foucault's Strata and Fields: An Investigation into Archaeological and Genealogical Science Studies*, Dordrecht: Kluwer.

Kusch, M. (1995a) *Psychologism: A Case Study in the Sociology of Philosophical Knowledge*, London: Routledge.

Kusch, M. (1995b) 'Recluse, Interlocutor, Interrogator: Natural and Social Order in Turn-of-the-Century Psychological Research Schools', *Isis* 86: 419–39.

Kusch, M. (1997) 'The Sociophilosophy of Folk Psychology', *Studies in History and Philosophy of Science* 28: 1–25.

Kusch, M. and H. Schröder (eds) (1989) *Text–Interpretation–Argumentation* (Studies in Textlinguistics), Hamburg: Buske.

Lagerspetz, E. (1995) *The Opposite Mirrors: An Essay on the Conventionalist Theory of Institutions*, Dordrecht: Kluwer.

Laudan, L. (1990) *Science and Relativism: Some Key Controversies in the Philosophy of Science*, Chicago: University of Chicago Press.

Lazarus, M. and H. Steinthal (1860) 'Einleitende Gedanken über Völkerpsychologie, als Einladung zu einer Zeitschrift für Völkerpsychologie und Sprachwissenschaft', *Zeitschrift für Völkerpsychologie und Sprachwissenschaft* 1: 1–73.

Leary, D. (1980) 'German Idealism and the Development of Psychology in the Nineteenth Century', *Journal of the History of Philosophy* 18: 299–317.

Lehmann, R. (1906) 'Der Rückgang der Universitätsphilosophie', *Die Zukunft* 54: 483–7.

Leslie, A. M. (1991) 'The Theory of Mind Impairment in Autism: Evidence for a Modular Mechanism of Development?', in A. Whiten (ed.) *Natural Theories of Mind: Evolution, Development and Simulation of Everyday Mindreading*, Oxford: Blackwell, 63–79.

Lewis, D. (1972) 'Psychophysical and Theoretical Identification', *Australasian Journal of Philosophy* 50: 249–58.

Lewis, D. (1983 [1970]) 'How to Define Theoretical Terms', in D. Lewis, *Philosophical Papers*, Oxford: Oxford University Press, 78–95.

Lichtenberg, G. (1958) *Aphorismen*, ed. M. Rychner, Zurich: Mannesse.

Lienhardt, R. G. (1961) *Divinity and Experience*, London: Oxford University Press.

Lindenfeld, D. (1978) 'Oswald Külpe and the Würzburg School', *Journal of the History of the Behavioral Sciences* 14: 132–41.

Lindworsky, J. (1916) *Das schlußfolgernde Denken: Experimentell-psychologische Untersuchungen*, Freiburg: Herder.

Lipps, G. F. (1906) Review of Ach (1905), *Literarisches Zentralblatt* 20 (12 May): 677–8.

Lipps, T. (1880) 'Die Aufgabe der Erkenntnistheorie und die Wundt'sche Logik I', *Philosophische Monatshefte* 16: 529–39.

Lipps, T. (1883) *Grundtatsachen des Seelenlebens*, Bonn: Cohen & Sohn.

Lipps, T. (1893) *Grundzüge der Logik*, Hamburg: Voss.

Lipps, T. (1901) 'Psychische Vorgänge und psychische Causalität', *Zeitschrift für Psychologie* 25: 161–203.

Lipps, T. (1903) *Leitfaden der Psychologie*, Leipzig: Engelmann.

Lück, H. E. and D.-J. Löwisch (eds) (1994) *Der Briefwechsel zwischen William Stern und Jonas Cohn: Dokumente einer Freundschaft zwischen zwei Wissenschaftlern*, Frankfurt am Main: Lang.

Lyons, W. (1986) *The Disappearance of Introspection*, Cambridge, MA: MIT Press.

Mach, E. (1886) *Die Analyse der Empfindungen*, Jena: Fischer.

MacKenzie, D. A. (1981) *Statistics in Britain, 1865–1930: The Social Construction of Scientific Knowledge*, Edinburgh: Edinburgh University Press.

Maier, H. (1908) *Psychologie des emotionalen Denkens*, Tübingen: Mohr.

Maier, H. (1914a) 'Logik und Psychologie', in *Festschrift für Alois Riehl, von Freunden und Schülern zu seinem siebzigsten Geburtstage dargebracht*, Halle: Niemeyer, 311–78.

Maier, H. (1914b) 'Psychologie und Philosophie', in F. Schumann (ed.) *Bericht über den VI. Kongress für experimentelle Psychologie*, Leipzig, Barth: 93–9, 144–6.

Malcolm, N. (1995 [1979]) 'Whether "I" is a Referring Expression', in N. Malcolm, *Wittgensteinian Themes: Essays 1978–1989*, ed. G.-H. von Wright, Ithaca, NY: Cornell University Press, 16–26.

Mandler, J. M. and G. Mandler (1964) *Thinking: From Association to Gestalt*, New York: Wiley.

Marbe, K. (1901) *Experimentell-psychologische Untersuchungen über das Urteil: Eine Einleitung in die Logik*, Leipzig: Engelmann.

Marbe, K. (1908) 'W. Wundts Stellung zu meiner Theorie der stroboskopischen Erscheinungen und zur systematischen Selbstwahrnehmung', *Zeitschrift für Psychologie* 46: 345–62.

Marbe, K. (1912a) 'Die Bedeutung der Psychologie für die übrigen Wissenschaften und die Praxis', *Fortschritte der Psychologie und ihrer Anwendungen* 1: 5–82.

Marbe, K. (1912b) Review of Dürr (1910) *Zeitschrift für Psychologie* 60: 114–27.

Marbe, K. (1913a) *Die Aktion gegen die Psychologie: Eine Abwehr*, Leipzig: Teubner.

Marbe, K. (1913b) *Grundzüge der forensischen Psychologie*, München: Beck.

Marbe, K. (1915) 'Zur Psychologie des Denkens', *Fortschritte der Psychologie und ihrer Anwendungen* 3: 1–42.

Marbe, K. (1916) *Die Gleichförmigkeit in der Welt: Untersuchungen zur Philosophie und positiven Wissenschaft*, vol. 1, München: Beck.

Marbe, K. (1919) *Die Gleichförmigkeit in der Welt: Untersuchungen zur Philosophie und positiven Wissenschaft*, vol. 2, München: Beck.

Marbe, K. (1926a) *Der Psycholog als Gerichtsgutachter im Straf- und Zivilprozess*, Stuttgart: Enke.

Marbe, K. (1926b) *Praktische Psychologie der Unfälle und Betriebsschäden*, München: Oldenbourg.

Marbe, K. (1927) *Psychologie der Werbung*, Stuttgart: Poeschel.

Marbe, K. (1945) *Selbstbiographie des Psychologen Geheimrat Prof. Dr. Karl Marbe in Würzburg*, Deutsche Akademie der Naturforscher, Halle a. d. Saale: Leopoldina.

Marbe, K. (1961) 'Autobiography', in C.A. Murchison (ed.) *A History of Psychology in Autobiography*, 7 vols, vol. III, New York: Russell & Russell, 181–213.

Martin, L. J. (1913) 'Quantitative Untersuchungen über das Verhältnis anschaulicher und unanschaulicher Bewusstseinsinhalte', *Zeitschrift für Psychologie* 65: 417–90.

Martius, G. (1896) Review of Külpe (1893), *Zeitschrift für Psychologie* 9: 23–45.

Mayer, A. (1903) 'Über Einzel- und Gesamtleistung des Schulkindes', *Archiv für die gesamte Psychologie* 1: 276–416.

Mayer, A. and J. Orth (1901) 'Zur qualitativen Untersuchung der Association', *Zeitschrift für Psychologie* 26: 1–13.

McCauley, R. N. (1993 [1986]) 'Intertheoretic Relations and the Future of Psychology', in S. M. Christensen and D. R. Turner (eds) *Folk Psychology and the Philosophy of Mind*, Hillsdale, NJ: Lawrence Erlbaum, 63–81.

McClelland, C. E. (1980) *State, Society and University in Germany, 1700–1914*, Cambridge: Cambridge University Press.

McGinn, C. (1983) *The Subjective View: Secondary Qualities and Indexical Thoughts*, Oxford: Clarendon Press.

McKeen Cattell, J. (1921) 'In Memory of Wilhelm Wundt', in G. S. Hall *et al.* (1921), 'In Memory of Wilhelm Wundt', *Psychological Review* 3: 153–88, here 155–9.

Meinong, A. (1902) *Über Annahmen* (*Zeitschrift für Psychologie* suppl. vol. 2), Leipzig: Barth.

Meinong, A. (1913 [1904]) 'Über Gegenstandstheorie', in A. Meinong, *Abhandlungen zur Erkenntnistheorie und Gegenstandstheorie*, Leipzig: Barth.

Merrill, K. (1979) 'Did Descartes Misunderstand the *Cogito*?', *Studia Cartesiana* 1: 5–42.

Merton, R. K. (1968) 'The Matthew-Effect in Science: The Reward and Communication Systems of Science', *Science* 159: 56–63.

Messer, A. (1899) *Die Wirksamkeit der Apperception in den persönlichen Beziehungen des Schullebens*, Berlin: Reuther & Reichard.

Messer, A. (1900) *Kritische Untersuchungen über Denken, Sprechen und Sprachunterricht*, Berlin: Reuther & Reichard.

Messer, A. (1906) 'Experimentell-psychologische Untersuchungen über das Denken', *Archiv für die Gesamte Psychologie* 8: 1–224.

Messer, A. (1907) 'Bemerkungen zu meinen "Experimentell-psychologischen Untersuchungen über das Denken"', *Archiv für die gesamte Psychologie* 10: 409–28.

Messer, A. (1908) *Empfindung und Denken*, Leipzig: Quelle & Meyer.

Messer, A. (1911) *Das Problem der Willensfreiheit*, Göttingen: Vandenhoeck & Ruprecht.
Messer, A. (1912a) *Das Problem der staatsbürgerlichen Erziehung historisch und systematisch behandelt*, Leipzig: Nemnich.
Messer, A. (1912b) 'Husserls Phänomenologie in ihrem Verhältnis zur Psychologie', *Archiv für die gesamte Psychologie* 22: 117–29.
Messer, A. (1914a) 'Die Bedeutung der Psychologie für Pädagogik, Medizin, Jurisprudenz und Nationalökonomie', in M. Frischeisen-Köhler (ed.) *Jahrbücher der Philosophie. Eine kritische Übersicht der Philosophie der Gegenwart*, vol. 2, Berlin: Mittler, 183–218, 232–7.
Messer, A. (1914b) 'Husserls Phänomenologie in ihrem Verhältnis zur Psychologie. (Zweiter Aufsatz)', *Archiv für die gesamte Philosophie* 32: 52–67.
Messer, A. (1914c) *Psychologie*, Stuttgart: Deutsche Verlags-Anstalt.
Messer, A. (1916) *Die Philosophie der Gegenwart*, Leipzig: Quelle & Mayer.
Messer, A. (1918) *Ethik: Eine philosophische Erörterung der sittlichen Grundfragen*, Leipzig: Quelle & Meyer.
Messer, A. (1920) *Natur und Geist: Philosophische Aufsaetze*, Osterwieck: Zickfeldt.
Messer, A. (1921) *Einführung in die Erkenntnistheorie*, 2nd edn, Leipzig: Meiner.
Messer, A. (1922) 'August Messer', in R. Schmidt (ed.) *Philosophie der Gegenwart in Selbstdarstellungen*, 7 vols, vol. 3, Leipzig: Barth (pages are not numbered consecutively).
Messer, A. (1924) *Glauben und Wissen: Geschichte einer inneren Entwicklung*, 3rd edn, München: Reinhardt.
Messer, A. (1925) *Ethik: Eine philosophische Erörterung der sittlichen Grundfragen*, 2nd rev. edn, Leipzig: Quelle & Meyer.
Messer, A. (1927) *Die Philosophie der Gegenwart in Deutschland*, 6th edn, Leipzig: Quelle & Meyer.
Meumann, E. (1894) 'Beiträge zur Psychologie des Zeitsinns. II', *Philosophische Studien* 9: 264–306.
Meumann, E. (1902) *Die Entstehung der ersten Wortbedeutungen beim Kinde*, Leipzig: Engelmann.
Meumann, E. (1903) *Die Sprache des Kindes*, Zürich: Zürcher & Furrer.
Meumann, E. (1904) Review of Wundt (1903b), *Archiv für die gesamte Psychologie* 2: 21–37.
Meumann, E. (1907a) Review of Wundt (1907b), *Archiv für die gesamte Psychologie* 10: 117–34.
Meumann, E. (1907b) 'Über Assoziationsexperimente mit Beeinflussung der Reproduktionszeit', *Archiv für die gesamte Psychologie* 9: 117–50.
Meumann, E. (1907c) *Vorlesungen zur Einführung in die experimentelle Pädagogik und ihre psychologischen Grundlagen*, 2 vols, vol. 1, Leipzig: Engelmann.
Meumann, E. (1908) *Intelligenz und Wille*, Leipzig: Quelle & Meyer.

Minsky, M. (1985) *The Society of Mind,* New York: Simon & Schuster.

Misch, G. (1935) 'Georg Elias Müller', *Nachrichten von der Gesellschaft der Wissenschaften zu Göttingen: Jahresbericht über das Geschäftsjahr 1934/35,* Berlin: Weidmann, 45–59.

Mischel, T. (1970) 'Wundt and the Conceptual Foundations of Psychology', *Philosophy and Phenomenological Research* 31: 1–26.

Mischel, T. (1977) 'Conceptual Issues in the Psychology of the Self: An Introduction', in T. Mischel (ed.) *The Self: Psychological and Philosophical Issues,* Oxford: Blackwell, 3–30.

Moog, W. (1918) 'Die Kritik des Psychologismus durch die moderne Logik und Erkenntnistheorie', *Archiv für die gesamte Psychologie* 37: 301–62.

Moog, W. (1919) *Logik, Psychologie und Psychologismus,* Halle: Niemeyer.

Moog, W. (1922) *Die deutsche Philosophie des 20. Jahrhunderts in ihren Hauptrichtungen und Grundproblemen,* Stuttgart: Enke.

Morrell, J. B. (1972) 'The Chemist Breeders: The Research Schools of Liebig and Thomas Thomson', *Ambix* 19: 1–46.

Morton, A. (1970) *Frames of Mind,* Oxford: Oxford University Press.

Moskiewicz, G. (1910a) Review of Messer (1908), *Zeitschrift für Psychologie* 54: 545–50.

Moskiewicz, G. (1910b) 'Zur Psychologie des Denkens (Erste Abhandlung)', *Archiv für die gesamte Psychologie* 18: 305–99.

Mühlhäusler, P. and R. Harré (1990) *Pronouns and People: The Linguistic Construction of Social and Personal Identity,* Oxford: Blackwell.

Müller, G. E. (1911) *Zur Analyse der Gedächtnistätigkeit und des Vorstellungsverlaufes, I. Teil* (*Zeitschrift für Psychologie* suppl. vol. 5), Leipzig: Barth.

Müller, G. E. (1913) *Zur Analyse der Gedächtnistätigkeit und des Vorstellungsverlaufes, III. Teil* (*Zeitschrift für Psychologie* suppl. vol. 8), Leipzig: Barth.

Müller, G. E. (1917) *Zur Analyse der Gedächtnistätigkeit und des Vorstellungsverlaufes, II. Teil* (*Zeitschrift für Psychologie,* suppl. vol. 9), Leipzig: Barth.

Müller, G. E. (1919) Review of Selz (1913), *Zeitschrift für Psychologie* 82: 102–20.

Müller, G. E. (1924) *Abriss der Psychologie,* Göttingen: Vandenhoeck & Ruprecht.

Müller, G. E. and A. Pilzecker (1900) *Experimentelle Beiträge zur Lehre vom Gedächtnis* (*Zeitschrift für Psychologie* suppl. vol. 1), Leipzig: Barth.

Müller, G. E. and F. Schumann (1894) 'Experimentelle Beiträge zur Untersuchung des Gedächtnisses', *Zeitschrift für Psychologie und Physiologie der Sinnesorgane* 6: 81–190, 257–339.

Müller-Freienfels, R. (1912) 'Vorstellen und Denken', *Zeitschrift für Psychologie* 60: 379–443.

Münsterberg, H. (1889) *Beiträge zur experimentellen Psychologie,* vol. 1, Freiburg i. Br.: Mohr.

Münsterberg, H. (1891) *Aufgaben und Methoden der Psychologie* Leipzig: Barth.

Myers, C. S. (1911) *A Text-Book of Experimental Psychology with Laboratory Exercises, Part II: Laboratory Exercises*, 2nd edn, Cambridge: Cambridge University Press.

Nanu, H. A. (1904) *Zur Psychologie der Zahlauffassung*, Diss. Würzburg, Würzburg: Becker.

Natorp, P. (1887) 'Über objective und subjective Begründung der Erkenntnis I', *Philosophische Monatshefte* 23: 257–86.

Natorp, P. (1901) 'Zur Frage der logischen Methode. Mit Beziehung auf Edm. Husserls "Prolegomena zur reinen Logik" ', *Kantstudien* 6: 270–83.

Natorp, P. (1912) 'Das akademische Erbe Hermann Cohens', *Frankfurter Zeitung* 12 October: 1–2.

Needham, R. (1972) *Belief, Language, and Experience*, Oxford: Blackwell.

Needham, R. (1981) 'Inner States as Universals: Sceptical Reflections on Human Nature', in P. Heelas and A. Lock (eds) *Indigenous Psychologies: The Anthropology of the Self*, London: Academic Press, 65–78.

Nelkin, N. (1986) 'Pains and Pain Sensations', *The Journal of Philosophy* 83: 129–48.

Newton-Smith, W. H. (1981) *The Rationality of Science*, London: Routledge.

Nipperdey, T. (1990) *Deutsche Geschichte, 1866–1918, Vol. 1: Arbeitswelt und Bürgergeist*, München: Beck.

Nipperdey, T. (1992) *Deutsche Geschichte, 1866–1918, Vol. 2: Machtstaat vor der Demokratie*, München: Beck.

Nisbett, R. E. and T. D. Wilson (1977) 'Telling More Than We Can Know: Verbal Reports on Mental Processes', *Psychological Review* 84: 231–59.

Normore, C. (1990) 'Doxology and the History of Philosophy', *Canadian Journal of Philosophy*, suppl. vol. 16: 203–26.

Oelze, B. (1991) *Wilhelm Wundt: Die Konzeption der Völkerpsychologie*, Münster: Waxmann.

Oesterreich, K. T. (1910) *Die Phänomenologie des Ich in ihren Grundproblemen*, Leipzig: Barth.

Olesko, K. M. (1993) 'Tacit Knowledge and School Formation', in G. L. Geison and F. L. Holmes (eds) *Research Schools: Historical Reappraisals, Osiris*, 2nd Ser., 8: 16–29.

Orth, J. (1901) 'Kritik der Associationseintheilungen', *Zeitschrift für pädagogische Psychologie und Pathologie* 3: 104–19.

Orth, J. (1903) *Gefühl und Bewußtseinslage: Eine kritisch-experimentelle Studie*, Berlin: Reuther & Reichard.

Ott, H. (1993) *Martin Heidegger: A Political Life*, tr. A. Blunden, New York: HarperCollins.

Paulsen, F. (1901a) *Einleitung in die Philosophie*, 7th edn, Berlin: Hertz.

Paulsen, F. (1901b) *Philosophia militans: Gegen Klerikalismus und Naturalismus*, Berlin: Reuther & Reichard.

Perner, J. (1991) *Understanding the Representational Mind*, Cambridge, MA: Bradford Books / MIT Press.

Perner, J., S. Baker, and D. Hutton (1994) 'Prelief: The Conceptual Origins of Belief and Pretence', in C. Lewis and P. Mitchell (eds) *Children's Early Understanding of Mind: Origins and Development*, Hove, East Sussex: Lawrence Erlbaum, 261–86.

Petersen, P. (1913) 'Referat über psychologische Literatur. Das Jahr 1912', *Zeitschrift für Philosophie und philosophische Kritik* 149: 193–221.

Petersen, P. (1914) 'Referat über psychologische Literatur. Das Jahr 1913', *Zeitschrift für Philosophie und philosophische Kritik* 153: 176–213.

Pfeiffer, L. (1908) *Über qualitative Arbeitstypen*, Diss. Würzburg.

Pintner, R. (1921) 'In Memory of Wilhelm Wundt', in G. S. Hall *et al.* (1921), 'In Memory of Wilhelm Wundt', *Psychological Review* 3: 153–88, here 186–8.

Popp, W. (1913) *Kritische Entwicklung des Associationsproblems*, Leipzig: Barth.

Putnam, H. (1975) 'The Meaning of "Meaning"', in K. Gunderson (ed.) *Language, Mind, and Knowledge: Minnesota Studies in the Philosophy of Science*, Vol. 7, Minneapolis: University of Minneapolis Press, 131–93.

Ramsey, W., P. S. Stich, and J. Garon (1996 [1991]) 'Connectionism, Eliminativism, and the Future of Folk Psychology', in P. S. Stich, *Deconstructing the Mind*, Oxford: Oxford University Press, 91–114.

Rangette, L. (1916) 'Untersuchung ueber die Psychologie des wissenschaftlichen Denkens auf experimenteller Grundlage', *Archiv für die gesamte Psychologie* 36: 169–254.

Reichwein, G. (1910) *Die neueren Untersuchungen über Psychologie des Denkens nach Aufgabestellung, Methode und Resultaten übersichtlich dargestellt und krititisch beurteilt*, Diss. Halle, Halle a. S.: Kaemmerer.

Revers, W. J. (1968) 'Külpe, Oswald', *International Encyclopedia of the Social Sciences*, vol. 8, New York: Macmillan, 467–8.

Richards, G. (1989) *On Psychological Language*, London: Routledge & Kegan Paul.

Richards, G. (1992) *Mental Machinery: The Origins and Consequences of Psychological Ideas*, London: Athlone Press.

Richards, G. (1996) *Putting Psychology in Its Place: An Introduction from a Critical Historical Perspective*, London: Routledge.

Rickert, H. (1904) *Der Gegenstand der Erkenntnis: Ein Beitrag zum Problem der philosophischen Transcendenz*, 2nd rev. edn, Tübingen: Mohr.

Rickert, H. (1913a) *Die Grenzen der naturwissenschaftlichen Begriffsbildung*, 2nd rev. edn, Tübingen: Mohr.

Rickert, H. (1913b) 'Zur Besetzung der philosophischen Profssuren mit Vertretern der experimentellen Psychologie', *Frankfurter Zeitung* 4 March: 1–2.

Roth, H. (1915) *Das sittliche Urteil der Jugend*, Diss. München.

Russell, J. A. (1991) 'Culture and the Categorisation of Emotions', *Psychological Bulletin* 110: 426–50.
Schanoff, B. (1911) *Die Vorgänge des Rechnens: Ein experimenteller Beitrag zur Psychologie des Rechnens*, Diss. Bonn, Leipzig: Nemnich.
Scheler, M. (1922) 'Die deutsche Philosophie der Gegenwart', in P. Witkop (ed.) *Deutsches Leben der Gegenwart*, Berlin: Wegweiser-Verlag, 127–224.
Schell, H. (1897) *Der Katholicismus als Princip des Fortschritts*, Würzburg: Göbel.
Schlesinger, A. (1908) *Der Begriff des Ideals*, Diss. Würzburg.
Schmidt, F. (1904) 'Experimentelle Untersuchungen über die Hausaufgaben des Schulkindes', *Archiv für die gesamte Psychologie* 3: 33–152.
Schneider, C. M. (1990) *Wilhelm Wundts Völkerpsychologie: Entstehung und Entwicklung eines in Vergessenheit geratenen, wissenschafts-historisch relevanten Fachgebietes*, Bonn: Bouvier.
Schrader, E. (1903) *Zur Grundlegung der Psychologie des Urteils*, Leipzig: Barth.
Schrader, E. (1905) *Elemente der Psychologie des Urteils, vol. 1: Analyse des Urteils*, Leipzig: Barth.
Schreiber, C. (1914) 'Die Erkenntnislehre des hl. Thomas und die moderne Erkenntniskritik', *Philosophisches Jahrbuch der Görres-Gesellschaft* 27: 488–520.
Schuhmann, K. (ed.) (1994a) *Edmund Husserl: Briefwechsel, vol. II: Die Münchener Phänomenologen*, Dordrecht: Kluwer.
Schuhmann, K. (ed.) (1994b) *Edmund Husserl: Briefwechsel, vol. V: Die Neukantianer*, Dordrecht: Kluwer.
Schultze, F. E. O. (1906) 'Einige Hauptgesichtspunkte der Beschreibung der Elementarpsychologie, I: Erscheinungen und Gedanken', *Archiv für die Gesamte Psychologie* 8: 241–338.
Searle, J. R. (1990) 'Collective Intentions and Actions', in P. R. Cohen *et al.* (eds) *Intentions in Communication*, Cambridge, MA: MIT Press, 401–15.
Searle, J. (1992) *The Rediscovery of the Mind*, Cambridge, MA: MIT Press.
Searle, J. R. (1995) *The Construction of Social Reality*, London: Allen Lane.
Segal, G. (1996) 'The Modularity of Theory of Mind', in P. Carruthers and P. K. Smith (eds) *Theories of Theories of Mind*, Cambridge: Cambridge University Press, 141–57.
Segal, J. (1908) 'Über den Reproduktionstypus und das Reproduzieren von Vorstellungen', *Archiv für die gesamte Psychologie* 12: 124–235.
Sellars, W. (1956) 'Empiricism and the Philosophy of Mind', in H. Feigl and M. Scriven (eds) *The Foundations of Science and the Concepts of Psychology and Psychoanalysis: Minnesota Studies in the Philosophy of Science*, vol. 1, Minneapolis: University of Minneapolis Press, 253–329.
Selz, O. (1910) 'Die experimentelle Untersuchung des Willensaktes', *Zeitschrift für Psychologie* 57: 241–70.
Selz, O. (1911) 'Willensakt und Temperament. Eine Erwiderung auf N. Achs Widerlegung', *Zeitschrift für Psychologie* 59: 113–22.

Selz, O. (1913) *Über die Gesetze des geordneten Denkverlaufs: Eine experimentelle Untersuchung, Erster Teil*, Stuttgart: Spemann.

Sentroul, C. (1911) *Kant und Aristoteles*, tr. L. Heinrichs, Kempten: Kösel.

Sganzini, C. (1913) *Die Fortschritte der Völkerpsychologie von Lazarus bis Wundt*, Bern: Francke.

Shapin, S. (1990) '"The Mind Is Its Own Place": Science and Solitude in Seventeenth-Century England', *Science in Context* 4: 191–218.

Shapin, S. (1994) *A Social History of Truth: Gentility, Credibility, and Scientific Knowledge in Seventeenth-Century England*, Chicago: University of Chicago Press.

Shapin, S. and S. Schaffer (1985) *Leviathan and the Air-Pump*, Princeton, NJ: Princeton University Press.

Sharpe, R. A. (1990) *Making the Human Mind*, London: Routledge.

Shweder, R. A. (1991) *Thinking through Cultures: Expeditions in Cultural Psychology*, Cambridge, MA: Harvard University Press.

Shweder, R. A. and E. J. Bourne (1991) 'Does the Concept of the Person Vary Cross-Culturally?', R. A. Shweder, *Thinking through Cultures: Expeditions in Cultural Psychology*, Cambridge, MA: Harvard University Press, 113–55.

Sigwart, C. (1889) *Logik*, vol. 1, 2nd edn, Freiburg i. Br.: Mohr.

Smith, C. and M. N. Wise (1989) *Energy and Empire: A Biographical Study of Lord Kelvin*, Cambridge: Cambridge University Press.

Smith, J. (1981) 'Self and Experience in Maori Culture', in P. Heelas and A. Lock (eds) *Indigenous Psychologies: The Anthropology of the Self*, London: Academic Press, 145–59.

Sokal, M., A. B. Davis, and U. C. Menzbach (1976) 'Laboratory Instruments in the History of Psychology', *Journal of the History of the Behavioural Sciences* 12: 59–64.

Specht, W. (1909) 'Das pathologische Verhalten der Aufmerksamkeit', in F. Schumann, *Bericht über den III. Kongress für experimentelle Psychologie*, Leipzig: Barth, 131–91.

Spengler, O. (1918) *Der Untergang des Abendlandes. Umrisse einer Morphologie der Weltgeschichte*, vol. 1: *Gestalt und Wirklichkeit*, München: Beck.

Spranger, E. (1905) *Die Grundlagen der Geschichtswissenschaft*, Berlin: Reuther & Reichard.

Sprung, L. and H. Sprung (1997) 'Georg Elias Müller (1850–1934): Skizzen zum Leben, Werk und Wirken', in G. Lüer and U. Lass (eds) *Erinnern und Behalten: Wege zur Erforschung des menschlichen Gedächtnisses*, Göttingen: Vandenhoeck & Ruprecht, 338–70.

Ssalagoff, L. (1911) 'Vom Begriff des Geltens in der modernen Logik', *Zeitschrift für Philosophie und philosophische Kritik* 143: 145–90.

Stern, W. and O. Lipmann (1908) 'Aufruf zur Sammlung psychologischer Fragebogen und Prüfungslisten', *Zeitschrift für angewandte Psychologie und psychologische Sammelforschung* 1: 575.

Stich, S. P. (1983) *From Folk Psychology to Cognitive Science*, Cambridge, MA: MIT Press.

Stich, S. P. (1994) 'What is a Theory of Mental Representation?', in S. P. Stich and T. A. Warfield (eds) *Mental Representation: A Reader*, Oxford: Blackwell, 347–64.

Stich, S. P. (1996) *Deconstructing the Mind*, Oxford: Oxford University Press.

Stich, S. P. and I. Ravenscroft (1996 [1994]) 'What *Is* Folk Psychology', in S. P. Stich, *Deconstructing the Mind*, Oxford: Oxford University Press, 115–35.

Stigler, J. W., R. A. Shweder, and G. H. Herdt (eds) (1990) *Cultural Psychology: Essays on Comparative Human Development*, Cambridge: Cambridge University Press.

Störring, G. (1908) 'Experimentelle Untersuchungen über einfache Schlussprozesse', *Archiv für die gesamte Psychologie* 11: 1–127.

Stumpf, C. (1906) 'Erscheinungen und psychische Funktionen', *Aus den Abhandlungen der könglichen Preussischen Akademie der Wissenschaften vom Jahre 1906*, Berlin: Verlag der königlichen Akademie der Wissenschaften.

Stumpf, C. (1910 [1896]) 'Leib und Seele', in C. Stumpf, *Philosophische Reden und Vorträge*, Leipzig: Barth, 65–93.

Stumpf, C. (1924) 'Carl Stumpf', in R. Schmidt (ed.) *Philosophie der Gegenwart in Selbstdarstellungen*, vol. 5, Leipzig: Barth (pages are not numbered consecutively).

Switalski, D. (1925) *Kant und der Katholizismus*, Münster: Aschendorff.

Tawney, G. A. (1921) 'In Memory of Wilhelm Wundt', in G. S. Hall *et al.* (1921), 'In Memory of Wilhelm Wundt', *Psychological Review* 3: 153–88, here 1178–81.

Taylor, C. (1977) 'What is Human Agency?', in T. Mischel (ed.) *The Self: Psychological and Philosophical Issues*, Oxford: Blackwell, 103–42.

Taylor, C. O. (1905/6) 'Über das Verstehen von Worten und Sätzen', *Zeitschrift für Psychologie und Physiologie der Sinnesorgane* 40: 225–51.

Thumb, A. and K. Marbe (1901) *Experimentelle Untersuchungen über die psychologischen Grundlagen der sprachlichen Analogiebildung*, Leipzig: Engelmann.

Tinker, M. A. (1932) 'Wundt's Doctorate Students and Their Theses, 1875–1920', *American Journal for Psychology* 44: 630–7.

Titchener, E. B. (1909) *Lectures on the Experimental Psychology of the Thought-Processes*, New York: Macmillan.

Tooby, J. and L. Cosmides (1992) 'The Psychological Foundations of Culture', in J. H. Barkow *et al.* (eds) *The Adapted Mind: Evolutionary Psychology and the Generation of Culture*, Oxford: Oxford University Press, 19–136.

Tooby, J. and L. Cosmides (1995) 'Foreword', in S. Baron-Cohen, *Mindblindness: An Essay on Autism and Theory of Mind*, Cambridge, MA: MIT Press, xi–xviii.

Trout, J. D. (1991) 'Belief Attribution in Science: Folk Psychology under Theoretical Stress', *Synthese* 87: 379–400.

Tuomela, R. (1995) *The Importance of Us: A Philosophical Study of Basic Social Notions*, Stanford, CA: Stanford University Press.

Turner, R. S. (1993) 'Vision Studies in Germany: Helmholtz versus Hering', in G. L. Geison and F. L. Holmes (eds) *Research Schools: Historical Reappraisals, Osiris*, 2nd Ser., 8: 80–103.

Turner, R. S. (1994) *In the Eye's Mind: Vision and the Helmholtz-Hering Controversy*, Princeton, NJ: Princeton University Press.

Ueberweg, F. and T. K. Oesterreich (1951) *Grundriss der Geschichte der Philosophie*, vol. 4, 13th edn, rev. by T. K. Oesterreich, Basel: Schwabe.

Ungerer, G. A. (1980) 'Wilhelm Wundt als Psychologie und Politiker', *Psychologische Rundschau* 31: 99–110.

Wartensleben, G. v. (1910) 'Beiträge zur Psychologie des Übersetzens', *Zeitschrift für Psychologie* 57: 89–115.

Watson, J. B. (1913) 'Psychology as the Behaviorist Views It', *Psychological Review* 20: 158–77.

Watt, H. J. (1905) 'Experimentelle Beiträge zu einer Theorie des Denkens', *Archiv für die Gesamte Psychologie* 4: 289–436.

Watt, H. J. (1906a) Review of Ach (1905), *Archiv für die gesamte Psychologie* 8: 80–9.

Watt, H. J. (1906b) 'Sammelbericht (II.) über die neuere Forschung in der Gedächtnis- und Assoziationspsychologie aus dem Jahre 1905', *Archiv für die gesamte Psychologie* 9: 1–21.

Weimer, W. B. (1974) 'The History of Psychology and Its Retrieval from Historiography, I: The Problematic Nature of History,' *Science Studies* 4: 235–58.

Wellek, A. (1968) 'Bühler, Karl', in S. L. Sills (ed.) *International Encyclopedia of the Social Sciences*, vol. 1, New York: Macmillan, 199–202.

Wellman, H. M. (1990) *The Child's Theory of Mind*, Cambridge, MA: MIT Press.

Wentscher, E. (1910) *Der Wille: Versuch einer psychologischen Analyse*, Leipzig: Teubner.

Wertheimer, M. (1906) 'Experimentelle Untersuchungen zur Tatbestandsdiagnostik', *Archiv für die Gesamte Psychologie* 6: 59–131.

Wertheimer, M. and J. Klein (1906) 'Psychologische Tatbestandsdiagnostik', *Archiv für Kriminal-Anthropologie und Kriminalistik* 15: 72–113.

Westphal, E. (1911) 'Über Haupt- und Nebenaufgaben bei Reaktionsversuchen', *Archiv für die gesamte Psychologie* 21: 219–434.

Whitley, R. (1984) *The Intellectual and Social Organisation of the Sciences*, Oxford: Clarendon Press.

Wilkes, K. (1984) 'Pragmatics in Science and Theory in Common Sense', *Inquiry* 27: 339–61.

Wilkes, K. (1991) 'The Long Past and the Short History', in R. J. Bogdan (ed.) *Mind and Common Sense: Philosophical Essays on Commonsense Psychology*, Cambridge: Cambridge University Press, 144–60.
Windelband, W. (1884) 'Kritische oder genetische Methode?', in W. Windelband, *Präludien*, Freiburg i. Br.: Mohr, 247–79.
Windelband, W. (1909) *Die Philosophie im deutschen Geistesleben des XIX. Jahrhunderts: Fünf Vorlesungen*, Tübingen: Mohr.
Wise, M. N. (1993) 'Mediations: Enlightenment Balancing Acts, or the Technologies of Rationalism', in P. Horwich (ed.) *World Changes: Thomas Kuhn and the Nature of Science*, Cambridge, MA: MIT Press, 207–56.
Wittgenstein, L. (1953) *Philosophical Investigations*, tr. G. E. M. Anscombe, Oxford: Blackwell.
Wundt, W. (1880) *Logik: Eine Untersuchung der Prinzipien der Erkenntnis und der Methoden wissenschaftlicher Forschung*, vol. 1, Stuttgart: Enke.
Wundt, W. (1887) *Zur Moral der literarischen Kritik: Eine Moralphilosophische Streitschrift*, Leipzig: Engelmann.
Wundt, W. (1888a) 'Selbstbeobachtung und innere Wahrnehmung', *Philosophische Studien* 4: 292–310.
Wundt, W. (1888b) 'Über Ziele und Wege der Völkerpsychologie,' *Philosophische Studien* 4: 1–27.
Wundt, W. (1889) 'Über die Eintheilung der Wissenschaften', *Philosophische Studien* 5: 1–55.
Wundt, W. (1892a) *Hypnotismus und Suggestion*, Leipzig: Engelmann.
Wundt, W. (1892b) 'Was soll uns Kant nicht sein?', *Philosophische Studien* 7: 1–49.
Wundt, W. (1893) *Logik: Eine Untersuchung der Prinzipien der Erkenntnis und der Methoden wissenschaftlicher Forschung*, vol. 1, 2nd edn, Stuttgart: Enke.
Wundt, W. (1894) 'Über psychische Causalität und das Princip des psychophysischen Parallelismus', *Philosophische Studien* 10: 1–124.
Wundt, W. (1896) 'Über die Definition der Psychologie', *Philosophische Studien*, 12: 1–66.
Wundt, W. (1897) *System der Philosophie*, 2nd rev. edn, Leipzig: Engelmann.
Wundt, W. (1903a) *Grundzüge der physiologischen Psychologie*, 5th edn, 3 vols, vol. 3, Leipzig: Engelmann.
Wundt, W. (1903b) *Naturwissenschaft und Psychologie*, Leipzig: Engelmann.
Wundt, W. (1904) 'Über empirische und metaphysische Psychologie: Eine kritische Betrachtung', *Archiv für die gesamte Psychologie* 2: 333–61.
Wundt, W. (1906) *Logik: Eine Untersuchung der Prinzipien der Erkenntnis und der Methoden wissenschaftlicher Forschung*, 1st vol., 3rd edn, Stuttgart: Enke.
Wundt, W. (1907a) *Grundriß der Psychologie*, 8th edn, Leipzig: Engelmann.
Wundt, W. (1907b) 'Psychologie', in W. Windelband (ed.) *Die Philosophie im Beginn des zwanzigsten Jahrhunderts*, Heidelberg: Winter, 1–55.

Wundt, W. (1907c) 'Über Ausfrageexperimente und über die Methoden zur Psychologie des Denkens', *Psychologische Studien* 3: 301–60.

Wundt, W. (1908a) 'Kritische Nachlese zur Ausfragemethode', *Archiv für die gesamte Psychologie* 9: 445–59.

Wundt, W. (1908b) *Logik: Eine Untersuchung der Prinzipien der Erkenntnis und der Methoden wissenschaftlicher Forschung*, 3 vols, vol. 3, Stuttgart: Enke.

Wundt, W. (1909) *Einleitung in die Philosophie*, 5th edn, Leipzig: Engelmann.

Wundt, W. (1910a) 'Das Institut für experimentelle Psychologie zu Leipzig', *Psychologische Studien* 5: 279–93.

Wundt, W. (1910b) 'Psychologismus und Logizismus', in W. Wundt, *Kleine Schriften*, 1st vol., Leipzig: Engelmann, 511–634.

Wundt, W. (1910c) 'Über reine und angewandte Psychologie', *Psychologische Studien* 5: 1–47.

Wundt, W. (1910 [1896]) 'Über naiven und kritischen Realismus', in W. Wundt, *Kleine Schriften*, Leipzig: Engelmann, 259–510.

Wundt, W. (1911) *Völkerpsychologie: Eine Untersuchung der Entwicklungsgesetze von Sprache, Mythus und Sitte, vol. 1: Die Sprache*, part 1, 3rd rev. edn, Leipzig: Kröner.

Wundt, W. (1912) *Ethik: Eine Untersuchung der Tatsachen und Gesetze des sittlichen Lebens*, 4th edn, 3 vols, Stuttgart: Enke.

Wundt, W. (1913a) *Die Psychologie im Kampf ums Dasein*, Leipzig: Engelmann.

Wundt, W. (1913b) *Elemente der Völkerpsychologie*, 2nd edn, Leipzig: Kröner.

Wundt, W. (1914) *Sinnliche und übersinnliche Welt*, Leipzig: Kröner.

Wundt, W. (1915) *Die Nationen und ihre Philosophie: Ein Kapitel zum Weltkrieg*, Stuttgart: Kröner.

Wundt, W. (1917) *Völkerpsychologie: Eine Untersuchung der Entwicklungsgesetze von Sprache, Mythus und Sitte, vol. 8: Die Gesellschaft*, part 2, Leipzig: Kröner.

Wundt, W. (1918) *Völkerpsychologie: Eine Untersuchung der Entwicklungsgesetze von Sprache, Mythus und Sitte, vol. 9: Das Recht*, Leipzig: Kröner.

Wundt, W. (1920a) *Erlebtes und Erkanntes*, Stuttgart: Kröner.

Wundt, W. (1920b) *Völkerpsychologie: Eine Untersuchung der Entwicklungsgesetze von Sprache, Mythus und Sitte, vol. 10: Kultur und Geschichte*, Leipzig: Kröner.

Wundt, W. (1923) *Völkerpsychologie: Eine Untersuchung der Entwicklungsgesetze von Sprache, Mythus und Sitte, vol. 6: Mythus und Religion*, part 3, 3rd edn, Leipzig: Kröner.

Ziehen, T. (1900) *Leitfaden der physiologischen Psychologie*, 5th edn, Jena: Fischer.

Index

Ach, N. 9, 18, 20–5, 27–8, 33–8, 43, 58–62, 67–8, 70, 72–4, 78, 80–4, 87–90, 93, 117–18, 125–6, 140, 146, 151, 160, 179–80, 196, 237, 369–70
act 39–42, 45, 57, 61, 63, 90, 128, 151, 157, 160, 213
actuality principle 133–4, 136–7, 139–41, 170, 200, 211–13, 218, 221–2, 130, 234
actus purus 76, 89–90, 92, 94, 200, 206, 221
alter-reference 246, 352–3, 360
Altschule, M. D. 374
Anschütz, G. 85–7, 126–7
Anscombe, E. 255, 360, 373
anti-Catholicism 196, 200, 206; *see also* Catholicism
anti-psychologism 89, 93, 147–63; *see also* psychologism
appearance 37–8, 90, 140
apperception 26, 34, 51–3, 59, 72, 98–9, 139, 146, 178–9, 207, 234, 237
Arbib, M. A. 322
Aristoteles 15, 17, 90, 94, 118, 195, 197, 201, 215, 222, 225–7, 347
Arps, G. F. 97
Ash, M. 151, 196, 369
association 11, 13, 35, 44–6, 50–1, 54–5, 58–66, 69, 71–2, 79, 80–3, 87, 98–9, 107–9, 111, 113, 135, 178–9, 181, 187, 207
associationism 10, 11, 43–6, 49, 62–4, 78, 82–4, 87, 89–90, 92–3, 107–8, 122–3, 130, 155, 178–9, 234, 253–4
associative tendency 34, 61–3, 68, 70, 82
Aster, E. von 91, 168
attention 77, 90, 92
attitude 60, 62–3
authority 101, 129, 202, 205, 210, 235–6, 246, 271, 352, 364
autonomy 103–4, 171, 202
Avenarius, R. 137–8, 216
awareness 10, 20, 33–40, 42–4, 56, 65–6, 68–9, 78, 83, 87–90, 117, 140

Baade, W. 88
Bacon, F. 203
Baerwald, R. 145
Baeumker, C. 89–90, 115, 126, 145, 217, 228
Bain, A. 107
Baker, L. R. 306–8, 327, 351–2
Balzer, W. 255
Barnes, B. 255–67, 270, 275, 366–7, 372
Baron-Cohen, S. 292
Beer, M. 26, 33, 369
belief 2, 4, 282, 285, 292–3, 297, 301, 303, 304, 306–7, 310–16, 320, 321–3, 329–30, 337–47, 354, 357–8
Bennett, J. 340
Berry, J. W. 331, 335
Betz, W. 91
Binet, A. 38
Bismarck, O. 195
Blackburn, S. 311, 314

Bloch, M. 367, 374
Bloor, D. 252, 255, 267–8, 275, 325, 342, 367, 369, 372
Blumenthal, A. 80, 99
Bourne, E. J. 333–4
Brentano, F. 20, 38–9, 45, 69, 74, 128, 154–61, 203–4, 206, 217
Bringmann, W. G. 142, 158, 371
Brock, A. 247
Brönner, W. 180–3, 191–3
Brown, J. R. 251
Brunner, J. 372
Brunswig, A. 90
Bühler, C. 228
Bühler, K. 9, 15, 17–18, 20, 25–8, 33, 39–43, 63–4, 67, 68–9, 72, 74, 77, 83–5, 88, 90–1, 114–15, 117–20, 124, 127–8, 140, 146, 159–61, 187–8, 191, 193–4, 206, 227–8, 231, 237, 241, 246–7, 250, 254, 369–70
Burke, E. 352
Burt, C. 247

capitalism 176–8, 193
card changer 23–5, 47, 59
Carruthers, P. 301, 373
Cassirer, E. 89, 93
Catholicism 12, 15–16, 90, 117, 184, 192, 194–8, 200–7, 218, 228–36
Caudle, F. M. 369
causality 73–4, 132–5, 137, 141, 176, 209, 284, 296
Celan, P. 330
Chapman, W. P. 359
Charcot, J. M. 17, 369
choice 51–3, 56–7, 70, 72, 171, 179, 207, 209, 221, 230
Chomsky, N. 295
Churchland, Patricia 308, 310, 372
Churchland, Paul 284–7, 297, 303, 310–11, 317, 319, 323, 351, 355, 372–4
Clark, A. 309, 338
class 3, 173–4, 180
Cogito 362–3
cognitive science 4, 303, 306, 312
Cohen, H. 148, 151
Cohn, J. 89
collective good 274–6, 327, 330, 347, 365
collectivism 15, 168–80, 182–3, 187, 193, 232–5, 279–80, 315, 318–19, 347, 350, 366–7
Collins, H. M. 240, 242, 369, 372
complex 65–6, 84, 93
concept application 255–6
Confucius 332
conscientialism 222–4, 230
consensus 270–3, 325
constellation 65, 82, 87
Cordes, G. 146, 369
Cosmides, L. 350
culture 174, 177, 179, 189

Danziger, K. 72–3, 138–9, 178, 371, 373
Darwin, C. 110, 289–90
Davidson, D. 252–4
Davies, M. 373
democracy 115, 175
Dennett, D. 250, 296–8, 313, 319, 324, 327, 340, 356
depictive 30, 34, 38, 77, 83, 86, 88–90, 133, 140, 247
Descartes, R. 177, 203, 297, 301, 362–3
Desjarlais, R. 373
determining tendency 10, 11, 56–63, 78, 81–2, 84, 86–8, 93, 122
determinism 177, 209, 212, 229
Deuchler, G. 85–6
Diamond, S. 170, 174, 245
Dilthey, W. 373
Diogenes 187–8
Dorsch, F. 145, 147
Double, R. 312
Douglas, M. 252, 353, 356
dream 42, 126, 128, 191
Driesch, H. 90–1
drillmaster 96, 104–14, 129, 238–9
Dubs, A. 153
Düker, H. 105
Dumont, L. 334
Dupré, J. 314
Durkheim, E. 251, 267, 325, 355, 363–4

Dürr, E. 26–7, 42–3, 69, 82, 84, 129, 146, 159, 161, 369

Ebbinghaus, H. 61, 75, 79–80, 138
Eckehart, *Meister* 205, 372
Edge, D. 248–9, 369
education 14, 113, 123, 129, 142–3, 145–7, 166, 188–9, 192
Egan, F. 298, 312
Ehrhard, A. 198
eliminativism 4, 252, 282, 302–15, 319–20, 338, 347, 351
Elsenhans, T. 88, 148, 369
empathy 19, 23, 25, 27, 142, 299
epistemology 16, 36, 148, 159, 166, 194, 198, 210–11, 214, 218, 220, 222–5, 229–30, 234
Erdmann, B. 37, 148
Ethemendy, J. 363
ethics 15, 172–3, 180, 189, 193, 198, 204
Eucken, R. 26, 74, 151, 198–9, 210, 237
Evans-Pritchard, E. E. 344
experiment: imperfect 75; interrogation 74–5, 92, 101–2, 121, 124, 127, 145, 160, 370; perfect 75, 92, 100–1, 238; reaction 21, 23, 24, 47, 52, 56–7, 62, 65–6, 72, 74, 88; sham 75, 78, 92
Experimenters' Regress 240, 372

Falckenberg, R. 216
false-belief task 291–2, 301–2, 342
Febvre, L. 367, 374
Fechner, G. T. 164, 211
feeling 10–11, 21–2, 30, 32–4, 37–40, 68, 71–2, 75–7, 83, 86–7, 97, 120, 128, 156, 159, 221, 223–4, 247
Feuchtwanger, A. 26, 33, 369
Feyerabend, P. 314
Fichte, J. G. 151
finitism 237, 273–4, 276, 320, 329–30
Flechsig, P. 126, 145
Fodor, J. 287–8, 306, 308, 338, 354
Foss, J. 314–15
Foucault, M. 251–2, 358, 364
fractioning method 28
Frede, M. 251
freedom 123, 192; of the will 209, 212, 218, 229–30
Frege, G. 148, 150
Freud, S. 126
Friedrich, J. 146
Frischeisen-Köhler, M. 88, 151, 197
Fröbes, J. 105
Fruton, J. S. 369
function 39, 42, 161, 181

Gadamer, H.-G. 330, 373
Galison, P. 369, 372
Galton, F. 17, 369
Gardner, S. 188, 228
Garfield, J. L. 373
Garfinkel, H. 373
Gassendi, P. 363, 374
Geertz, C. 333
Geison, G. L. 248, 349, 369
Gelman, S. A. 373
Gemelli, A. 226
Geulincx, A. 220
Geuter, U. 126
Geyser, J. 89–90, 153, 213–14
Gigerenzer, G. 249–50
Gillett, G. 359
Glover, J. 373
Goldman, A. 299–301
Goldstein, D. 250
Gopnik, A. 288–93
Gordon, R. 299–301
Grabmann, M. 89–90, 218, 225–7
Gracia, J. 251
Graf, F. W. 196, 209
Graham, G. 373
Grahek, N. 359
Grane, L. 195–6
Greenwood, J. 311, 313, 373
Groos, K. 216
Grünbaum, A. A. 28, 56, 72, 125–6, 369
Grünholz, E. 213–14, 225
Gutberlet, C. 89–90, 127, 197, 211–16, 372
Guttenplan, S. 372
Güttler, K. 150

Habrich, J. 146

Hacker, F. 42, 126–7
Hacker, P. 293, 295–6, 306–7, 323
Hacking, I. 353–4, 372–3
Haering, T. 27, 369
Hall, G. S. 97–8
Hammer, S. 116, 160
Harnack, A. von 196
Harré, R. 332, 356–7, 359, 361, 373
Harris, P. 299, 302
Harwood, J. 369
Haugeland, J. 255, 267–8
Haupt, E. 111, 371
Hawkins, R. D. 359
Heal, J. 299
Heelas, P. 331–2, 356–7
Hegel, G. W. F. 190–1, 201, 280
Heidegger, M. 153, 196, 228, 364
Heil, J. 312–13
Hellpach, W. 97–8, 115
Helmholtz, H. von 88, 155, 245
Henning, H. 87–8
Hensel, P. 196–7
Herbart, J. F. 53, 107, 109–10, 155, 168, 177, 204
Hering, E. 88
Hesse, M. 322
heuristics of discovery 248–50
Hewson, C. M. 373
Heymans, G. 148
Hilbert, D. 370
Hintikka, J. 374
Hipp chronoscope 23, 25, 27, 47, 68
Hirschfeld, L. A. 373
Höffding, H. 158
Holmes, F. L. 249, 369
Homes, H. 228
Hönigswald, R. 89, 93
Horgan, T. 310–12
Horst, S. 313
Horwicz, A. 150
Howell, S. 332
humanities 11, 13–14, 132, 137, 140–3, 146, 165, 167, 233
Hume, D. 267, 308, 325
Humphrey, G. 18, 31
Hunter, J. S. 330
Husserl, E. 15, 38, 40–1, 55, 69, 74, 87, 128–9, 148–54, 156–7, 159–62, 196, 206, 226, 233–4
hypnosis 19, 23, 28, 56–8, 72, 75, 101–2, 125–6, 145

I 184, 208, 213, 359–65
idealism 172, 183, 197–9, 204, 222, 245
illusion 75–6, 262
individualism 15, 168, 175–87, 190, 193, 203, 232–5, 250–1, 279, 280–1, 315–20, 325, 365–7
innateness 292–3, 309, 317, 347
institution 1–5, 12, 16, 235–8, 240–6, 253–76, 279–81, 316–17, 320–7, 337, 346, 348–9, 356, 359, 361, 364–8; performative theory of 5, 255–76
intellectualism 15, 154–6, 176, 178, 193, 199, 202, 204, 206, 221, 233
intention 34, 41, 55, 227, 285, 291
intentionality 43, 69, 91, 268, 349
interlocutor 96, 114–24, 232–3, 238–9
internationalism 177, 195
interrogator 96, 124–9, 238–9
interruption method 28, 88, 107, 125, 129
introspection 9, 11, 18, 38, 57, 72, 75–7, 80, 85, 88, 119, 141, 169, 175, 178, 186, 192–3, 233, 240–2, 300, 328, 332
Introspectionists' Regress 241
introspective report 24, 26, 29–30, 45, 48, 54, 57, 65, 72, 75–6, 80–1, 85, 89, 93, 115, 124, 126, 140, 241, 328
intuition 280, 321, 347–52, 362, 365
Itkonen, E. 255

Jaensch, E. 151
James, W. 33
Jardine, N. 373
Jasanoff, S. 249
Jaspers, K. 129
Jerusalem, W. 148
Jesus 202
Jodl, F. 158
Joël, C. 211
Johnson, D. M. 356–8, 373–4
Jones, C. M. 359
Judd, C. H. 97–8

judgement 10–11, 20, 32, 45–7, 53–5, 67, 72, 77, 86–7, 121, 148–9, 152, 154, 156, 158–9, 187, 193, 241

Kandel, E. R. 359
Kanerva, L. 373
Kant, I. 26, 53, 89, 97, 151, 189, 196–202, 204, 207, 210–11, 213, 216, 218, 222, 223–5, 227, 286, 308
Katz, D. 105, 112
Katz, R. 105–6, 112
Kennedy, J. F. 283
Kepler, J. 289
Kierkegaard, S. 364
Kiesow, F. 115
Kim, U. 331
kinds 245–8, 352–9; artificial 245–6, 248, 321, 352–6; human 352–4; natural 245–7, 256–8, 352–4; social 245–7, 256, 352, 354
Klein, F. 370
Knorr Cetina, K. 372
knowledge 34, 36, 41–3, 63, 65–6, 69–70, 83, 91, 93, 234
Kobes, B. W. 356
Koch, A. 146
Koehler, W. 86
Koffka, K. 33, 42, 62–3, 82–3, 369
Kraepelin, E. 126
Krebs, E. 228
Kroh, O. 104–5, 111–13
Krüger, F. 372
Kuhn, T. 302, 314, 328
Külpe, O. 9–11, 14, 16, 21, 23, 26–7, 29–31, 36, 38–9, 41–3, 47, 56, 62, 69, 73–4, 82, 84, 87, 90, 96, 110, 114–24, 127, 129, 131, 135, 137–41, 145–7, 160–1, 163–6, 181, 188–94, 199, 211, 213, 216–27, 229–31, 241, 243, 369, 370–2

Lagerspetz, E. 255, 264
Lassalle, W. 176
Laudan, L. 251
law 283–4, 314, 322–3, 327; of thought 30, 43, 148, 153
Lazarus, M. 168
Lehmann, R. 150–1
Leibniz, G. W. 53
Leo XIII 195, 201
Leslie, A. M. 292
Lewis, D. 283–4, 310, 323
Lichtenberg, G. 363
Lienhardt, G. 331–2
Lindenfeld, D. 117, 218
Lindworsky, J. 27, 42, 63, 369
Lipmann, O. 145
Lipps, G. F. 84
Lipps, T. 74, 141, 148, 158, 216, 369
Lock, A. 331, 356
Locke, J. 225
logic 9, 14, 37, 89, 91, 131, 142, 147–63, 198
logicism 154–6, 204, 206, 233
Lotze, H. 211
Löwisch, D.-J. 151, 196
Lück, H. E. 151, 196
Luther, M. 200, 202, 204–5, 235, 372
Lyons, W. 328–9, 332, 373

McCauley, R. 314
McGinn, C. 373
Mach, E. 36, 137–9, 216, 255
McKeen Cattell, J. 97
MacKenzie, D. 369
McKinley, W. 329
Maier, H. 88
Malcolm, N. 360
Malebranche, N. 220
Mandler, G. 18, 31
Mandler, J. M. 18, 31
Marbe, K. 9, 15, 18, 20–1, 26–7, 31–7, 40, 42–7, 53–4, 67–72, 74, 83–7, 89, 115–17, 121, 128, 147, 151, 159–60, 170, 180–1, 183–8, 191–4, 227–8, 231, 237, 369, 371
Martius, G. 216, 370
Marx, K. 176
Marxism 176–7, 180, 192–3
materialism 90, 176–7, 193, 211, 213, 220, 305
Mayer, A. 31, 44, 68, 71, 146, 369
medicine 14, 117, 127, 145–6, 163, 165
Meinong, A. 38, 158, 160–1
memory 22, 26, 36, 49, 64–5, 68, 75–6, 80–1, 84, 111, 138, 214; drum 79, 106

Merrill, K. 374
Merton, R. 110
Messer, A. 9, 16, 18, 28–9, 33, 37–8, 40–1, 54, 67–9, 82, 85–6, 115, 122, 140–1, 146–7, 159–62, 164, 188–9, 193–4, 199, 211, 228–9, 231, 369
metaphysics 15–16, 90, 112, 166, 194, 203, 207–9, 215, 218
Meumann, E. 75, 85, 122, 135, 142, 157, 369–71
Mill, J. S. 107
Minsky, M. 250
Misch, G. 111, 371
Mischel, T. 74, 133, 356
modernism 195, 203, 217
modularity 292–3, 304, 312, 347, 350–1
monarchy 115, 175, 191–2
monism 199, 209, 213, 219–21
Moog, W. 153
Morrell, J. 369
Morse key 24–5, 57
Morton, A. 283
Moskiewicz, G. 87
Mühlhäusler, P. 361
Mulkay, M. 248–9, 369
Müller, *Frau* 105–6, 112–13
Müller, G. E. 10–11, 13–14, 17, 22, 71, 74, 78–84, 87–8, 92–3, 96, 104–14, 116, 118, 121–3, 125–6, 129–31, 141, 151, 166, 234, 236, 241, 247, 250, 370–1
Müller-Freienfels, R. 90
Müller-Lyer, F. C. 75
Münsterberg, H. 135, 141
Myers, C. S. 369

Nanu, H. A. 33, 369
national liberalism 192, 196
nationalism 170, 173, 185
Natorp, P. 148, 150–1, 153, 196
natural science 9, 13–14, 73, 132–4, 138–44, 146–7, 150, 163, 166–7, 220, 222, 225, 234
Needham, R. 332, 343–7, 357, 374
Nelkin, N. 359
neo-Kantianism 71, 89, 93, 148, 151, 158, 210–11, 217, 222, 234
Neoscholasticism 16, 89–90, 194, 203–4, 225, 227, 230–1, 233; *see also* Scholasticism, neo-Thomism
neo-Thomism 15, 71, 90, 92–4, 194, 197, 201–2, 204, 210–20, 225–30, 372; *see also* Neoscholasticism, Scholasticism
Neumann, J. von 359
neuroscience 4, 250, 287, 303–4, 313–14, 338
Newton-Smith, W. 251
Nietzsche, F. 139, 237, 251–2, 364
Nipperdey, T. 174, 192, 195, 372
Nisbett, R. E. 328
non-depictive 11, 33, 38–9, 68–9, 81, 83–5, 89, 92–3, 129–30, 140, 161, 206, 222, 233–4, 241, 247
nonsense syllable 47, 61, 79–80, 106, 122
normativity 255, 267–71, 275, 294, 296–7, 313, 316–17, 319–20, 324–6, 346–8, 351, 365–6
Normore, C. 251

Oelze, B. 170, 371
Oesterreich, K. T. 99, 129, 226, 369
Olesko, K. M. 249
organism 170–1, 174, 176, 179
Orth, J. 31–3, 44–5, 68, 71–2, 83, 369
Oswald, L. H. 283

pain 33, 45, 318, 348–9
pattern matching 255, 265, 258, 270, 273, 351, 366
pattern recognition 256, 258–60, 265, 270
Paul, H. 186
Paulsen, F. 198–202, 204, 206, 210–11, 219–21, 229
Perner, J. 288–9, 330
perseveration 11, 13, 22, 68, 70, 78, 80–1, 108–9, 111–13, 122–3
perseveration tendency 78, 92, 108
personality 104, 170–1, 174, 179–80, 196, 209, 221
Petersen, P. 91
Pfeiffer, L. 146
phenomenology 20, 38, 91, 152, 156, 161–3, 226, 363
physiology 14, 105, 111, 131–2,

134–8, 140–1, 150, 155–7, 163, 167, 211, 338
Pilzecker, E. 22, 79, 106–7, 111–12
Pilzecker, *Frau* 106
Pintner, R. 97
Pius X 195, 229
Planck, M. 239
platitude 284–5, 288, 299, 307, 317, 320, 323–7, 330, 339, 347, 349
Plato 115, 118, 201, 223, 250, 262, 339
Popp, W. 89–90
positivism 138–9, 217, 222
presentation 10–11, 21–2, 30–1, 33–5, 37–41, 43–5, 48–51, 55, 58–9, 62–3, 68, 76–80, 83, 86–90, 93, 98, 107, 109–10, 119–20, 128–9, 133–4, 140, 152, 155–6, 159, 207, 214–15, 223–4, 247
priming 259–60, 264
privileged access 286, 318, 362–4
promiscuist 131, 138–42, 232–4
Protestantism 12, 15, 194–211, 216, 218, 227, 230, 232–3, 235
proto-normative system 267–9, 275, 366
psychiatry 14, 27, 29, 111, 126–9, 145–6, 163, 166
psychologism 147–63, 198, 213, 266
psychology: animal 186; applied 13–14, 92, 103, 125, 129, 131, 142–7, 163, 233–4; collective 11, 13, 73, 77, 92, 99–100, 120, 143, 166–80, 184, 186–7, 208, 215, 233, 243; cross-cultural 334–7; cultural 333–4; descriptive 161; developmental 288–93, 301, 318, 341–3; educational 14, 122, 142–6; folk 1–4, 77, 143–4, 154–5, 157, 255, 275, 279–368; indigenous 331–2, 346, 356; individual 169, 178–9, 233, 251; normal 145–6; physiological 136–7; practical 142, 166; pure 103–4, 111, 125, 142–7, 162–3; social 304, 328
psychopathology 127–9, 145, 163
psychophysical dualism 137
psychophysical interactionism 214–15, 219–20
psychophysical materialism 135–6, 140, 170, 178, 233
psychophysical parallelism 74, 134–7, 170, 200, 211, 213–14, 219–20, 229–30, 233–4
purist 131–7, 152–8, 232–3
Putnam, H. 304–5

quale 20, 45–6, 53, 55, 67, 70, 88, 158, 241, 301
questioning method 22, 24–7, 68, 74, 76, 80–1, 92–3, 107
Quine, W. V. O. 348

Ramsey, W. 304
Rangette, L. 42, 369
rationality 244–5, 280, 294, 296, 313, 319, 324–5, 327–8, 338
Ravenscroft, I. 298–9, 313
realism 211, 214, 222–7, 230–1, 234
recluse 96–104, 117, 232–3, 238–9, 245
Reichwein, G. 85–6
reproduction 48–9, 107
reproductive tendency 36, 49, 52–4, 62
research school 2, 12, 17, 96, 234, 248–9
Richards, G. 369
Rickert, H. 89, 93, 148, 151, 164
Riehl, A. 151
Ritschl, A. 196
Roth, H. 146
rule 40–1, 237, 296, 341
Russell, J. A. 335–7
Ryle, G. 347

St Augustine 90, 227
St Paul 345
St Thomas Aquinas 90, 94, 195, 197–8, 201, 203, 222, 225–7, 254
sanctioning 268–70, 275, 317, 328, 342–3, 366
Sartre, J.-P. 364
saturation 261, 264
Schaffer, S. 369, 372
Schanoff, B. 42, 63, 369
Schell, H. 229
Schleiermacher, F. 373

Schlesinger, A. 145
Schneider, C. M. 371
Scholasticism 15, 76–7, 90, 92, 159–60, 162, 183, 202–3, 206, 215, 221, 226–7, 231; *see also* Neoscholasticism, neo-Thomism
Schrader, E. 88, 369
Schreiber, C. 225
Schröder, H. 361
Schuhmann, K. 196
Schultze, F. E. O. 27–8, 37, 68, 126, 369
Schumann, F. 106–7
Searle, J. 255, 264, 266, 310–11, 314, 349, 367, 372–4
Segal, G. 292, 369
self-observation 72, 76–7, 93, 102, 125, 129, 182; occasional systematic 28; retrospective 10, 18–30, 67–8, 80, 92; simultaneous 19–20, 74; systematic experimental 18–30, 67–8, 73, 75, 78, 85, 87–8, 92, 117, 121, 126, 145, 233–4; *see also* introspection
self-perception 19–20, 72, 103, 121, 221
self-reference 1, 235–6, 242, 246, 256, 258, 263–6, 269, 275, 280, 322–6, 331, 340, 352, 360, 362–5
self-validation 256, 265–6, 326, 359, 362
Sellars, W. 284–5, 303, 323
Selz, O. 9, 18, 27, 42–3, 63, 65–7, 69–70, 78, 84, 369
sensation 10, 21, 30, 33–4, 36–7, 39, 41, 55, 73, 76, 119–20, 129, 162, 176, 210, 221, 223–4, 247, 285, 291, 358–9
sensualism 40, 241
Sganzini, C. 181, 371
Shapin, S. 280, 369, 372
Sharpe, R. A. 356, 373
Shweder, A. 333–4
Sichler, A. 371
Sigwart, C. 20, 45
simulation theory 299–301, 318, 349–50
situation of consciousness 10, 21, 30–4, 36–9, 43, 45, 49, 69, 78, 83, 87–8, 81, 117, 140
Smith, C. 369
Smith, J. 332
Smith, P. K. 373
sociability 12–13
social democracy 176–7, 180, 185, 188–9, 192–3, 195
socialism 177, 184
society 173–80, 192, 358
sociologism 244, 251–4
sociology of scientific knowledge 5, 17, 96, 232, 249, 252, 281, 368
sociophilosophy 4, 279, 281, 316, 346, 366, 368
Sokal, M. 369
solipsism 188, 318
solitude 13, 86–7, 96–7, 100–1, 103–4, 129, 180, 233
soul 90, 177, 181, 184, 187–8, 190–1, 213–15, 219, 230, 233–4; of a people 170–1, 177, 179, 182, 184–8, 192–3
Specht, W. 90
Spencer, H. 204
Spengler, O. 204, 371
Spinoza, B. 213
Spranger, E. 153
Sprung, H. 371
Sprung, L. 371
Ssalagoff, L. 153
state 12, 16, 169, 173–6, 178–80, 189–92, 233, 236
status 264, 266, 337
Steinthal, H. 168
Stern, W. 145
Stevens, G. 188, 228
Stich, S. 283–4, 298–9, 303–6, 309–10, 313, 320, 323, 329, 340–1, 343, 355, 373
Stigler, J. W. 333
stimulus error 42–3
Stölzle, J. 216
stopwatch 25, 28, 236
Strawson, P. 311
Stumpf, C. 10, 17, 71, 74, 90, 110, 128, 156, 160, 161, 196, 206, 217–20, 228
suggestion 23, 60, 76, 85, 125, 241

supervenience 254, 305
Switalski, D. 235

task 10, 25–6, 30, 47–56, 60–2, 66–9, 75, 81–2, 86, 90–1, 118, 120, 122, 159, 225, 234
Tawney, G. A. 97
Taylor, C. 356
Taylor, C. O. 26, 33, 36, 369
theory 1–4, 242–4, 281–97, 302–3, 309–11, 314–15, 317, 319, 323, 347–51; bedrock 309, 350; change 291–2, 302; 'theory theory' 283–93, 298, 316, 318, 320, 323, 347, 349
thought 10, 25, 30, 33, 36–44, 63–4, 69, 76–7, 86, 88–9, 91, 117, 120, 122, 128, 140, 177–80, 246–8, 253
threshold of consciousness 34–5, 79, 82, 107, 109–10, 156
Thumb, A. 44–5, 71, 183, 187
Titchener, E. 17, 31, 43
Tooby, J. 350
training 81, 93, 102, 107
Troeltsch, E. 196
Trout, J. D. 306, 308, 373
Tuomela, R. 255, 264
Turner, R. S. 245, 369

Ueberweg, F. 226
unconscious 54, 59, 63, 73, 81, 133–4, 140–1, 155–6, 213–14, 234, 304
Ungerer, G. A. 142, 158, 174, 371
uniformity 180–7
universalism 190, 316
utilitarianism 176–8, 193

voice key 24–5, 27, 68
volition 30–1, 45, 55–7, 61, 72, 83, 207–8; *see also* will
Volkmann, J. 216–17
voluntarism 15–16, 199, 202, 206–7, 211, 214–15, 220–1, 230, 232–4

Wartensleben, G. von 26, 33, 369
Watson, J. B. 241
Watt, H. J. 9, 18, 28, 33, 47–56, 62, 67–70, 72, 85, 140–1, 159–60, 179–80, 369
Weber, M. 196
Weimer, W. B. 99
Wellman, H. 288–93
Wentscher, E. 88
Wertheimer, M. 33, 146, 369
Westphal, E. 29, 56, 129, 369
Whitley, R. 369
Wilkes, K. 293–6, 310–11, 319
will 55–7, 61, 63, 85, 88, 98, 109, 170–1, 179, 190, 206–9, 214–15, 221; *see also* volition
Wilson, T. D. 328
Windelband, W. 148, 151, 153
Wise, N. 369, 372
Wittgenstein, L. 267, 293, 296, 318, 325
Woodward, J. 310–12
Wundt, M. 165
Wundt, W. 10–11, 13–17, 20, 30, 32–3, 40, 44–5, 51, 53, 56–7, 68–78, 84–5, 87–90, 92, 94, 96–104, 106, 109–10, 114–16, 118–27, 129–37, 139–45, 148, 150–62, 165–224, 227, 229–39, 241, 243–5, 247, 250–1, 254, 369–72

Ziehen, T. 20, 45